THE ENCYCLOPEDIA OF EARTHQUAKES AND VOLCANOES

THE ENCYCLOPEDIA OF EARTHQUAKES AND VOLCANOES

by David Ritchie

AN INFOBASE HOLDINGS COMPANY

To St. Mamas, from one cat person to another.

The Encyclopedia of Earthquakes and Volcanoes

Facts On File, Inc.
460 Park Avenue South
New York NY 10016

Library of Congress Cataloging-in-Publication Data

Ritchie, David, 1952 Sep. 18–
The encyclopedia of earthquakes and volcanoes / David Ritchie.
p. cm.
Includes bibliographical references and index.
ISBN 0-8160-2659-9
1. Earthquakes—Encyclopedias. 2. Volcanoes—Encyclopedias.
I. Title.
QE521.R58 1994
551.2′03—dc20 93-7670

Text design by Grace M. Ferrara
Printed in the United States of America

10 9 8 7 6 5 4 3 2

This book is printed on acid-free paper.

Contents

Acknowledgments

For extraordinary patience and helpful advice, the author would like to acknowledge Elizabeth Ryan, literary agent; and Phil Saltz, editor on this project. Also, special, posthumous thanks to Father Seraphim Rose (1934–1982), for giving a wandering Protestant a push toward Holy Orthodoxy.

Preface

This book is intended to serve as a single-volume general reference for the nonexpert on the subject of earthquakes and volcanoes. To discuss every major earthquake in history and to describe the eruptive history of every known volcano on earth would be beyond the scope of such a work. Therefore, the list of topics has been restricted to some of the most famous volcanoes and earthquakes, along with a selection of interesting but lesser-known examples of these phenomena (especially active calderas) and a selection of pertinent geological terms and concepts. Because an eruptive history does not always convey the "flavor"—that is, the human aspect—of an earthquake or eruption, some contemporary accounts of various events, such as the eruption of Peleé have been added. Short biographies of prominent scientists in seismology and volcanology are also included.

The work is cross-referenced to provide the reader with additional information. In a given entry, the reader may find terms in small capital letters in the body or at the end. These terms refer readers to other articles that provide pertinent information to the given entry. Every entry in the book is listed in the index.

aa A Hawaiian word (pronounced "AH-ah"), aa is a particular kind of LAVA flow with an irregular, jagged surface. (See also PAHOEHOE.)

acoustics Various noises are associated with earthquakes and volcanic eruptions. Earthquakes are often accompanied by a deep, audible, grinding or rumbling noise. The noise often is compared to that of thunder or of heavy traffic passing in the streets. In one instance, noises associated with an area of occasional earthquake activity have become something of a tourist attraction. This case involves the "Moodus noises" in the state of Connecticut. These are mysterious sounds similar to gunfire that have been reported in the vicinity of East Haddam. The name "Moodus" is derived from the Native American word *Mackimoodus,* meaning "meeting place."

The acoustic effects of volcanic eruptions can be surprising. The noise that accompanied the explosion of the volcano Krakatoa in 1883, for example, was heard some 3,000 miles away on the island of Rodriguez in the Indian Ocean. This is said to be the greatest distance at which the noise of a natural event has been heard within historic times without the aid of electronic communications. In some eruptions, the noise may be audible hundreds of miles away yet go unheard in areas much closer to the point of eruption. When Mount Katmai in Alaska erupted on June 6, 1912, for example, the sound of the eruption was heard some 800 miles away but reportedly was not distinct at Kodiak, only about 100 miles from the volcano.

A flow of aa lava at Mount Etna, 1974 (*Earthquake Information Bulletin*/U.S. Geological Survey [USGS])

active volcano An active volcano is considered to be one that has shown activity within historical times, or the past several thousand years. A historically active volcano, however, may be inactive at present and indeed may have shown no activity for hundreds of years. Approximately 500 volcanoes around the globe are thought to be active, but this figure may be a serious underestimate because of some submarine volcanoes whose activity has not been observed and reported.

Aeolian Islands See LIPARI ISLANDS.

Africa The African continent has areas of strong seismic and volcanic activity. Much volcanism is concentrated along Africa's GREAT RIFT VALLEY, which extends through the eastern central portion of the continent and contains numerous volcanoes and calderas, including Duguna, Erta Ale, FANTALE, Groppo, K'ONE (Gariboldi), ASAWA and the BARRIER. (See also DEIRBA.)

Agadir, Morocco The earthquake of February 29, 1960, struck the community of Agadir (population almost 50,000) at the foot of the Atlas Mountains at 11:45 P.M., killing some 12,000 people and destroying some 70% of all structures in the city. The earthquake measured 6.25 on the Richter scale of magnitude and was preceded by two milder shocks. The earthquake was also accompanied by a tsunami that reached almost a hundred yards inland from the sea. Effects of the earthquake included ruptured sewers, from which large numbers of rats were reportedly released into the city. The city's fire-fighting capability was effectively neutralized by the earthquake, with the result that fires burned unchecked. The dome of a mosque collapsed upon a group of praying Muslims, and the Jewish community of Agadir was devastated; of some 2,200 Jews in Agadir, approximately 1,500 were said to have died in the earthquake. Corpses were so numerous in Agadir after the earthquake that most of the 12,000 dead were simply buried in common graves.

Agnano, volcano, Italy The Agnano volcano formed a crater in the Phlegraean Fields near Naples that once was filled with a lake but later was drained and converted into a race track.

Agua de Pau, caldera, Azores The stratovolcano Agua de Pau has a record of historical activity extending back to 1563, when an eruption of pumice reportedly covered the nearby island of São Miguel. Strong earthquakes preceded and continued during the eruption and destroyed most of the community of Ribeira Grande, several miles north of the Agua de Pau caldera. Basalt lava extruded from the volcano following the eruption from the main vent. Another, less powerful eruption took place in the caldera the following year. In October 1952, destructive earthquakes preceded an eruption in which fissures opened at the foot of Agua de Pau. This eruption lasted one week and produced a lava flow and a small cone. Very small earthquakes occurred at Agua de Pau during the 1980s.

Agung, volcano, Bali, Indonesia Agung is best known for its powerful eruption in 1963, which killed between 1,200 and 2,000 people and sent large amounts of ash into the upper atmosphere. This airborne material is thought to have caused spectacular atmospheric effects in the following weeks, such as brilliant red sunsets and halos around the moon and sun. The high-altitude cloud from this eruption of Agung was also implicated in a sharp decrease in starlight as measured at observatories. Average temperatures at the Earth's surface dropped measurably for three years after this eruption.

Many of the fatalities occurred at a religious festival that was in progress near the volcano at the time of the eruption. Clouds of lethal gas swept down from the volcano and killed large numbers of participants in the religious rites. Lava overwhelmed the villages of Sebih, Sebudi and Sorgah. A combination of heat, ash and poisonous gases is said to have killed animals for miles around the volcano. Huge boulders cast out from the volcano during the eruption landed in the village of Subagan. (See also CLIMATE, VOLCANOES AND.)

Aira, caldera, Japan The Aira caldera lies a few miles north of the Ata caldera in southern Japan, in the region of Kagoshima Bay. The bay itself is thought to be a GRABEN formed by volcanic and tectonic activity. Uplift of the bay floor has also occurred on occasion, and the Aira caldera has been cited to show that volcanically related uplift and subsidence can affect the whole area of a caldera even when an active volcano is located at the caldera's very edge. The Aira caldera is famous for the violent eruptions of the SAKURA-ZIMA volcano, although volcanic activity occurs at other points in the caldera as well.

An eruption from 1779 to 1781 began with a series of strong earthquakes in early September. Changes in water level (sometimes involving energetic spouting) were observed at water wells on Sakura-zima on the morning of November 8, 1779, at about the same time clouds of steam started rising from the summit of the volcano. On the afternoon of November 8, a major eruption started. After several days, small islands began to emerge from the waters near Sakura-zima. These islands are thought to have been formed partly through underwater eruptions but also in part through uplift of the bay floor.

Changes in hot springs in the vicinity also were observed; two new hot springs emerged, and another stopped flowing. The area around Sakura-jima appears to have subsided in the decades following this eruption, because the waters encroached on low areas along the shore, flooding parts of the city of Kagoshima. In some areas, floods covered the land to a depth of perhaps 10 feet or more. Local authorities tried building embankments to bar the rising waters but were unsuccessful. Some communities along the shore were wiped out by the rising waters. Uplift affected other areas around Sakura-zima about this time. A seacoast road on the southern shore of Sakura-zima rose several feet until it lay more than 100 feet inland from the waters. Uplift also affected the northwest shore of Kagoshima Bay so that trade in the harbor at one community had to be conducted using wagons rather than boats.

Another major eruption began in 1913, when earthquake activity to the north in May and June signaled the beginnings of renewed volcanism.

The nearby Kirishima volcano erupted in late 1913 and early 1914. Strong earthquakes occurred near Kagoshima in late June. Dramatic changes also were observed in the activity of local hot springs. On the eastern shore of Sakura-zima, hot springs stopped flowing in the spring of 1913, and shortly afterward, other hot springs on the south side of the volcano became too hot for bathing when the tide was low. Changes in the water table occurred early in 1914; a pond on the southern side of Sakura-zima dried up, killing the fish in it, while the water table dropped, and some of the island's water wells went dry. On the morning of January 12, 1914, a spring at a beach on the north side of Sakura-zima emitted a gush of cool water, while water spouted to a height of several feet from hot springs on the other side of the island and extremely hot water poured out from the ground at several locations. On the same morning, a powerful eruption of Sakura-zima began. This eruption was preceded by strong earthquake activity over more than 24 hours. Earthquakes were especially frequent on Sakura-zima itself. A particularly strong earthquake occurred several hours after the eruption began. Since the 1913–15 eruptions, numerous small eruptions have been recorded.

In 1935 earthquakes felt on the southern side of Sakura-zima in the middle of the year were followed by eruptions of ash beginning in September. Occurring only a few years after the violent events of 1913–15, these eruptions convinced several hundred residents of the area to evacuate. Ash was deposited to a depth of several inches on the southern and eastern sides of the volcano, and some damage to crops occurred. An eruption may have occurred underwater on March 13, 1938, when waters about 1,000 feet offshore rose abruptly while a roar was heard. This phenomenon was repeated soon afterward, some distance away. Sakura-zima itself started erupting again two weeks later. Minor eruptions took place between 1939 and 1942. In 1946 the volcano exhibited explosive activity and extruded lava. Minor explosions also occurred over the next eight years. Starting in 1955, explosions concentrated on the summit of the volcano. In the middle to late 1980s, explosive activity appeared to become more frequent after a comparatively quiet period. Earthquake activity at the Aira caldera has not always been related clearly to eruptions, although in some cases, eruptions plainly had earthquakes as precursors. The character of slippage along faults has been seen to change during periods of eruptive activity. When the mountain is not erupting, earthquakes are characterized by strike-slip (or predominantly horizontal) movement, which changes to oblique-slip normal faulting in the initial stages of eruptions and then to oblique-slip normal or reverse faulting when eruptive activity is at its height. As the eruption subsides, movement along faults returns to strike-slip or reverse.

As noted earlier, the Aira caldera is noted for the dramatic uplift and subsidence it has displayed on occasion. After the eruption of 1914, the caldera and adjacent areas displayed dramatic subsidence, almost 20 feet in some locations. A few months later, uplift started again, and has continued through the 1980s. Measurements of uplift at various points on and around Sakura-zima indicate that a reservoir of magma under the caldera has expanded at an average rate of perhaps 30 million cubic feet per year at an estimated depth of perhaps four miles. The caldera does not appear to show deformation in a uniform pattern; some scientists have suggested that there is more than one source of uplift within Aira. In addition to deformation patterns observed in the caldera as a whole, Sakura-zima may exhibit comparatively brief and shallow deformations.

In the 20th century, the Aira caldera has shown some curious phenomena related to heat flow. After the major eruption in 1914, the temperature of the soil began rising near the northwestern shore of Sakura-zima. Fruit trees and other flora died. Eventually, trees were killed within an area some 500 feet wide, and benzene and chlorine fumes were detected there. Some months earlier, similar emissions of fumes at a spot on the mainland, near a line of vents passing through the summit of Sakura-zima, reportedly killed an ox and made several humans sick. By the spring of 1915, soil dug up at the heat-affected area was too hot to hold in one's hands. After 1915, this unusual concentration of heat diminished.

Akan, caldera, Japan The Akan caldera is located in northern Japan near the southern end of the Kuril Islands. Lake Akan occupies part of the caldera. Several cones are also found in the caldera; these include Furebetsu, Fuppushi, O-akan and Me-akan, the last of which has been active within historical times. Strong earthquakes were felt in the vicinity of Akan caldera in the late 1920s and early 1930s, and in late 1937, a cloud of vapor was seen rising from the foot of Me-akan. Marked seismic activity increased in the early 1950s, and small quantities of ash may have been released during this period, although it is not known if there was any direct relationship between eruptive and seismic activity at that time. An eruption began in November 1955, and the following year, observations of earthquakes showed a rise in activity of earthquake swarms several days before an explosion on June 15. Seismicity increased for approximately three weeks preceding an explosion in 1959. Earthquakes accompanied eruptions of ash in 1988. The Akan caldera has been studied intensively to examine the relationship between earthquakes and a magmatic system. Tectonic (as opposed to volcanic) earthquakes have also been studied for their relationship to earthquake swarms at Akan. On one occasion, a major tectonic earthquake followed changes in temperature at hot springs in Akan.

Alabama, United States The state of Alabama varies geographically in its degree of seismic risk. The southern portion of the state is characterized by low seismic risk, whereas the degree of risk generally increases as one moves northward toward the Tennessee border. There have been several notable earthquakes in the history of Alabama, including the earthquakes of February 4 and 13, 1886, in Sumter and Marengo counties, where perceptible movement of the earth was reported along the Tombigbee River. An earthquake on May 5, 1931, in northern Alabama was felt in Birmingham and caused minor damage at Cullman; the Mercalli intensity was V–VI, and the affected area was about 6,500 square miles. On April 23, 1957, an earthquake in the area of Birmingham, estimated at intensity VI on the Mercalli scale and affecting an area of about 2,800 square miles, caused minor damage in Birmingham; loud noises were associated with this earthquake in some locations. The August 12, 1959, earthquake along the border of Alabama and Tennessee caused minor damage and was estimated at intensity V; the earthquake affected an area of some 2,800 square miles. An earthquake on February 18, 1964, along the border of Alabama and Georgia measured Mercalli intensity V and Richter magnitude 4.4.

Alaska, United States The largest and northernmost state of the United States, Alaska also is one of the most seismically and volcanically active parts of the country. Earthquakes in Alaska are concentrated in two belts, one extending along the southeastern coast and another reaching from the interior near Fairbanks southwestward along the ALEUTIAN ISLANDS. The 1964 GOOD FRIDAY EARTHQUAKE, one of the most powerful and destructive earthquakes of the 20th century, occurred along the southern coast of Alaska and, along with the tsunami associated with it, caused extensive destruction as far south as Crescent City, California.

Volcanism in Alaska has been both frequent and destructive throughout history. A familiar case in point is the eruption of KATMAI in 1912. This eruption created a caldera some three miles wide and laid down a plain of fumaroles later named the VALLEY OF TEN THOUSAND SMOKES. The volcanic arc in Alaska extends more than 1,000 miles, from Cook Inlet in the east to Buldir Island near the tip of the Aleutian chain in the west. More than 70 volcanoes exist in the Aleutian Islands and on the Alaska Peninsula. The Alaskan volcanoes are thought to be an expression of activity along a SUBDUCTION ZONE marked by the Aleutian Trench south of the Alaska Peninsula and the Aleutian Islands. The Aleutian Trench reaches depths greater than 20,000 feet. North of the Aleutian Islands, in the Bering Sea, lie the Pribilof Islands, which were formed by eruptive activity but do not constitute part of the Aleutians.

A history of earthquakes and volcanism in Alaska would occupy an entire volume, and all an article of this length can do is present a few examples.

A very strong earthquake accompanied the eruption of PAVLOV volcano on the Alaska Peninsula in 1786. A tsunami, or seismic sea wave, reportedly flooded land on Sanak Island, the Shumagin Islands and the Alaska Peninsula on July 27, 1788, with considerable loss of human life and of livestock. In May 1796 an earthquake with frightening noises affected Unalaska Island, and BOGOSLOV volcano cast out rocks as far away as Umnak Island. In 1812 an eruption of Sarycheff volcano was accompanied by powerful earthquakes on Atka Island. Umnak Island underwent a strong earthquake in April of 1817 when Yunaska volcano erupted. Sometime in 1818, an earthquake near Makushin volcano and Unalaska Island is said to have caused great alterations in the landscape. Unalaska Island experienced two earthquakes in June of 1826, but details are unavailable. An earthquake described as "severe" struck the Pribilof Islands on April 14, 1835; and in April of 1836, the Pribilofs were subjected to shocks so powerful that they knocked people off their feet. An earthquake on September 8, 1857, was very powerful but apparently caused no damage. A minor earthquake on May 3, 1861, at St. George Island in the Pribilof Islands was accompanied by noise from underground. On August 29, 1878, an entire town on Unalaska Island appears to have been destroyed by a tsunami and earthquake. AUGUSTINE volcano erupted on October 6, 1883; a very powerful earth-

Rails buckled by earthquake in Alaska, March 27, 1964 (*Earthquake Information Bulletin*/USGS)

quake and a tsunami occurred in connection with this eruption. An earthquake in the area of Prince William Sound in May of 1896 was so violent that people who were standing had trouble remaining on their feet.

The Yakutat Bay earthquakes of September 3 and 10, 1899, were estimated at Mercalli intensity XI and at Richter magnitudes 8.3 and 8.6 respectively. The epicenter was located near Cape Yakataga. The first of these earthquakes was felt with tremendous violence at Cape Yakataga, but the second earthquake was the one that caused major changes in topography. A U.S. Geological Survey expedition to the region six years after the earthquakes found widespread evidence of topographic changes. Beaches had been raised, and barnacles and other aquatic organisms were lifted out of the water. On the west shore of Disenchantment Bay, an uplift of more than 47 feet was measured—approximately the height of a five-story building. Over a wide area, uplift of 17 feet or more was observed. In some areas, depressions of several feet occurred. A tsunami thought to have been perhaps 35 feet high occurred in Yakutat Bay, and tsunamis were reported at other locations along the coast of Alaska as well. There were reports of volcanic eruptions associated with these earthquakes, but the "eruptions" are presumed to have been merely large clouds of snow released in slides caused by the earthquakes. Strong aftershocks occurred over several months following these earthquakes. No loss of life was attributed to the earthquakes because the area was not yet settled; a small number of Native Americans and prospectors, however, witnessed the earthquakes firsthand.

On September 21, 1911, an earthquake of Richter magnitude 6.9 on the Kenai Peninsula and Prince William Sound broke cables, caused great rockslides and killed large numbers of fish; water at Wells Bay was reportedly disturbed greatly. Cables broke also in another earthquake on January 31, 1912, in the vicinity of Prince William Sound; this earthquake, which was measured at Richter magnitude 7.25, appears to have been centered west of Valdez and was felt in Fairbanks. Very strong shocks occurred at Kanatak, Nushagak and Uyak on June 4–5, 1912, and were felt more than 100 miles away from Mount Katmai, although the earthquakes may have been unaffiliated with the June 6 eruption of Mount KATMAI. An earthquake of Richter magnitude 6.4 at Cook Inlet on June 6, 1912, coincided with a bright display of light from Katmai, and the shock was recorded at many distant locations, including Ottawa, Ontario and Irkutsk in Russia. Very strong earthquakes were reported on the night of June 6 at Kodiak, and on June 7, a strong earthquake struck Kanatak, together with rockslides and a powerful rumbling noise.

An earthquake near Seward on January 3, 1933, measured at Richter magnitude 6.25, was felt very strongly at Anchorage and caused alarm at Seward; the ground cracked in numerous places in the vicinity of Seward, notably for a distance of 20 miles along a road running north from the city. On April 26, 1933, an earthquake northwest of Anchorage severed telegraph lines and broke plate-glass windows and was felt also in Fairbanks and in the Aleutian Islands. Houses were displaced from their foundations at Old Tyonek. The principal shock measured Richter magnitude 7.0. Old Tyonek experienced further damage several weeks later when an earthquake of magnitude 6.25 occurred there on June 13, 1933. The May 14, 1934, earthquake on Kodiak Island measured magnitude 6.5 and was felt strongly on Whale and Kodiak Islands; plaster was cracked, and roads were blocked by landslides. An earthquake of magnitude 6.75 in southern central Alaska was strong enough to break plate glass in Anchorage on August 1, 1934.

Tsunami damage was remarkable in the magnitude 7.4 earthquake of April 1, 1946; centered about 90 miles southeast of Scotch Cap Lighthouse, the earthquake produced a tsunami that demolished the lighthouse and caused damage at widely separated locations in and around the Pacific Basin, along the Pacific coasts of North and South America, in the Aleutian Islands, and in Hawaii, where 173 persons drowned and property damage was estimated at $25 million.

The earthquake of March 9, 1957, measured Richter magnitude 8.3 and was one of the greatest natural calamities in Alaskan history. The earthquake, which involved hundreds of aftershocks and affected an area approximately 700 miles in length along the southern border of the Aleutian Islands between Amchitka Pass and Unimak Island, was accompanied by a tsunami 40 feet high that struck the shore at Scotch Cap, and a 26-foot tsunami that caused extensive damage at Sand Bay. On the islands of Kauai and Oahu in Hawaii, the waves destroyed two villages and caused several million dollars in damage. The tsunami was 10 feet high along the coast of Japan, and a wave six feet high was reported in Chile.

The earthquake of July 9, 1958, is famous for the dramatic effect it produced at Lituya Bay, on the Gulf of Alaska in the southeastern part of the state. A tremendous rockslide at the head of the bay produced a giant wave—more than 1,700 feet high—that swept outward through the mouth of the bay and is thought to have killed two people who were caught in the wave. A fishing boat with two occupants was carried out of the bay by the wave front and reportedly cleared the spit at the mouth of the bay by at least 100 feet. The wave also wiped the rim of the bay clean of trees. Otherwise, little damage was reported from this earthquake, except that underwater communication cables were broken in the vicinity of Skagway, and Yakutat experienced damage to bridges, a dock and oil lines. Great landslides reportedly occurred in the mountains, and fissures and sand blows were reportedly widespread on the coastal plain near Yakutat.

The Good Friday earthquake of March 27, 1964, is covered in detail elsewhere in this volume.

Among the well-known volcanoes of Alaska are Katmai, Augustine, Pavlov and Shishaldin. Numerous calderas, indicative of eruptive activity followed by collapse, are found at locations such as ANIAKCHAK, Emmons Lake, FISHER, Katmai, LITTLE SITKIN, OKMOK, SEMISOPOCHNOI, VENIAMINOF and the WRANGELL MOUNTAINS.

Alban Hills, volcanic structures, Italy The Alban Hills are located near Rome and are believed to have originated through a combination of explosive and effusive eruptions. A period of predominantly effusive eruptive activity is thought to have produced a stratovolcano that developed a collapse CALDERA, from which a new central cone arose later. The record of activity at the Alban Hills in historical times is uncertain. Eruptions are reported in 114 B.C. and possibly several centuries earlier, but there is some question whether these events were volcanic in nature or represent other natural phenomena such as fires and falls of hail. An ashfall was reported in nearby Rome in 540 B.C. The volcano Albano is located in the Alban Hills area, and Lago Albano occupies an eccentric crater just west of the rim of the inner caldera. The Via Appia Nuova, Via Appia Antica and Via Tuscolana traverse the Alban Hills.

Alcedo, volcano, Galapagos Islands, Ecuador Alcedo is one of several volcanoes on Isabela Island in the Galapagos. A caldera is present. A lava flow, identified from aerial photos, appears to have occurred on the southeast side of the volcano between 1945 and 1961. Radial fissures account for many of the lava flows on the volcano. Volcanic activity is suspected (although this has not been proven) as the cause of an uplift of a short length of shoreline on the western side of the island, possibly in 1954. A large amount of coral reef was lifted above sea level, probably at the same time.

Aleutian Islands, Alaska, United States The volcanic Aleutian Island chain extends westward from the southern shores of Alaska. The islands are part of a range of volcanic mountains, the Aleutian Range, extending more than 1,600 miles, from the Alaska Peninsula to a point just east of the International Date Line. The Aleutian Range contains dozens of recently active volcanoes, including AUGUSTINE, BOGOSLOV, KATMAI, Novarupta, PAVLOV, REDOUBT and TRIDENT. The tallest volcanoes (up to 11,000 feet) occur at the northeastern end of the range. Summit elevations generally diminish as one moves southwestward along the Aleutians. Several kinds of volcanoes occur in the Aleutian Range. Some are shield volcanoes made up of numerous thin flows of lava. Other Aleutian volcanoes are composite volcanoes with steep sides. In some places, these composite cones occur atop older, shield volcanoes, resulting in a structure much like that of the Cascade Mountains in the northwestern United States and the Canadian province of British Columbia. Volcanic domes may also be seen where viscous lava has emerged. A notable CALDERA formed from the collapse of Katmai during its 1912 eruption. The Aleutian volcanoes are associated with an offshore subduction zone marked by the presence of the Aleutian trench, a deep undersea trough located to the south of the Aleutians and the Alaska Peninsula. The Trench becomes shallower and eventually vanishes as one approaches the mainland. The progressive shallowing of the trough is thought to be due to a buildup of sediment.

The Aleutian Range has been the site of powerful earthquakes, such as the GOOD FRIDAY EARTHQUAKE of 1964, which caused great destruction in the vicinity of Anchorage. Ground subsidence destroyed much of Anchorage's main street. Approximately 75 homes in a residential neighborhood on Turnagain Bluff were wrecked when the land on which they rested underwent a sudden slump. The earthquake also demolished the airport control tower and killed the controller on duty when the structure collapsed. Alaskan earthquakes have been accompanied by powerful tsunamis on several occasions in this century. The tsunami that accompanied the Good Friday Earthquake, for example, caused tremendous damage along the southern coast of Alaska, wiped out much of the state's commercial fishing fleet and carried destruction as far south as Crescent City, California. More than 200 persons were killed as a result of tsunamis originating in Alaskan waters between 1946 and the Good Friday earthquake.

In recent years, areas along the Aleutian Range have emerged as cause for concern as potential sites of major future earthquakes. One of these areas is the "Commander gap," an area near the western end of the Aleutian chain where no major earthquake has occurred since the mid-19th century. The "Shumagin gap" near the western tip of the Alaska Peninsula also has been identified as a prospective source of powerful earthquakes because others have occurred there in 1788 and 1947. (Another strong earthquake in 1903 may have originated in this area.) The potential for destructive tsunamis from earthquakes in the Shumagin gap is also considerable. A third "gap" along the southern Alaska coast, the "Yakataga gap," lies near the northern tip of the Alaska panhandle and has been an area of concern for the U.S. Geological Survey, which expects that strain in the Yakataga gap may manifest itself in the near future in the form of earthquakes of magnitude 8.0 or stronger.

Alexandria, Egypt An earthquake on July 21 in A.D. 365 shook much of the Mediterranean basin and appears to have caused widespread destruction. Among the most notable casualties of this earthquake was reportedly the great lighthouse at Alexandria in Egypt. Said to have been some 600 feet high, the lighthouse was reduced to a ruin that remained in place for the next five centuries. More than 50,000 people in Alexandria were reportedly killed in this earthquake, which was accompanied by tsunamis. Edward Gibbon, in his history *The Decline and Fall of the Roman Empire,* describes the effects of this earthquake on the shores of the Mediterranean:

> In the second year of the reign of Valentinian and Valens, on the morning of the twenty-first day of July, the greatest part of the Roman world was shaken by a violent and destructive earthquake. The impression was communicated to the waters; the shores of the Mediterranean were left dry by the sudden retreat of the sea; great quantities of fish were caught with the hand; large vessels were stranded on the mud; and a curious spectator [evidently the historian Ammanius, whose accu-

racy Gibbon questions in a footnote to the work] amused his eye, or rather his fancy, by contemplating the various appearance of valleys and mountains which had never, since the formation of the globe, been exposed to the sun. But the tide soon returned with the weight of an immense and irresistible deluge, which was severely felt on the coasts of Sicily, of Dalmatia, of Greece, and of Egypt; large boats were transported and lodged on the roofs of houses, or at the distance of two miles from the shore; the people, with their habitations, were swept away by the waters; and the community of Alexandria annually commemorated the fatal day on which fifty thousand persons had lost their lives in the inundation.

The psychological impact of this earthquake on the Romans appears to have been considerable. Gibbon continues:

> This calamity, the report of which was magnified from one province to another, astonished and terrified the subjects of Rome, and their affrighted imagination enlarged the real extent of a momentary evil. They recollected the preceding earthquakes, which had subverted the cities of Palestine and Bithynia; they considered these alarming strokes as the prelude only of still more dreadful calamities; and their fearful vanity was disposed to confound the symptoms of a declining empire and a sinking world.

Amatitlán, caldera, Guatemala The Amatitlán caldera is located several miles south of Guatemala City and includes the volcano Pacaya, which has a long history of unrest within historical times. Large deposits of TEPHRA in central Guatemala have been traced back to the Amatitlán caldera. Strong earthquakes were associated with reported explosive eruptions in 1565 and 1651. Explosive eruptions continued through the next three centuries, in 1664–74, the 1690s, 1775 and 1846. Activity from fumaroles in the Los Humitos area of Pacaya started in 1891 and continued for years, although no actual eruption accompanied this activity. Explosive eruptions resumed in 1961 and continued through the late 1980s. Lava flows were associated with eruptive activity in this last period. Increases in seismic activity have been noted before some eruptions of Pacaya.

Ambrym, volcano, Vanuatu Situated at the intersection of the Vanuatu archipelago and the D'Entrecasteaux Fracture Zone near Loyalty Island, the volcanic island Ambrym has exhibited numerous explosive eruptions and lava flows over the past two centuries. Roughly triangular in shape, Ambrym is approximately 30 miles long and 20 miles wide at its broadest point. A large CALDERA occupies the summit. Two cones inside this caldera, Mount Marum and Mount Benbow, show nearly constant activity. Several small volcanoes (Rahoum, Tower Peak and Tuvio) also are found on the island. The historical record of activity on Ambrym is brief but colorful. Emissions resembling smoke were reported in 1774, and in 1888 a flow of lava was observed from a rift on the southeastern side of the island. An eruption in 1894 was characterized by large numbers of earthquakes. Dates of eruptive activity between 1912 and 1915 are not entirely certain, but Mount Benbow and various other sites on the island appear to have shown eruptive activity. Strong earthquakes accompanied eruptions in 1913 along fissure lines running east to west across the island, and reports mention a large eruption cloud and emissions of flames. A hospital was destroyed in this set of eruptions. Strong earthquakes accompanied another eruption in late March of 1937, and marked seismic activity was noted before the lengthy eruption of 1950 through 1954. An explosion in 1972 did not show any precursive earthquake activity. Ambrym remained in nearly continuous eruption between 1964 and 1980. Acid rain generated by emissions of sulfur dioxide harmed crops in February 1979. Eruptions are thought to have occurred around the end of 1985 and in early 1986. In February 1988 the crew of an aircraft flying near the island observed an eruption from Mount Benbow.

Anak Krakatoa See KRAKATOA

Anatolia, Turkey On October 16, 1883, an earthquake at Anatolia in Asia Minor, now part of Turkey, killed perhaps 1,000 people and left some 20,000 homeless. Great fissures are said to have opened and shut in the earth during this earthquake. Starvation and cold temperatures reportedly killed several hundred more residents of Anatolia before assistance could arrive.

Andes Mountains The Andes mountain range is part of the Andean cordillera, which runs roughly north-south along the western edge of South America. The Andes are believed to have been formed by an ongoing collision between the crustal plate bearing South America and the plate underlying the Pacific Ocean. Volcanic and hydrothermal activity along the Andean cordillera has generated hot springs and numerous commercially viable deposits of ores of various metals, among them copper and gold. Tsunami activity also has been associated with earthquakes along the Andean cordillera.

A curious magnetic anomaly has been reported along the Andean cordillera. A reversal of change in the vertical geomagnetic field has been accompanied by an intensified change in the horizontal magnetic field. This pattern indicates the existence of internal induction currents at a depth of perhaps 40 miles, running along a high-conductivity zone beneath the mountains.

Earthquakes occur frequently in and near the Andean cordillera. An earthquake in CHILE in 1822, for example, reportedly killed some 10,000 persons and raised the shoreline by several feet. Along the Pacific shore, earthquakes are sometimes accompanied by tsunamis. (See also BOLIVIA; CONCEPCIÓN; PERU; PLATE TECTONICS.)

andesite One of the most commonplace volcanic rocks, andesite is widely distributed around the Pacific Basin (the "RING OF FIRE"), where chains of andesitic volcanoes form the "andesite line" that has been used to mark the boundary of the Pacific basin. Andesitic volcanoes also are found in other regions of the world, including Europe's Carpathian Mountains and the Colorado Plateau of the United States. Andesite varies in composition but is generally characterized as lighter in color than basalt and intermediate in silica content. The composition of andesite appears to have little relationship to local lithology and tectonics. This indicates that andesite must derive its chemical makeup from conditions in the mantle below the continents, not from the influence of rock in the continental crust, although in some cases andesite is believed to assimilate components of crustal rock as magma rises to the surface. The characteristics of andesite output vary greatly from one volcano or cluster of volcanoes to another. In some areas, a group of volcanoes may emit andesite of remarkably consistent composition, whereas a single volcano elsewhere may put out a variety of types. Many volcanoes release two main types of rocks, a principal andesitic series and another group made up largely of basalt with either RHYOLITE or DACITE mixed in. Composite andesitic volcanoes, widely found around the Pacific Ocean basin, are composed of TEPHRA and flows of andesite and rhyolite and commonly feature CALDERAS formed by explosive eruption and the collapse of a cone into a depleted magma chamber beneath the mountain, as in the case of Crater Lake in Oregon.

andesite line See ANDESITE.

Aniakchak, caldera, Aleutian Islands, Alaska, United States The Aniakchak caldera is located in the eastern Aleutian Island arc near Bristol Bay. Surprise Lake occupies part of the crater. Several cones and necks are found on the floor of the caldera, including Vent Mountain. Powerful explosions occurred at Aniakchak in 1931, possibly from a cinder cone. A dome formed in the vent late in the eruption.

Antarctica Although the historical record of seismic and volcanic activity in Antarctica is not as extensive as the record for more densely settled portions of the world, much is known about earthquake and volcanic activity on the Antarctic continent. The volcano Mount EREBUS was discovered by Captain James Clark Ross of Britain on an expedition to reach the south magnetic pole. Erebus reportedly was erupting at the time of Ross's visit. The volcano and a nearby crater were named Erebus and Terror respectively after the two ships on Ross's expedition. A later expedition under the command of Ernest Shackleton climbed Erebus in the first ascent of a mountain in Antarctica. Other sites of volcanic activity on or near Antarctica include DECEPTION ISLAND, Hampton/Whitney, Takahe, Thule Island and Waesche.

Antioch, Syria The ancient city of Antioch was the site of two notable earthquakes, one in A.D. 115 and the other in A.D. 526. The first earthquake coincided with a celebration in honor of the Roman emperor Trajan, who reportedly had to leave a building by crawling through a window. The latter earthquake occurred on May 20 of 526, again during a great public observance, the festival of the Ascension. More than a quarter of a million persons are said to have been killed in this earthquake, which appears to have destroyed all but a few buildings in Antioch.

Aoba, volcano, Vanuatu A shield volcano, Aoba is basaltic and is situated along a fissure system that has given the island an elongated shape. Although the volcano appears to have grown through outpourings of fluid lava, there is evidence of explosive activity in the island's history as well, signified by pyroclastic materials. Two nested calderas occupy the summit of the island, and a line of SPATTER CONES accompanies the fissure system along its trace from southwest to northeast. Several craters less than a mile in diameter are located at the extreme northeastern and southwestern ends of the island. Dates of recent eruptions are inexact, but effusive and explosive eruptions are thought to have occurred several hundred years ago, possibly involving the inner caldera's collapse. An explosive eruption approximately a century ago cast out large quantities of ash, and lahars reportedly wiped out villages on the southeastern side of the island. Emissions of steam from the summit caldera increased and then subsided in 1971, and fumarole activity may have been responsible for discoloring a lake on the summit in 1971.

Apoyo, caldera, Nicaragua The Apoyo caldera is located in the Nicaraguan Depression near the town of Granada. Lake Apoyo occupies much of the caldera. The collapse of the volcano, forming the caldera, is thought to have occurred following great eruptions that expelled perhaps a third of a cubic mile of magma. Apoyo caldera is noted for a long history of earthquake activity that appears to have started in the 16th century, although some of this earthquake activity may have been tectonic rather than volcanic in origin and involved the whole region, not merely this caldera. Although no actual eruptions have been observed at Apoyo, earthquake swarms, along with changes in the temperature of Lake Apoyo and in its sulfate content, indicate that some of the disturbances at Apoyo are due to volcanic processes. Several domes (El Cerrito, Cerritito, Lomo Poisentepe and Apoyoito) are located near the caldera, as is a line of cinder cones along a fault running roughly along a north-south line immediately to the east of the caldera, between Apoyo and the shore of nearby Lake Nicaragua.

Arabian crustal plate A plate of the crust adjacent to the African plate, the Arabian plate is separated from Africa by the Red Sea and also borders on the Indo-Australian plate and the Iranian plate. The Arabian plate is colliding actively with the Indo-Australian plate. (See also PLATE TECTONICS.)

Arizona, United States Located in a region of moderate seismic risk, the state of Arizona experiences earthquakes that originate within its own territory as well as vibrations from earthquakes centered in neighboring states, notably CALIFORNIA. A very powerful earthquake, estimated at Mercalli intensity VIII–IX, occurred near Fort Yuma on November 9, 1852; fissures opened in the desert along the Colorado River, and the earthquake knocked down parts of Chimney Peak. Shocks were reported on almost a daily basis for months.

There is abundant evidence of volcanic activity in Arizona. One interesting example is Vulcan's Throne, a cinder cone on the northern rim of the Grand Canyon. Output from its eruptions is thought to have blocked the canyon time and again, but on each occasion, the river formed a new channel. The Kitt Peak Observatory near Tucson is built atop a mountain that formed as intrusive igneous rock, and hydrothermal activity in Arizona deposited ores that made the state a major source of copper. The spectacular San Francisco Peaks near Flagstaff are also volcanic in origin, as is nearby Sunset Crater, which is thought to have erupted in the 11th century. Arizona also has one of the most famous IMPACT STRUCTURES on earth, namely Meteor Crater, an impact crater near Flagstaff.

Arkansas, United States Although the state of Arkansas has seldom experienced powerful earthquakes, one series of such earthquakes was the strongest in United States history: the New Madrid, Missouri earthquakes of 1811–1812, which altered the topography of northeastern Arkansas considerably. A less destructive, but nonetheless powerful, earthquake occurred in Arkansas on October 22, 1882; this earthquake was estimated at Mercalli magnitude VI–VII and affected an area of some 135,000 square miles, although the epicenter was difficult to ascertain because reports from the affected area were so few. The October 28, 1923, earthquake at Marked Tree was remarkably strong (Mercalli intensity VII) and affected some 40,000 square miles; the earthquake was felt in Arkansas and in nearby states, caused considerable damage to buildings and disturbed the surface of the St. Francis River. On November 16, 1970, an earthquake of Mercalli intensity VI and Richter magnitude 3.6 in northeastern Arkansas was felt over some 30,000 square miles and resulted in minor damage.

Asamayama, volcano, Japan The volcano Asamayama on Honshu, the central and largest island of Japan, underwent its most famous eruption in 1783 when the volcano cast out large numbers of hot rocks that landed on nearby communities. The eruption is said to have killed about 5,000 people. One rock expelled in this eruption reportedly measured 120 feet by more than 260 feet and formed an island where it landed. Together with the Icelandic volcano SKAPTAR JÖKUL, Asamayama was implicated by Benjamin Franklin in the unusually low temperatures that affected the northern hemisphere that year and in a curious "dry fog" seen hanging over the land. Ash cast out by the two volcanoes may have been responsible for the apparent fog and the drop in temperature.

Asawa, volcanic complex, Ethiopia The Asawa volcanic complex is located in the central part of the Main Ethiopian Rift valley near Lakes Abaya and Shalla. Also known as Asawa/Corbetti, the Asawa complex includes the CALDERAS Aluto, Awasa, Corbetti, Duguna, Gadamsa, Gademota, Hobicha, Shalla and Wonchi and the Wagebeta caldera complex. The area is characterized by flows of basalt, obsidian and scoria, as well as by layered pumice and lava domes. The Asawi caldera adjoins the Corbetti caldera, which is thought to have formed through the eruption of large amounts of material from fissure vents. The historical record of activity at Asawa/Corbetti is brief. There are reports of eruptions dating back to the early 20th century, but the accuracy of these reports is questioned. In recent years FUMAROLES have been active here. In 1984, earthquakes damaged buildings and caused the evacuation of a school.

Ascension Island Ascension Island is the summit of a volcanic mountain along the mid-Atlantic Ridge approximately midway between Africa and South America.

ash Ash is finely divided solid material ejected from a volcanic eruption. Volcanic ash differs in color and composition but is usually gray. Four eruption processes give rise to volcanic ash. One is magmatic. In this process, gas bubbles form in magma as pressure on the molten rock diminishes on its way to the surface. The bubble-filled magma then fragments in the vent of the volcano and is expelled as finely divided solid material. In the second process, the hydrovolcanic process, magma mixes with ground water or surface water in an explosive manner. The phreatic process involves fast expansion of steam and/or hot water and fragments of COUNTRY ROCK. The fourth process is abrasion, which occurs when grains of ash collide with one another. The shape of ash particles depends on the conditions in which they were formed. Where large bubbles form in the vent of the volcano during decompression of magma, for example, resulting bits of ash may occur as thin sheets of glass formed when the bubbles solidified and broke apart. Hollow "needles" may be found in ash where the flow of magma within the vent elongated gas bubbles in the molten rock.

Many ash particles fall out of the air soon after being ejected from the volcano that gave rise to them, but extremely fine ash may rise into the upper atmosphere and block incoming sunlight, causing a drop in surface temperatures. Another observed effect of volcanic ash in the upper atmosphere is extremely colorful sunsets, caused by the high-level ash cloud's tendency to block short wavelengths of solar radiation and let through only the longer wavelengths, notably orange and red. Such vivid sunsets followed the 1883 eruption of the volcano Krakatoa, for example.

Ash in the upper atmosphere may remain there for years. In the lower atmosphere, ash from eruptions may fall and cover the land or sea in layers many feet thick. One of the most famous ashfalls, from Vesuvius in A.D. 79, preserved the entire cities of Pompeii and Herculaneum and hid them from discovery for some 18 centuries. Archeologists exploring the buried cities found curious cavities, or lacunae, in the ash. These lacunae turned out to be the preserved outlines of persons killed by the eruption. The ash buried them where they fell, and the lacunae remained after their bodies decayed. Plaster casts have been made of some of these lacunae and offer a striking glimpse of the human aspect of the two cities' destruction. The 1991 eruption of Mount Pinatubo in the Philippine Islands deposited so much ash on nearby Clark Air Base that buildings collapsed. During an eruption, ash clouds may interfere with the navigation of aircraft. Pilots flying near Mount St. Helens in Washington State during its 1980 eruptions reported that very fine ash from the volcano made its way into their aircraft. Airborne ash also caused severe damage to an airliner flying through the cloud from an eruption of Alaska's Redoubt volcano in 1989. (See also AVIATION AND VOLCANOES; ISOLATION; TEPHRA; "YEAR WITHOUT A SUMMER.")

ash-flow eruptions In an ash-flow eruption, a NUÉE ARDENTE or similar phenomenon lays down a deposit of very hot ash, which may range in thickness from only several feet to perhaps a thousand feet or more. In portions of the western United States, particular ash-flow deposits may extend for 100 miles. Material from these eruptions covers large areas of the western United States, Mexico, New Zealand and other parts of the world. Deeper levels of such a deposit may become denser and resemble lava-like rock, which may include pieces of obsidian. This increase in density is thought to result from intense heat acting on the deeply buried layers of material. The heat fuses the ash particles together and in some places turns them into obsidian, or volcanic glass.

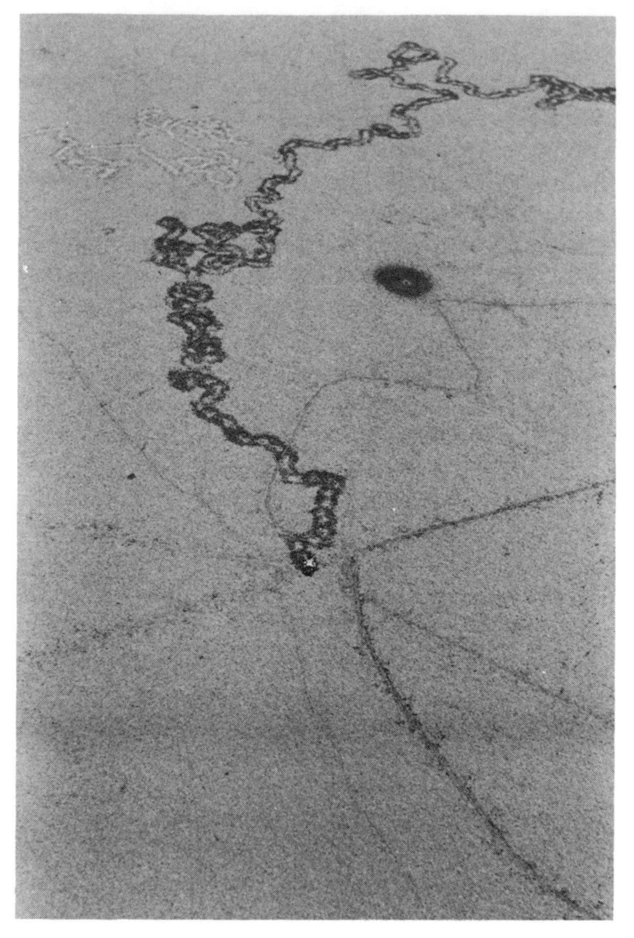

Beetle on volcanic ash, Cerro Negro, Nicaragua, 1947. Note beetle's trail (R. E. Wilcox, USGS)

Ash-flow eruptions commonly lay down deposits very quickly, even over wide areas. The rapidity of this process is reflected in the fusing of ash particles, which were deposited so quickly that the initial heat had little or no chance to dissipate. In Alaska, an eruption of Mount Katmai produced in less than 24 hours the VALLEY OF TEN THOUSAND SMOKES, a famous ash-flow deposit that endured for decades as a plain of fumarole. A similar phenomenon occurred during the violent eruption of Bezymianny volcano in Russia. That eruption produced an ash-flow deposit later named Valley of Ten Thousand Smokes of Kamchatka.

Askja, caldera, Iceland Prehistoric in origin and located in the rift zone that runs through the middle of Iceland, the Askja CALDERA was the site of an eruption in 1875 that created the Oskjuvatn caldera. The Askja caldera itself appears to have formed when magma moved underground into a nearby fissure swarm. A comparable set of circumstances is thought to have produced the Oskjuvatn caldera, which is notable for having emerged during an episode of rifting. A major volcanic eruption took place in 1874 and 1875 along a fracture zone about 60 miles north-northwest of the Askja caldera. These events were preceded and accompanied by earthquakes, including one episode that shook all of northern Iceland continually for the last week of December 1874. Eruptions (both effusive and explosive) began at Askja on January 1, 1875. A violent eruption took place at Askja on March 28 and 29. Eruptions continued through April at Askja and through October at the fissure swarm to the north. Magma moved underground from the vicinity of Askja toward the fissure swarm, and the collapse of Oskjuvatn caldera started in February 1875, before the great eruption of March 28–29. The collapse was largely completed by July.

There may have been a minor eruption in 1919, but the only sign of it is a layer of tephra in an ice core. Lava flowed from the rim of Oskjuvatn between 1921 and 1926, and around 1929 (the exact date has not been determined), a fissure eruption took place to the south of Askja. An eruption in 1961 was heralded by earthquakes and began with emergence of geysers and solfataras at Askja in early October, along with numerous fissures on the floor of the caldera. Hot water flowed from the main fissure. Very energetic geyser activity occurred for several days in mid-October. In an eruption beginning on October 26, basaltic lava emanated from a fissure running east to west across the floor of Askja caldera. The water level in a caldera lake dropped several feet, and fissures formed in the walls of the caldera.

Aso, caldera, Japan The Aso caldera is located on the southern Japanese island of Kyushu and is situated on the Oita-Kumamoto fault zone, which runs northeast to southwest through the center of Kyushu. Several great eruptions of magma from Aso are thought to have occurred in prehistoric times, and one of these eruptions appears to have left a layer of ash all over Japan. The Nakadake volcano within the caldera has been active within historical times. Activity at Nakadake has not been confined to one area of the crater but rather has migrated over the last few decades. Lava emanating from Nakadake is basaltic ANDESITE. Several vents occupy the caldera. Hot springs also occur there.

Aso caldera is noted for the earthquake activity associated with eruptions there. There is evidence from studies of S-waves and P-waves that magma is present in large amounts under the eastern and central portions of the caldera, several miles down.

The Aso caldera has exhibited frequent seismic activity in the latter half of the 20th century. Several types of earthquakes have been observed and associated with eruptive activity at the caldera. One type is characterized by surface waves with periods of about one second and is thought to be produced by internal volcanic gas explosions. A second type of earthquake is characterized by surface waves lasting for slightly longer periods (up to eight seconds), resulting from vibration by the magma chamber and possibly also from explosions. The third type consists of body waves averaging about half a second long and produced by explosions in the magma during eruptions. The fourth type, associated with periods of eruptive activity, is characterized by body waves with a period of approximately one-fifth of a second. (During one eruption of Nakadake in 1958, vibrations lasting a much longer period, 40 seconds or longer, were noted for two days before the eruption but ceased about an hour before the eruption began.

Tilt or inclination measurements of the ground conducted since 1931 at Aso caldera have provided information that has helped scientists understand the relationship between tilt and eruptive activity. Measurements made in the 1950s revealed inflation for some months before major eruptions, and deflation following eruptions. Tilt measurements made in the middle 1960s also showed inflation occurring during a period of eruptive activity. Temperature changes and gas emissions have also been studied as possible precursors of eruptions in the Aso caldera. Temperatures in the crater of Nakadake, monitored in the late 1950s, went up sharply in the days just prior to an eruption, and the level of dissolved carbon dioxide in hot springs in the vicinity rose for several months before an eruption in 1979. Temperature measurements and fluctuations in water level in a pond in Nakadake's crater were used to anticipate explosive activity and emissions of ash that began in late 1984 and continued through early 1985. There also appears to be a relationship between emissions of sulfur dioxide and eruptive activity at Aso caldera: Sulfur dioxide output rises before and during periods of explosive activity but then diminishes as explosions subside. (See also SEISMOLOGY.)

Assam, India One of the most powerful earthquakes ever recorded, the Assam earthquake of 1897 killed more than 1,000 people. A similarly powerful earthquake on August 15, 1950, at Assam reportedly killed more than 1,000 people and demolished about 2,000 homes. Damage estimates exceeded $25 million. The earthquake reportedly continued over five days and was accompanied by geyser action, the appearance of large fissures and landslides that blocked rivers. The natural dams created by these slides caused water to collect behind them; when the dams eventually failed, they released the water as floods that caused considerable destruction downstream.

asthenosphere The layer of the earth's mantle immediately below the crust. (See also EARTH, INTERNAL STRUCTURE OF; PLATE TECTONICS.)

Ata, caldera, Japan The Ata caldera is located in southern Japan near the end of the Satsuma Peninsula and under the waters of Kagoshima Bay. The caldera extends from the tip of the Satsuma Peninsula to the tip of the Osumi Peninsula several miles across the water. This has been an area of intense volcanic activity in recent centuries, notably at SAKURA-ZIMA volcano in the AIRA caldera and at Kaimon-dake in the nearby Ibusuki volcanic field. The small Ikeda and Yamakawa calderas are thought to be nested inside Ata. Strong eruptions occurred in the first and ninth centuries. Pronounced earthquake activity occurred in the vicinity of Ata caldera in the late 1960s and 1970.

Atitlán, caldera, Guatemala The Atitlán caldera is believed to have formed during an eruption of several cubic miles of magma. Lake Atitlán occupies a portion of the caldera. Three stratovolcanoes—Atitlán, San Pedro and Toliman—have formed in the caldera since its origin, along the caldera's southern edge. Atitlán volcano reportedly erupted in 1469, then again in 1717 and 1721. Eruptions continued at intervals of several years through the first half of the 19th century. Of these eruptions, only one was strong, in May of 1853. Activity at Atitlán volcano subsided following an eruption in 1856.

Atlantic Ocean Although earthquake and volcanic activity is less pronounced in the Atlantic Ocean basin than in and around the PACIFIC OCEAN, the Atlantic contains numerous features of earthquakes and volcanism. Most prominent on a physiographic map of the Atlantic basin is the MID-OCEAN RIDGE that runs down the middle of the Atlantic, roughly equidistant from the Americas and from Europe and Africa. The AZORES, a group of islands off the western coast of Africa, consist of several mountains along this ridge, which was discovered during World War II by military ships making depth measurements. ICELAND is another island on the Atlantic mid-ocean ridge. Exploration of the Atlantic mid-ocean ridge in the late 1970s revealed the existence of hydrothermal vents and large colonies of animals living in the vicinity of the vents. Among the animals found there were mussels with shells approximately one foot long and polychaete worms up to six feet long. Numerous SEAMOUNTS are found in the Atlantic Ocean. Bermuda occupies the summit of one seamount.

Atlantis A volcanic eruption in the Mediterranean may have had something to do with the origin of the Atlantis legend. The story of Atlantis, as related by Plato in writings dated around 400 B.C., concerns a continent that disappeared beneath the sea within 24 hours, taking with it an advanced civilization. Plato attributes the story to Critas, an Athenian politician. Critas in turn heard the story from his father, who was a friend of Solon, who lived in the sixth and seventh centuries B.C. and is considered the founder of democracy in Athens. In a period of exile, when his political career was at a low, Solon visited Egypt and heard there the story of a gigantic island, located somewhere west of the Straits of Gibraltar. The island was called Atlantis, and (according to legend) it sank beneath the sea in a single day and night after an alliance led by the Athenians had defeated the Atlanteans in battle some 9,000 years ago. The legend of Atlantis has exerted a powerful fascination on the Western imagination, and there have been efforts to locate an actual land or natural catastrophe that might account for the Atlantis story. Eventually, speculation focused on the island of Santorini (Thera) in the Aegean Sea. Santorini once had been the hub of the advanced Minoan culture, which vanished suddenly and mysteriously from the Mediterranean in approximately 1400 B.C. It is now widely presumed that the Minoan culture perished in an eruption that resulted in the collapse of the volcano and the formation of a caldera, accompanied by a tsunami that emanated from the island and spread destruction through nearby portions of the Mediterranean basin. The destruction of Thera and Minoan civilization appears to have been recorded, in an exaggerated form, as the legend of Atlantis.

Augustine volcano, Alaska, United States An island stratovolcano in the Cook Inlet area of southern Alaska, Augustine has undergone explosive eruptions on several occasions since the early 19th century. The volcano has the potential to produce economically devastating ashfalls and/or tsunamis.

Australia Unlike its neighbor New Zealand, Australia is not especially noted for earthquake activity in modern times, although several powerful earthquakes have occurred there in the past century, including quakes in Adelaide in 1897 and 1954, southeastern Queensland in 1918, New South Wales in 1961 and Victoria in 1966.

avalanche A large mass of unconsolidated material, such as rock, soil, ice or snow, falling because of the influence of gravity. Earthquakes may set off avalanches, which are especially hazardous in seismically active areas where population centers are located close to mountains.

aviation and volcanoes Although volcanic eruptions are infrequent compared to other phenomena that pose dangers to aircraft, such as thunderstorms, clouds of ash from volcanoes have the potential to do tremendous damage both to aircraft on the ground and to those in flight. This danger is especially great to aircraft in flight because clouds of volcanic ash do not show up on airplane radar—a result of the limitations on the sensitivity and power/aperture of these radars. The fine airborne ash can cause a wide variety of damage to aircraft that encounter it in flight. Abrasive action from the ash can damage engines, landing lights, control surfaces and windows and windshields. The windshield of a jet aircraft passing through an ash cloud may become opaque. Jet engines may cease operating, leaving an aircraft in a powerless descent. Damage to a single aircraft may amount to tens of millions of dollars and leave the airplane unusable without extensive repair.

A well-documented case of aircraft damage from a volcanic ash cloud occurred during the 1989 eruption of Redoubt volcano in Alaska. A Boeing 747-400 aircraft entered the cloud from Redoubt at approximately 26,000 feet while descending for landing at Anchorage. The aircrew tried at once to gain altitude and escape the cloud, but all four engines died after climbing only some 3,000 feet. Compressor erosion and other damage were considerable. Ash that melted and resolidified on the stage one turbine nozzle guide vanes was found later to be the primary cause of loss of engine thrust. Eight minutes after the engines stalled, after the airplane had descended approximately 13,000 feet, the aircrew succeeded in restarting the engines. All engines, nose cowls and thrust reverses had to be replaced. The sandblasting effect of passing through the ash cloud caused heavy damage to the pilots' windshields,

This 700-ton block of rock was carried along and deposited by an avalanche in Peru in 1970 (USGS)

some cabin windows, and landing light covers and required their removal and replacement. Sandblasting also forced the removal and replacement of leading edges on the wing, stabilizer and vertical fin. Ash contaminated the whole interior of the aircraft, so that all seats, side walls and other interior furnishings had to be taken out and either cleaned or replaced. Contamination from ash extended to the electronic equipment. The fuel, oil, hydraulic and drinking-water systems also were contaminated and required draining and cleaning. Total damage to the aircraft exceeded $80 million.

Before this incident, other aircraft had encountered clouds of volcanic ash over Alaska on several occasions and undergone various degrees of damage. The 1986 eruption of Augustine volcano had extensive effects on aircraft operation in the vicinity of Cook Inlet. Air carrier service at Anchorage all but ceased during the eruption, and the U.S. Air Force moved most of its aircraft out of Anchorage for several days. No disabling incidents involving aircraft were reported during this eruption. During an earlier eruption of Augustine, in 1976, two F-4E Phantom jets passed through the volcano's ash plume and had their canopies scoured. Sandblasting removed some paint from the wings, and fine material penetrated many portions of the planes' interiors. A DC-8 passing through the ash cloud on the way to Tokyo had its center windshield scoured, and the windshield had to be replaced. Large amounts of ash stuck to the plane, and some abrasion was reported on landing gear and other external parts of the aircraft. Two other passenger aircraft had ash adhering to them, but damage was less extensive than to the DC-8. A brief eruption of Mount Spurr near Anchorage in 1953 resulted in sandblasting damage to three aircraft that flew through the ash plume. In this eruption, the Air Force's 5039th Air Transport Squadron evacuated more than 20 of its aircraft to Laird and Eielson Air Force Bases. Three big military cargo aircraft that remained on the ground, exposed to the ashfall, needed 10 days to be cleaned of ash.

Other eruptive activity in various parts of the world has provided further information on airborne ash and its effect on aircraft. Eruptions of the volcano Galung Gung in Indonesia in 1982 affected two aircraft that flew above the volcano on June 24 and July 13. The planes had to make emergency landings at Jakarta. The 1986 eruption of the Lascar volcano in Chile also had a bearing on aviation safety, although no incidents involving aircraft were reported. The eruption of Lascar occurred at a remote location and lasted only several minutes, but its ash plume traveled quickly over populated areas served by commercial air travel and illustrated how even a largely ignored volcano such as Lascar (it was not considered dangerous before its eruption) could pose hazards to air travel.

The Boeing Commercial Airplane Group has reported that flights have observed many peculiar conditions during flights through volcanic ash, including heavy discharges of static, a glow in the engine inlets, and false cargo fire warnings. Ash also may enter the cockpit, accompanied by an odor of acid.

Azores The Azores island group off the western coast of Africa includes several volcanoes, notably AGUA DE PAU, FURNAS and SETE CIDADES. Five of the volcanoes have undergone eruptions within historical times, and submarine eruptions have been numerous. Minor eruptions occurred at the eastern end of the Azores in 1638 and 1881 but did not form any lasting additions to the island chain. (See also HOT SPOT.)

B

Baikal, Lake, Russia Lake Baikal, in the seismically active southeastern corner of Russia, has been the site of many notable earthquakes, including events recorded in 1828, 1839, 1862, 1869, 1871 and 1959. Many extinct volcanoes are located in the vicinity of the lake. A rift passing through Lake Baikal has been linked tentatively to an oceanic ridge passing into the continental landmass of eastern Asia. Earthquakes around Lake Baikal are mostly restricted to a narrow zone around the rift.

Baja California, Mexico The elongated strip of land running approximately north to south along the western coast of MEXICO, Baja California occupies a small plate of crustal rock that is pushing northward into California. This movement is part of the tectonic environment that gives rise to frequent strong earthquakes in southern California, as the northward-moving plate strikes and grinds against the deeply rooted mass of the Sierra Nevada. (See also LOS ANGELES.)

Baker, Mount, Washington, United States One of the peaks of the volcanic Cascade Mountains, Mount Baker has been active on numerous occasions over the past few centuries. Many eruptions were observed in the last 200 years, notably 1792, 1843 and 1880. The 1843 eruption deposited ash over a large area. Although Mount Baker appears to have emitted lava in the recent geologic past, it is considered more likely to expel ash and fragments of rock. Mount Baker stands just north of Seattle and is heavily glaciated. It was the site of an interesting phenomenon in 1975: the rapid formation of a crater lake. The lake appeared as a result of increased activity at Sherman Crater, a vent several hundred feet from the mountain's summit. Ice in the crater broke apart and melted, forming a lake more than 150 feet wide and 200 feet long. The lake is thought to have been buried under an avalanche of ice in 1977.

Banda Api, caldera, Banda Sea, Indonesia Banda Api is located in the Banda Sea near the islands of Buru and Ceram and is part of the Banda group, which includes 10 islands. Two of these islands (Banda Besar and Pisang) are situated along the rim of the Banda Api caldera. Another island, Neira, stands on the eastern rim of a smaller caldera nested inside the one marked by Banda Besar and Pisang. Banda Api appears to have formed in much the same pattern as Krakatoa, with two or more episodes of caldera-building, each one involving the growth and destruction of a stratovolcano.

Banda Api has been active frequently in the 19th and 20th centuries. One frightening eruption occurred without warning in June of 1820 and scared the inhabitants into fleeing by boat. Clouds of ash and smoke emanated from the volcano in 1824 during the formation of a new crater. Sulfur vapor from the volcano caused a thick mist to form in late 1835, and a subsequent strong earthquake was followed by still more shocks that continued for several days.

Less powerful earthquakes occurred in late 1853, and the volcano put out unusually large amounts of steam. Noises from underground were heard in December of 1855, and a minor earthquake occurred in January of 1856. More subterranean noises were heard later that month. The volcano reportedly emitted large amounts of smoke during these times of rumbling. Five earthquakes, one of them accompanied by a tsunami, occurred in June and July of 1859. Earthquakes continued in August. One especially strong earthquake on September 25 was associated with a dramatic rise in the sea along the southern coast of Neira, although this disturbance of the sea appears to have caused no significant damage. More strong earthquakes occurred on October 18, November 7–8 and December 29. In 1859 and 1860, Banda Api emitted unusually large amounts of smoke.

Several notable earthquakes occurred in 1860, including one that was accompanied by a strange underground roaring noise that reportedly appeared to move from southeast to northeast. Neira experienced frequent earthquakes between early January and late June of 1877, although the volcano did not erupt. A fairly strong earthquake in 1887 that appeared to be centered near Banda Api was followed by an increase in steam from the volcano, although no such rise in activity was observed in connection with other, strong shocks that followed several months later. A very strong earthquake with pronounced subterranean rumbling shook Banda Api on August 12, 1890, followed by an even stronger earthquake on November 23. Aftershocks continued for days. Damage to buildings was widespread. A fresh active crater with three fissures emanating from it appeared at this time.

Earthquakes and detonations occurred on May 18, 1901, and a glow was seen near the summit of Gunungapi. A minor earthquake occurred in March 1902, and more subterranean rumbling was heard. A strong earthquake in June of 1987, approximately 90 miles southeast of Banda Api, preceded the release of a steam plume from the volcano. Earthquake activity increased dramatically in early May before an eruption on May 9.

Bandai-San, volcano, Japan The July 15, 1888, eruption of Bandai-San is one of the worst volcanic disasters in Japanese history. Although the mountain had displayed no signs of activity for perhaps a thousand years beforehand, Bandai-San blew up in a series of approximately 20 steam-blast explosions in a single day. Most of the destruction occurred within several minutes, when the most powerful explosion set off a huge landslide that sent the entire north slope of the mountain rolling into the Nagase Valley. More than 400 people were reported killed, but only 116 bodies were found. According to one estimate, the eruption expelled some 1,600,000,000 cubic yards of solid material into the valley below the volcano. One volcanologist put this volume of material into more easily understandable terms when he explained that if the material from this eruption were formed into ships of 15,000 tons each, the resulting line of ships would span the Pacific Ocean from Japan to California. This eruption was one of several (including BOGOSLOV and VULCANO) that occurred about the same time and were implicated in a sharp drop in the amount of solar radiation reaching Earth in 1890–91. (See CLIMATE, VOLCANOES AND.)

Bandelier Tuff, southwestern United States The Bandelier Tuff in New Mexico is made up of volcanic ash deposited over a wide area by an eruption of the Valles caldera.

Bardarbunga, volcano, Iceland Buried under the ice of the Vatnajökull, Bardarbunga and its surrounding area are thought to have been the site of several notable eruptions since the late ninth century. A large eruption occurred along fissures to the southwest of Bardarbunga around 900 and again around 1480. Layers of TEPHRA have been associated with eruptions in 1477 and 1717. An eruption from a fissure some 20 to 25 miles southwest of Bardarbunga in the 1860s released a substantial lava flow and apparently some tephra as well. Earthquake activity around Bardarbunga showed a marked increase in the mid-1970s. Between 1974 and 1986, several moderately strong earthquakes occurred. Activity at Bardarbunga may be related to activity at KRAFLA, some 70 miles north of Bardarbunga. It has been suggested that inflation at Krafla accompanies deflation at Bardarbunga, and that changes in pressure are transmitted through a layer of partly molten subcrustal rock.

Barrier, The, caldera, Ethiopia The Barrier is located south of Lake Turkana and separates the lake from the Sugata Valley to the south. The Barrier has an outer and an inner, younger caldera. Although there are reports of eruptions in the mid- to late-19th century and in 1921–22, details are unavailable.

basalt A dark volcanic rock, basalt is widespread on the earth's surface (it forms much of the floor of the Pacific Ocean basin) and varies greatly in chemical composition. "Plateau basalts" are found where large quantities of basalt have poured from fissures, sweeping over the surface and overwhelming landforms in their path. Plateau basalts may contain hundreds of thousands of cubic miles of rock. Examples of plateau basalts are found in the Pacific Northwest of the United States, specifically the Columbia River Plateau and Snake River Plain. Other examples of plateau basalts are found in India's Deccan Plateau and along Manchuria's borders with Korea and Russia, notably around the Baekdoo-San volcano on the Korean border. Yet

Columnar jointing is often seen in basalt (H. E. Malde, USGS)

another plateau basalt, possibly made up of several plateaus, stretches from Scotland to Greenland. Plateau basalts are noted for their absence of pyroclastic material. Individual lava flows in a plateau may be up to approximately 150 feet thick and may extend for 10 miles or more. The geological record indicates these flows were not accompanied by explosive eruptive activity. Basalt sometimes exhibits a phenomenon called "columnar jointing," in which the rock forms a vertical prismatic structure reminiscent in some ways of a honeycomb. A spectacular example of a basalt formation exhibiting columnar jointing is Giant's Causeway in the British Isles. Basalt is believed to form from magma originating in the upper mantle. Several processes have been suggested as the principal cause of melting in the upper mantle that gives rise to basaltic magma, including heat from radioactivity and a cooling and fractionating process associated with convective activity in the mantle.

basal wreck The bowl-like structure formed when the peak of a volcano is destroyed in an eruption. A basal wreck is characterized by a roughly circular shape, a raised rim and a central depression.

basin and range province, western United States The basin and range province in the United States extends from the vicinity of Klamath Falls, Oregon and Bear Lake, Idaho southward through Nevada and Utah and occupies large portions of southeastern California, southern Arizona and southwestern and central New Mexico. The province covers some 300,000 square miles, or roughly 8% of the United States. More than 200 mountain ranges occupy the basin and range province, which shows widespread evidence of recent faulting (notably along the Wasatch Mountains in Utah, at the base of which one finds fresh exposures of unweathered rock indicating movement along a fault there in the very recent geologic past).

The basin and range province consists of a collection of parallel mountain ranges running north-south. The ranges enclose and are separated by long depressions. Geologists believe the basin and range province represents a region of thinning and splitting of the earth's crust, where crustal rock has split into many blocks, some of which rose through motion along faults to become mountains, while other blocks sank and became depressions.

The basin and range province is noted for strong earthquake activity. A recent example was the Dixie Valley, Nevada earthquake of 1952 (magnitude 7.1). One of the greatest earthquakes recorded in the basin and range province occurred in 1872 and involved more than 20 feet of vertical displacement between Owens Valley and the Sierra Nevada. Seismic activity is generally most intense along the eastern and western edges of the province. The powerful Kern County, California earthquake of 1952 took place just beyond the western boundary of the basin and range province, and the Owens Valley earthquake of 1872 took place on that boundary at the Southern Sierra Gap. On the opposite boundary, a 1959 earthquake near Hegben Lake, Montana, had a magnitude of 7.1 and was felt over an area of more than half a million square miles. This earthquake set off an avalanche that dammed the Madison River, formed a lake and killed more than 20 persons. Another earthquake of approximately the same magnitude struck Idaho in 1983, caused two deaths and resulted in more than $12 million in damage to property. Other strong earthquakes in the basin and range province affected Nevada in 1869, 1903, 1915, 1932, 1934 and 1954. The GREAT BASIN, which includes DEATH VALLEY, is part of the basin and range province. Volcanic activity also occurs along the boundaries of the province, a notable example being the LONG VALLEY CALDERA.

Evidence of volcanism is also abundant within the basin and range province. Lavas and other products of eruptions have buried older rock along the northern edge of the Great Basin. Volcanic rocks extending across the Great Basin include huge volumes of tuff, or pyroclastic material welded together. Large flows of basalt are found on the northwestern and southeastern edges of the Great Basin, along the COLUMBIA PLATEAU and SNAKE RIVER PLAIN and the southern edge of the Colorado Plateau respectively. Ubehebe Crater is a famous product of relatively recent volcanic activity in the basin and range province.

Mines and mining have played an important part in the development of the basin and range province during the past two centuries. Copper, gold, silver, lead, zinc, antimony, iron, mercury, magnesium, manganese and beryllium have all been mined in large quantities from deposits in the basin and range province. (See also NORTH AMERICA; NORTH AMERICAN CRUSTAL PLATE.)

batholith A very large pluton (body of intrusive igneous rock) that may have great economic importance when associated with valuable ore deposits. Batholiths are prominent in the landscapes of the western United States and Canada. The northern Rocky Mountains contain tremendous batholiths, but it is uncertain what combination of processes created these batholiths. The Idaho Batholith, roughly 250 miles long and 100 miles across at its broadest point, contains the Salmon River and Clearwater mountains. The Bitterroot Range along the eastern edge of the batholith runs along the border between Montana and Idaho. A break along the northern part of the Idaho Batholith contains the Coeur d'Alene district, famous for its lead, zinc and silver mines. The nearby Boulder Batholith in Montana is also a rich source of metals and has yielded about a third of the copper mined in the United States. Granite batholiths form large portions of the northern CASCADE MOUNTAINS, the Coast Mountains in British Columbia, southern Alaska and the SIERRA NEVADA. The granite in these batholiths may be very coarse-grained, indicating that the magma which formed the rock cooled slowly and at great depth. (Fine-grained igneous rocks generally cooled more rapidly, at or near the surface, so that coarse grains did not have an opportunity to develop.) The chemical composition of these granitic rocks may vary. Sections of COUNTRY ROCK may separate individual plutons, and entire blocks of country rock may be included with plutons, having become separated from the walls or roof of a pluton and fixed in the intruding magma.

Batur, caldera, Bali, Indonesia Batur is situated on the island of Bali and undergoes frequent explosive eruptions. Two calderas, one inner and one outer, have been described at Batur. Inside the inner caldera, a small stratovolcano (also called Batur) has arisen and is active. Dark clouds were seen emerging from Batur in a minor eruption in 1854. In 1888, numerous earthquakes were felt in Bali in late May, days before an eruption of Batur. The Batur volcano also may have been involved in detonations and earthquakes perceived from October 9 to October 11 of 1902. A lava flow from Batur invaded a local village in 1904 or 1905. Although strong earthquakes occurred in the vicinity of Batur in 1917, these are thought to have been tectonic rather than volcanic in origin. Batur erupted again from 1921 to 1926. (It is not certain whether this resumption of eruptive activity was related to the earthquakes of 1917.) The volcano expelled incandescent pieces of rock, and detonations shook the ground. There was a brief diminution in eruptions at the summit of the volcano as lava started flowing on April 9 and 10. Powerful explosions began again, however, in mid-April. Lesser eruptions took place between 1922 and 1925, followed by a more powerful eruption with accompanying earthquakes in 1926. Lava emerged from a fissure on the southwestern side of Batur, and soon afterward another fissure opened on the southeast side of the volcano, releasing a lava flow that overran a nearby village. Lava tubes formed during the 1926 eruption. In 1959 emissions of hydrogen sulfide increased from the northern part of the caldera lake. Small earthquakes were observed on the rim of the caldera during the eruption of AGUNG volcano, several miles southeast of Batur, on September 5, 1963. Batur started erupting later that day. Lava flowed from several vents on the west side of Batur during this eruption. Minor explosions occurred every few days at Batur between August and December of 1965. Explosions occurred again on a frequent basis in early 1968, and lava flowed from Batur. Bubbles of gas, smelling of sulfur, rose from the lake in 1969, and small explosions took place in 1970.

Bayonnaise, caldera, Japan A mostly submerged caldera in southern Japan, Bayonnaise has a record of submarine eruptions dating back to 1896. An eruption of the volcano Myozin-Syo, a lava dome in the middle of the caldera, in 1952 destroyed a research ship that crossed over a vent and killed more than 30 people on board.

Benioff zone A zone of deep earthquake activity associated with a SUBDUCTION ZONE, a Benioff zone extends perhaps 400 miles below the surface and is named after Hugo Benioff (1899–1968), an American seismologist who mapped such zones in the 1940s and 1950s.

Bermuda The island of Bermuda occupies the summit of a volcanic seamount that rises some 6,000 feet from the ocean floor. At one time, waters covered the peak, and coral formed on the summit. The coral now constitutes the surface of the island. A well-drilling project showed that almost 400 feet of limestone overlie the basaltic rocks that make up the volcanic portion of the seamount.

Bezymianny, volcano, Kamchatka, Russia Part of the KLYUCHEVSKAYA complex of volcanoes, Bezymianny erupted with great violence in 1955–56. In many ways, this eruption closely resembled that of Mount ST. HELENS in Washington in the United States in 1980. On March 30, 1956, a tremendous explosion destroyed the summit of Bezymianny and produced a crater one mile by one and one-fourth miles wide. The eastern side of the mountain was blown open. A lava dome formed in the crater. The March 30 eruption alone is believed to have cast out about 2.4 billion tons of rock. An ash flow laid down by the volcano produced an area of fumaroles similar to the VALLEY OF TEN THOUSAND SMOKES created by the eruption of Mount Katmai in Alaska; this site came to be known as The Valley of Ten Thousand Smokes of Kamchatka.

Bible, earthquakes in the The Bible contains numerous references to earthquakes, in both the Old and New Testaments. In the first book of Kings (I Kings 19:11–12), the prophet Elijah experienced an earthquake:

> . . . And, behold, the Lord passed by, and a great and strong wind rent the mountains, and brake in pieces the rocks before the Lord; but the Lord was not in the wind; and after the wind an earthquake; but the Lord was not in the earthquake: and after the earthquake a fire; but the Lord was not in the fire . . .

Isaiah also described God's power in terms of an earthquake (Isaiah 29:6):

> Thou shalt be visited of the LORD of hosts with thunder, and with earthquake, and great noise, with storm and tempest, and the flame of devouring fire.

Amos, who lived around 750 B.C., used a famous earthquake to set the date of his prophecies (Amos 1:1):

> The words of Amos, who was among the herdmen of Tekoa, which he saw concerning Israel in the days of Uzziah king of Judah, and in the days of Jeroboam the son of Joash king of Israel, two years before the earthquake.

This same earthquake, apparently, is mentioned in Zechariah 14:5:

> And ye shall flee to the valley of the mountains; for the valley of the mountains shall reach unto Azal; yea, ye shall flee, like as ye fled from before the earthquake in the days of Uzziah king of Judah: and the Lord my God shall come, and all the saints with thee.

Matthew wrote that Christ predicted His return would be preceded by earthquakes (Matthew 24:7):

> For nation shall rise up against nation, and kingdom against kingdom: and there shall be famines, and pestilences, and earthquakes, in divers places.

Mark repeated this same warning in similar language (Mark 13:8):

> For nation shall rise against nation, and kingdom against kingdom: and there shall be earthquakes in divers places, and there shall be famines and troubles: these are the beginnings of sorrows.

Luke's version of this prophecy (Luke 21:11) reads as follows:

> And great earthquakes shall be in divers places, and famines, and pestilences; and fearful sights and great signs shall there be from heaven.

Matthew also reported that an earthquake accompanied the crucifixion of Christ and greatly impressed onlookers (Matthew 27:54):

> Now when the centurion, and they that were with him, watching Jesus, saw the earthquake, and those things that were done, they feared greatly, saying, Truly this was the Son of God.

Matthew added (Matthew 28:2) that another earthquake occurred when the stone was rolled away, by angelic action, from Christ's tomb:

> And behold, there was a great earthquake: for the angel of the Lord descended from heaven,

and came and rolled back the stone from the door, and sat upon it.

In Acts 16:26, an earthquake released the apostles Paul and Silas from prison:

And suddenly there was a great earthquake, so that the foundations of the prison were shaken: and immediately all the doors were opened, and every one's hands were loosed.

Revelation, the final book of the New Testament, abounds in references to earthquakes in connection with prophecies of the final days before Christ's return. The first-person narrator was the author, the apostle John:

And I beheld when he had opened the sixth seal, and, lo, there was a great earthquake; and the sun became black as sackcloth of hair, and the moon became as blood. *[6:12]*

And the angel took the censer, and filled it with the fire of the altar, and cast it into the earth: and there were voices, and thunderings, and lightnings, and an earthquake. *[8:5]*

And the same hour was there a great earthquake, and the tenth part of the city fell, and in the earthquake were slain of men seven thousand; and the remnant were affrighted, and gave glory to the God of heaven. *[11:13]*

And the temple of God was opened in heaven, and there was seen in his temple the ark of his testament; and there were lightnings, and voices, and thunderings, and an earthquake, and great hail. *[11:19]*

And there were voices, and thunders, and lightnings; and there was a great earthquake, such as was not since men were upon the earth, so mighty an earthquake, and so great. *[16:18]*

Bishop Tuff, western United States A pyroclastic flow, the Bishop Tuff covers almost 600 square miles of Nevada and central California. The tuff is thought to have been laid down in the eruptions associated with the formation of the LONG VALLEY CALDERA in California. The Bishop Tuff extends through Arizona, Colorado and Utah and has been identified as far east as Kansas. (See also TEPHRA.)

Bogoslov, volcano, Aleutian Islands, Alaska, United States Bogoslov volcano is one of the most famous in the world because of its record of self-destruction and reemergence over the past few centuries. It is visible as an island that is the summit of the volcano based on the seabed some 6,000 feet below. The summit of the volcano keeps building itself up and tearing itself apart in explosive eruptions or being destroyed by wave activity. The island actually is a set of lava domes that have appeared at various locations. An island at the location of Bogoslov was reported first in 1768 and was named Ship Rock. Castle Rock, another island, appeared in 1796 during a loud eruption that frightened natives on a neighboring island. By 1826 Castle Rock was about two miles long, less than a mile wide and more than 300 feet high. A great mass of volcanic rock arose from the waters near Ship Rock, on the opposite side from Castle Rock, in 1883. This new island was called New Bogoslov. A bar of clastic, or loose and fragmented, material linked New Bogoslov with Ship Rock and Castle Rock. Wave activity soon produced a channel, however, and thus cut the temporary, single island into two. Two new domes, Metcalf Cone and McCulloch Peak, formed in the waters between the two islands in 1906 and 1907. Metcalf Cone appeared first but was partially destroyed by an explosive eruption before McCulloch Peak formed. McCulloch Peak was itself short-lived; only 10 months after it appeared, the peak was demolished by an explosion, leaving behind a heated lagoon almost a mile wide. This eruption is said to have been spectacular, with a thick black cloud and lightning reported. Further eruptions were reported in 1910, 1926 and 1931. The 1926 eruption produced a new dome. An eruption of Bogoslov in the 1890s was one of several eruptions that occurred around the same time (BANDAI-SAN in 1888 and VULCANO in 1888–90).

Bolivia Located in the seismically and volcanically active RING OF FIRE that encircles the Pacific Ocean basin, Bolivia is situated in northwestern South America. A volcanic mountain in Bolivia, Cerro Rico, played an important part in the economic exploitation of the Incas by the Spanish empire. The mountain contained rich veins of silver ore and yielded great quantities of silver to be shipped to Europe. (See also ANDES MOUNTAINS.)

bombs Fluid masses of lava that are cast out from volcanoes and solidify in mid-air, volcanic bombs tend to be streamlined but may occur in many different shapes.

Bonneville Slide, Columbia River Gorge, Oregon and Washington, United States The

These volcanic bombs were cast out from Lassen Peak, California (J. S. Diller, USGS)

Bonneville Slide is a huge quantity of soil and rock that fell from the sides of the Columbia River gorge near Crown Point, east of Portland, Oregon, some centuries ago, possibly around A.D. 1100. Almost half a cubic mile of solid material was involved in the Bonneville Slide, which may have occurred as a result of an earthquake engendered by the CASCADIA SUBDUCTION ZONE offshore to the west. The northern side of the gorge is especially susceptible to landslides because the uplift of the Cascade Mountains has tilted the gorge toward the south and because easily eroded sedimentary material underlying lava flows along the northern side of the gorge undermines the land. Landslides along the gorge are believed to have buried more than 10 square miles of the gorge and created a huge natural dam perhaps 200 feet high and more than 5,000 feet wide. Other slides in the vicinity of the Bonneville Slide include the Fountain Slide, Ruckel Slide, Skamania Slide, Wind Mountain Slide and Washougal Slide.

Boqueron, volcano, El Salvador Located near the city of San Salvador, Boqueron erupted explosively on June 6, 1917. Though brief, the eruption destroyed most of San Salvador and carried destruction over an area of some 20 miles around the volcano. A lake that had formed inside the crater of Boqueron spilled out in the form of a hot flood that overwhelmed more than a dozen communities. Lava from this eruption collected to depths of 160 feet in some locations, and ash accumulated in layers several feet in depth. One curious effect of this eruption was hair loss among residents of the area; evidently acid in an ash fall was responsible for this occurrence. More than 400 people were reported killed in this eruption of Boqueron. On more than a dozen previous occasions, eruptions had destroyed the city of San Salvador.

breccia A coarse sedimentary rock, a breccia is made up of angular clasts, or particles, and may indicate that the source of the rock fragments is nearby. In the case of volcanoes, a breccia may also be defined as a rock made up of angular clasts from the vent of a volcano.

British Columbia, Canada The Canadian province of British Columbia is located on the Pacific coast within the circum-Pacific RING OF FIRE, the belt of earthquake and volcanic activity that encircles the Pacific basin. Although not as noted for earthquake and volcanic activity as the coastal United States (CALIFORNIA, OREGON and WASHINGTON) to its south, British Columbia has a history of considerable seismic and volcanic activity. Part of this activity is thought to be due to the subduction of the Juan de Fuca crustal plate along the Cascadia subduction zone beneath the Pacific Ocean to the west of the province. This subduction zone is believed to have produced the volcanoes of the CASCADE MOUNTAINS, which reach into southern British Columbia and include Mount Garibaldi, a volcano north of Vancouver.

Bulusan, volcano, Philippine Islands Also known as Irosin, the stratovolcano Bulusan is located in the southern part of Luzon Island. Recorded explosive eruptions date back to 1852 and have occurred every few years since then. Seismic records from 1978 to 1983 indicate that earthquake activity increased before eruptions, and temperatures at hot springs in the vicinity during this period have shown marked fluctuations. Earthquake swarms, however, have not always been followed by eruptions. Earthquakes and hot-spring activity did not provide any warning of a small eruption of ash in 1988, although an earthquake swarm was noticed some hours following the eruption.

C

Calabria, Italy The earthquakes that began in Calabria in southern Italy in early February of 1783 and continued for the rest of that year were felt over a wide area of the Italian peninsula and Sicily. The earthquakes began with a powerful shock on February 5 and continued with another strong event on March 28. The earthquakes appear to have been centered near the town of Oppido and to have caused extensive destruction for several miles around. Scarcely any structure in Oppido was said to remain standing after the first earthquake. Large chasms opened in the ground and reportedly swallowed numerous houses. One chasm opened in these disturbances measured some 500 feet long and 200 feet deep. Landslides occurred along the shores of the Straits of Messina and allegedly destroyed numerous villas.

Following the February 5 earthquake, residents of the area recognized the danger of remaining on the coast and evacuated the shoreline, either moving slightly inland to elevated ground or putting out to sea on fishing boats. These safety measures proved ineffective, however, for the night after the evacuation, a strong earthquake dislodged a large mass of earth from Mount Jaci; part of the falling material landed in the sea and generated a disturbance that sent a wave rolling over the ground on which the refugees had gathered. The wave is also said to have caused the destruction of the fleet of boats. More than 1,400 people were reported killed. Much of the city of Messina was destroyed in the earthquakes by the shocks themselves and by subsequent fires.

The earthquakes in February and March are believed to have killed some 40,000 people, not counting an estimated 20,000 who died later from exposure, disease and other causes. Reportedly, thermal springs burst forth from fissures in the ground. One spectacular account says that some individuals fell into these fissures and then were cast out again in outbursts of boiling water; although survivors of these ordeals were reportedly rare, some did live through them but were said to have been crippled by burns for the remainder of their lives.

In another great earthquake at Calabria on December 16, 1857, whole villages reportedly fell into huge fissures in the earth, and more than 10,000 people are thought to have been killed. According to one estimate, some 111,000 people in the vicinity of Calabria were killed in earthquakes between the years 1783 and 1857—an average of 1,500 deaths per year.

Calbuco, volcano, Chile The stratovolcano Calbuco has erupted several times since 1837. A 1929 eruption buried forests in the vicinity under thick layers of ash.

caldera A large, roughly circular depression commonly formed by the explosion or collapse of a volcanic peak. A caldera may be miles in diameter and harbor lakes containing volcanic islands. A famous example of a caldera is CRATER LAKE in Oregon. A caldera is much wider than the volcanic vent or vents involved in its formation. A caldera complex consists of the rock assemblage underlying a caldera and may involve a wide variety of constituents, including, but not limited to, TUFF, BRECCIA, DIKES, STOCKS and SILLS. A "resurgent" caldera is one in which volcanic activity resumes after the caldera has formed. Crater Lake again is a familiar example of a resurgent caldera. Others include Creede Caldera in Colorado and Valles caldera in New Mexico.

The formation of a caldera is a spectacular event. A tremendous explosion or series of explosions may be involved, as in the case of the destruction of KRAKATOA, where explosions demolished much of the island and left behind a fresh caldera. Such an event may affect areas hundreds or even thousands of miles away, through mechanisms

Model of Kilauea caldera, with Mauna Loa in background (H. T. Stearns, USGS)

such as ashfalls and tsunamis, or seismic sea waves. The opposite process—collapse, rather than explosion—also may be responsible for forming calderas. Collapse may occur when magma underneath a volcanic peak drains away through underground conduits, or perhaps through a parasitic vent on the flank of the volcano; as the molten rock below it is removed, the peak loses its underpinnings, so to speak, and falls inward upon itself. This process appears to have created Crater Lake in Oregon, one of the most accessible and scenic calderas in the world. Calderas are not always so easy to identify as those at Krakatoa and Crater Lake; subsequent volcanic activity and/or erosion may cover or erase many traces of a caldera. Over many years, several calderas may form at a given location and form a complex pattern of overlapping, roughly circular features.

Activity at calderas may take many forms. Ground deformation may occur. In some cases, ground deformation is so slight that it can be measured only by using special, sensitive instruments. In other cases, ground deformation may be dramatic, measured in inches per day. In rare cases, ground deformation has lifted structures at harbors many feet out of the water. Giant shield volcanoes, such as those in Hawaii, exhibit a special pattern of deformation. The entire summit inflates until lava emanates from the volcano. Sometimes uplift at calderas can elevate the surface more than 1,000 feet, as happened during an eruption at TOYA caldera in Japan. A caldera floor may also drop suddenly by hundreds of feet under the appropriate conditions. Perhaps the most famous case in point occurred at FERNANDINA caldera in the Galapagos Islands in 1968, when a portion of caldera floor subsided up to 1,000 feet or more in little more than a week. As a rule, however, uplift and subsidence at calderas tend to be more gradual.

Temperature changes are commonly seen when activity occurs at calderas. Fumaroles, lakes, hot springs and the soil itself may show marked increases in temperature. The water level in crater lakes may change, as may the colors of lakes. Chemical changes in ground water, hot springs and fumaroles also may be observed. Another phenomenon observed in and near calderas in connection with eruptions is change in the earth's magnetic field. Little is understood about these changes, although they are thought to be linked to tectonic forces or to demagnetization of COUNTRY ROCK. Changes in gravity have been noted at calderas and have been attributed to uplift or subsidence, changes in density of rock beneath the surface and movement of magma or ground water.

Unrest at a caldera may be brief or prolonged, depending on circumstances at the site. Some calderas have been active almost continually in the recent past, whereas others have been quiet for many centuries. There is also a great range of intensity in caldera unrest. Activity at some calderas has been extremely violent, whereas other calderas have a history of only mild activity. The most evident unrest may be confined to the area of an active vent, although it is commonplace for unrest at a caldera to involve the entire caldera in one way or another; deformation and other signs of unrest may occur all through a caldera during an episode of unrest.

Among the more exotic factors that appear to influence activity at calderas are "ocean loading and unloading," meaning the increase and decrease respectively of the weight from water piled atop the area of a caldera. Increased ocean loading is thought to contribute to eruptions in some situations by exerting pressure on a magma reservoir underground and forcing magma toward the surface. This process is believed to operate at PAVLOV volcano in Alaska from time to time. On the other hand, ocean unloading also is suspected of creating conditions favorable to eruptions; a case in point is RABAUL, where an eruption in 1983 and 1984 is believed to have been connected with ocean unloading.

As a rule, earthquakes in or near calderas are small, although there have been notable exceptions to this pattern, such as the powerful earthquakes, greater than magnitude 7 on the Richter scale, that have occurred in the vicinity of KILAUEA in Hawaii on occasion. Earthquakes in a caldera setting have various causes. Some are tectonic in origin, meaning they result from the readjustment of shifting blocks of the earth's crust rather than directly from volcanic processes. (A very strong earthquake in the vicinity of a caldera stands a good chance of being tectonic rather than volcanic in origin.) Some earthquakes at calderas are associated with intrusion of molten rock underground, whereas others may result from processes including tectonic activity and the draining away of molten rock from beneath the caldera. "Volcanic tremor" is the expression for earthquakes that are thought to result from flow of magma and/or gases through narrow space underground or other mechanisms such as the collapse of bubbles in magma. There is no single pattern of seismic activity associated with eruptions at calderas, although one pattern often observed is for an eruption to occur as earthquakes reach a maximum of energy release after a quick increase; another pattern sometimes seen is for an eruption to begin in the "lull" after seismic energy has increased and then diminished sharply. There appears to be no connection between the intensity of earthquake activity before an eruption at a caldera and the power of the eruption itself. Eruptions at calderas may be classified as normal explosions, involving magma and gases released from it, and phreatic explosions, involving ground water turned into vapor. Phreato-magmatic explosions, as they are sometimes called, are combinations of the previous two varieties. Eruptive phenomena may include ash falls, lava flows, lava lakes, domes, cryptodomes, landslides, cinder cones, nuées ardentes, avalanches, and submarine and subglacial eruptions.

California, United States California is known as the "earthquake state" because it has been the site of some of America's most powerful and destructive earthquakes, including the famous SAN FRANCISCO earthquake of 1906. The great SAN ANDREAS FAULT runs along the western shore of California, but it is not the only active fault in the state. Numerous faults, inactive and active, are found in California, and no part of the state is completely free from seismic disturbances. Well-known faults in California include the HAYWARD FAULT, Calaveras Fault Zone, Imperial Fault, Manix Fault and Garlock Fault. California is so active from a seismic standpoint that a complete list of all earthquakes in its history is impossible in the space available here. What follows is a partial list of major California earthquakes over the past two centuries.

Earthquake History of California An earthquake on July 28, 1769, in the region of Los Angeles involved four very strong shocks followed by many others over the next few days and possibly into the following year. In October of 1800, an earthquake at San Juan Bautista left every dwelling there uninhabitable, and carts outdoors were used for sleeping; the date of the principal shock is not known, but it reportedly was accompanied by an extremely loud noise. An earthquake in June of 1838 in the area of San Francisco is thought to have been comparable in power to the catastrophe that destroyed the city in 1906; the earthquake generated effects of Mercalli intensity X, and large displacements were reported along the San Andreas Fault. The shock was especially strong at Yerba Buena, and heavy damage was reported at the missions in San Francisco, San Jose and Santa Clara, as well as at the Presidio in San Francisco.

The October 21, 1868, earthquake at Hayward caused maximum damage there and at other locations along the Hayward Fault. Virtually every building in Hayward experienced heavy damage, and some were ruined completely. Civic buildings were destroyed in San Leandro. San Francisco again experienced damage to buildings on landfill, and property damage appears to have exceeded $350,000. Some 30 persons are thought to have been killed in this earthquake, which was followed by aftershocks continuing into the following month. The great San Francisco earthquake of April 18, 1906, is covered in detail elsewhere in this volume.

The Santa Barbara earthquake of June 29, 1925, caused an estimated $8 million in damage in Santa Barbara, where most buildings on the principal street in the commercial district were damaged. Thirteen people were killed in the earthquake, and movement of approximately one foot was seen along a boulevard built on a beach. An earth dam at a reservoir failed, but the freed water did little harm. Much of the exterior of a hotel collapsed. An office building failed, and the roof of a church collapsed. This earthquake served as a reminder that sound design and construction can preserve a building even in earthquake-prone territory and on less than stable soil; one concrete building, well-designed and given a good foundation, went through the earthquake with little or no damage, even though the building was constructed in a marshy area. Reinforced concrete proved resistant to the earthquake, except where construction was faulty. Strong aftershocks were reported on July 3. Another earthquake at Santa Barbara on June 29, 1926, killed a child and knocked down some chimneys.

The Long Beach earthquake of March 10, 1933, was not one of the most powerful in California's history (magnitude of only 6.3), but it caused extensive damage because it occurred in a heavily settled area. More than 100 people were killed, and hundreds more were injured. Damage was estimated at approximately $40 million. Most damage occurred in an area characterized by water-rich soil and structurally unsound buildings. The most strongly affected area extended from southern Los Angeles to Manhattan Beach, Anaheim and Laguna Beach. Destruction was especially notable in such places as Compton, Huntington Park and Long Beach. Severe damage to school buildings occurred. No fault displacement was noted.

An earthquake series in August 1975 rotated this gravestone on its base in Oroville, California (Earthquake Information Bulletin/USGS)

One of the most famous and powerful earthquakes in California history was the Imperial Valley earthquake of May 18, 1940. The earthquake was estimated at magnitude 7.1, generated effects of Mercalli intensity X and had its epicenter southeast of El Centro. Horizontal displacement up to 15 feet and vertical displacement of as much as four feet were reported. Damage was widespread in the Imperial Valley and was estimated at $6 million on the United States side of the border alone, not counting indirect effects such as those on agricultural irrigation systems. Surface ruptures occurred over some 40 miles. About four-fifths of the buildings in Imperial were damaged, and the city's water tanks were ruined. Almost half the buildings in nearby Brawley were damaged, and a city water tank collapsed at Holtville. The earthquake also was blamed for a fire at a hotel in Mexicali, Mexico.

The Kern County earthquake of July 21, 1952, was one of the most powerful and destructive in the history of California and the strongest in the United States since the great San Francisco earthquake of 1906. Estimated at Richter magnitude 7.7, the earthquake generated effects of up to Mercalli intensity XI and caused an estimated $50 million in property damage. The earthquake resulted in extensive damage along the Southern Pacific Railroad near Bealville; tunnels built of reinforced concrete with walls 1.5 feet thick were twisted and cracked and even caved in. Mercalli intensities of up to VIII were reported in cities. At one facility alone, the earthquake caused several million dollars in damage from fire. Twelve people died in this earthquake, nine of them in a single incident involving the fall of a brick wall at Tehachapi; 18 people required hospitalization, and injuries numbered in the hundreds. This shock was felt in Nevada and Arizona as well as in California. More than 100 aftershocks of magnitude 4 or greater were recorded, 47 of those on July 21 alone.

The SAN FERNANDO EARTHQUAKE of February 9, 1971, is described in detail elsewhere in this volume, as is the 1989 Loma Prieta earthquake in the San Francisco area.

Tsunamis in California California, having a long coastline, is subject to occasional tsunamis, or seismic sea waves. These waves may be generated by earthquakes along the California coast or elsewhere; the 1964 GOOD FRIDAY EARTHQUAKE in Alaska, for example, produced a tsunami that caused extensive damage in CRESCENT CITY, California near the Oregon border. One of the most remarkable tsunamis in California history occurred on December 17, 1896, when a wave that may have been seismic in origin struck Santa Barbara and carried away a large portion of a boulevard that had been reinforced especially to resist wave action. A large hill of sand between the boulevard and the usual high-tide mark was also carried away by the wave. There is a possibility, however, that this wave was produced by a storm rather than by seismic activity because there appears to have been no strong earthquake in the vicinity on this date.

Because of the earthquake potential of southern California, the "tsunamigenic," or tsunami-making, potential of faults there has received considerable attention. Generally speaking, faults along the southern California coast do not appear to pose a great tsunamigenic threat, but faults thought to be capable of producing tsunamis are found in some locations off the coast of southern California, notably in the vicinity of Point Arguello. In and near the Santa Lucia Bank, an elevated area on the continental slope near the California shore, faults are suspected of being able to produce powerful earthquakes and the tsunamis associated with them. Another offshore area with suspected tsunamigenic potential is located in the area of Santa Barbara Channel and the nearby Transverse Ranges. Powerful, destructive earthquakes have occurred in this area (the December 21, 1812, earthquake that caused such heavy damage to missions onshore, for example, is thought to have involved fault rupture under the Santa Barbara Channel), and future earthquakes of high magnitude might result in tsunamis of considerable destructive power. Tsunamis produced by submarine slides are another possibility along the southern California coast, although this particular hazard is thought to be small. The California coast remains vulnerable, however, to tsunamis involved with seismic disturbances elsewhere around the Pacific Basin, particularly in Alaska, where certain areas contain the potential for very powerful earthquakes that might produce large tsunamis.

California also has a history of recent volcanic activity, some of it within the 20th century. LASSEN PEAK, a volcano in northern California, erupted with great violence during World War I. Volcanic formations may be seen also at LONG VALLEY CALDERA. GEOTHERMAL ENERGY plays a small but interesting part in providing electric power for California. (See also LOS ANGELES.)

Campi Phelagraei See PHLEGRAEAN FIELDS.

Campo Bianco, volcano, Lipari Islands Campo Bianco (White Field) is a volcanic cone, also known as Monte Pelato, on the northern end of the island of Lipari, just north of the island of Vulcano. The cone was named for the remarkable whiteness of the pumice deposits there. Prior to World War II, these deposits were an important source of pumice for industrial and other applications, but the unreliability of ocean transport during the war forced users to develop other sources for pumice, and the economic importance of Campo Bianco diminished. The crater of Campo Bianco has steep sides several hundred feet high and has been breached on the east, from which side a broad stream of lava extends down to the ocean. The lava flow is called Rocche Rosse because oxidation has given it a red coating of iron oxide. A submarine cable running southeast from the community of Lipari, on the eastern side of the island, was broken on several occasions by eruptions associated with Vulcano between September of 1889 and December of 1892.

Canada Though not commonly associated in the public mind with earthquakes and volcanoes, Canada has both. The volcanic CASCADE MOUNTAINS extend northward from the United States into Canada and include Mount GARIBALDI, and both eastern and western Canada have zones of pronounced seismic activity. British Columbia has its share of earthquakes as a result of North America's collision with the Juan de Fuca crustal plate along the CASCADIA SUBDUCTION ZONE off Canada's Pacific shore. Eastern Canada also is known for earthquakes, especially along the ST. LAWRENCE VALLEY separating Canada from the northeastern United States. Seismic activity is so commonplace in this region that it has become a leading natural "laboratory" for studying earthquakes and techniques for predicting them.

Canadas, Las, caldera, Canary Islands The Las Canadas caldera is located on the island of Tenerife and is the site of the Pico Viejo and Pico de Tiede volcanoes. An eruption is believed to have occurred in 1492, and there is evidence of earlier eruptions around 1341, the 1390s and 1444. Eruptive activity also occurred between 1704 and 1706. More than a year of seismic activity preceded an eruption in 1909.

Canary Islands The volcanic Canary Islands are located on the continental shelf of western Africa. Volcanoes active in recent times in this group include Chahgorra, on Tenerife, which erupted in 1798 for more than three months and emitted ashes, lava and stones; some of the stones ejected in this eruption are thought to have reached an altitude of more than half a mile before falling back to earth. Other eruptions on Tenerife in 1704 and 1706 are said to have destroyed harbor facilities on the island. On the island of La Palma is a great caldera approximately a mile deep.

The island of Lanzarote was the site of a great series of eruptions that began on September 1, 1730, produced a large hill of volcanic material overnight and proceeded to emit lava that overran several villages. On September 7, a very large, solid rock was thrust up with a loud noise from the bottom of a stream of lava. The rock changed the course of the lava so that the flow ran through the town of St. Catalina and several villages. When the lava reached the ocean four days later, fish were killed in large numbers. Later, three new craters developed on the site of St. Catalina itself and cast out large volumes of ash, sand and stones. All the cattle on the island were killed on October 28 by vapors from the volcano that reportedly condensed from the air and fell as a lethal rain. A fierce storm also beset the island at this time. On January 10, 1731, an eruption produced a large hill of volcanic material that reportedly collapsed on that same day into the crater from which it arose and that was followed by several streams of lava, which then flowed to the sea. Between January and March, several new cones appeared and emitted lava. Altogether, approximately 30 new cones grew on the island. Another fish kill occurred in June, and there was apparently a submarine eruption, because smoke and fire were reported rising from the waters near the shore. The eruptions in this series continued through 1736 and required the evacuation of many residents to safer islands. An eruption on the island of Lanzarote in 1824 was preceded by very strong earthquakes and appears to have resembled the eruption that produced Monte Nuovo near Naples.

capable fault As defined for use in selecting sites for nuclear power facilities in the United States, "capable fault" means a fault that is capable of movement in the "near" future. This, in turn, signifies a fault along which movement has occurred within a few thousand years.

Cape Ann, Massachusetts, United States A major earthquake that damaged buildings over a wide area of eastern Massachusetts in 1755 is believed to have been centered off Cape Ann. Ground motions were reportedly so strong that they knocked standing people off their feet.

Cape Verde Islands The volcanic Cape Verde Islands are located west of Africa, and the volcano Fuego is noted for an eruption in 1847 in which no fewer than seven openings developed and sent out streams of lava.

capillarity A set of conditions in which a liquid, water being the most familiar example, is drawn up by surface tension into small tubes or interstices in rock. The drawing-up process is known as capillary action. The capillary fringe in soil represents a boundary zone between the water table and the dryer "zone of aeration" above it, in which air as well as water fills spaces between soil particles. Water in the capillary fringe is known as capillary water. Capillary migration (also known as capillary flow and capillary movement) refers to water movement through capillarity.

Caracas, Venezuela The March 26, 1812, earthquake in Caracas destroyed about 90% of the city and is thought to have killed some 10,000 people in the city itself, plus another 5,000 or so in the surrounding area. The number of casualties rose to approximately 20,000 by the middle of April, partly because of disease spread by contaminated water. The earthquake apparently was most destructive in the northern portion of Caracas, where two churches collapsed on worshipers (the earthquake occurred on Holy Thursday).

Caribbean Sea The Caribbean islands have been a hotbed of volcanic activity, notably the highly destructive eruptions of Mount PELÉE and SOUFRIÈRE in 1902. Volcanic islands make up the Lesser Antilles, a chain of islands stretching some 500 miles across the Caribbean. The Lesser Antilles have two separate branches at their northern end: the Volcanic Caribbees, including active volcanoes, and just to their east, the Limestone Caribbees, made up of older volcanic lands crowned with carbonate rock. Earthquake activity has also been prominent in the Caribbean. (See also PORT ROYAL.)

Cartago, Costa Rica The nation of Costa Rica is extremely vulnerable to earthquakes because of its position in seismically active Central America. One of the most destructive earthquakes in the history of Costa Rica occurred on August 27, 1841, killing some 4,000 people out of a population of fewer than 20,000. Another earthquake on May 4, 1910, killed more than 200 people and caused extensive damage.

Cascade Mountains, United States and Canada The Cascades are the location of some of the most beautiful volcanic mountains in the world, as well as the recently active volcano Mount ST. HELENS. Volcanoes in the Cascades include Mount Garibaldi, Three Sisters, Mount RAINIER, CRATER LAKE, MEDICINE LAKE volcano, Mount BAKER, Mount HOOD and LASSEN PEAK. About 100 to 150 miles from the Pacific Ocean, the Cascades stretch from British Columbia through the states of Oregon and Washington and into northern California. Three rivers run from east to west through the Cascades: the Fraser, Klamath and Columbia.

Cascadia subduction zone, Pacific Ocean This area is located off the Pacific coast of the northwestern United States and British Columbia, where the NORTH AMERICAN CRUSTAL PLATE, moving westward, is thought to be overriding the JUAN DE FUCA PLATE immediately to its west. The Cascadia subduction zone is believed to have generated volcanism that produced the volcanoes of the Cascade Mountains, including Mount HOOD, Mount BAKER, Mount RAINIER and Mount ST. HELENS.

Catania, Sicily, Italy This community is known for its early attempt to divert a flow of lava. During an eruption of Mount ETNA in 1669, a party of men covered themselves with wet hides for protection against the heat and made an attempt to divert the lava flow that was approaching Catania. At first, this tactic appeared successful. The new flow of lava moved toward the town of Paterno, several hundred of whose residents chased the men of Catania away from their lava-diversion project. The breach in the wall of the lava flow was short-lived; solidified lava filled the opening, and the lava flow continued toward Catania as before. The lava overran and destroyed a large portion of Catania, and large quantities of the same lava flow are still visible in the city.

Catena A chain of craters, of suspected volcanic origin, on Mars.

Central America Located along the "RING OF FIRE," the belt of intense earthquake and volcanic activity encircling the Pacific Ocean basin, Central America is a narrow land bridge connecting North and South America and has numerous volcanoes as well as a history of highly destructive earthquakes. In Guatemala, for example, the earthquake of June 7, 1773, destroyed the city of Santiago and killed almost 60,000 people. Volcanic events in Guatemala's history include the April 13, 1902 eruption of Tacona, which buried the city of Retalbulen and is thought to have caused approximately 1,000 deaths.

San Salvador, capital of El Salvador, has been struck repeatedly by major earthquakes, notably on May 3, 1965, when an earthquake of Richter magnitude 7.5 killed 100 people, injured several hundred and left thousands homeless. An eruption of the volcano San Miguel in 1844 killed several hundred.

The city of Cartago in Costa Rica was destroyed by an earthquake on May 4, 1910, with the loss of more than 200 lives. Volcanic activity in Costa Rica in recent years includes the July 30, 1968, eruption of Arenal, with damage estimated at more than $40 million. A 1623 eruption of the volcano Irazu killed hundreds and destroyed several villages.

Nicaragua's history of earthquakes and volcanic eruptions is especially violent and tragic. The Dec. 23, 1972, earthquake in Managua measured 6.25 on the Richter scale of earthquake magnitude and killed some 11,000 people, leaving thousands more homeless. The January 1835 eruption of the volcano Coseguina lasted three days and killed hundreds. An example of 20th-century volcanism in Nicaragua is the Jan. 12, 1947, eruption of Cerro Negro, which destroyed almost 300 square miles of territory and caused dozens of deaths, as well as extensive damage to Leon City. The seismic and volcanic history of Nicaragua was one of the factors that influenced the decision to construct an interocean canal in Panama, where geologic hazards are less extreme, rather than in adjacent Nicaragua.

Cerro Azul, caldera, Galapagos Islands, Ecuador A small shield volcano on Isabela (Albemarle) Island, Cerro Azul has erupted in 1932, 1940, 1943, 1948, 1949 (uncertain), 1951, 1959 and 1979.

Cerro Negro Nicaragua This cinder cone, whose name means "Black Hill," arose in 1850 and has erupted frequently since then.

Cerro Rico, Bolivia The remains of an ancient volcano, laden with rich ores of silver and tin, Cerro Rico (Hill of Silver) yielded vast quantities of silver for the Spanish during their occupation of the former Inca empire.

Cha, caldera, Cape Verde Islands The Cha caldera is located on Fogo Island. Although there is disagreement on the dates of eruptions in the historical record, it is said to have erupted continuously between 1680 and 1713, then been quiet until 1798. An earthquake killed individuals in 1847, and powerful earthquakes and volcanic tremors occurred before an eruption in June 1951.

Chaos Craigs, California, United States A collection of volcanic domes immediately to the north of Lassen Peak, Chaos Craigs, which includes four or more domes, formed approximately 1,000 to 1,200 years ago, according to one estimate based on radiocarbon dating of charcoal produced by pyroclastic flows. Each dome is about one mile wide and some 1,800 feet high. Large accumulations of debris surround the domes. Tremendous rockfalls at one of the domes produced a pile of debris known as Chaos Jumbles. An explosion some 300 years ago is thought to have given rise to Chaos Jumbles.

Aerial view of erupting Cerro Negro volcano, Nicaragua. Streaks radiating from the crest of the volcano were produced by bombs rolling down the slopes (Earthquake Information Bulletin/USGS)

Charleston, South Carolina, United States The earthquake that struck Charleston on August 31, 1886, is believed to be the most destructive earthquake that has occurred on the Atlantic coast of the United States in historical times. The shock was felt over an area of almost three million square miles, although damage was confined largely to the immediate vicinity of Charleston. Approximately 100 buildings were destroyed, and most brick structures were damaged. The earthquake apparently was most intense 12 miles west of Charleston, where large amounts of sand and water spewed from fissures in the ground. One such outburst of water reportedly left behind a crater more than 20 feet wide. The poorly consolidated soil under Charleston was responsible for much of the damage from the earthquake. Relatively minor earthquakes occurred earlier in

Results of 1886 Charleston, South Carolina earthquake (J. K. Hillers, USGS)

the summer of 1886 but do not appear to have caused any great concern in Charleston. More notable shocks occurred on August 27 and 28, and several minutes before 10 P.M. on August 31, the major earthquake began, accompanied by a noise that later was compared to the sound of steam escaping from a boiler or of fast-moving street traffic at close range. The initial shock is believed to have lasted about 35 to 40 seconds, but recollections of the event do not agree on whether this shock occurred singly or consisted of several separate movements. The earthquake killed more than 20 people directly, and a greater number reportedly died later from exposure or from injuries received in the earthquake. Although some structures came through the earthquake with less damage than others and few buildings were shaken completely to the ground, cracks appeared in many stone structures, and safety required them to be demolished. Damage to railway tracks provided evidence of the ground motions: The rails showed both lateral and vertical displacement, and tracks were displaced toward the southeast. A study conducted soon after the earthquake concluded that the disturbance appeared to have two centers separated by approximately 13 miles. The main focus of the earthquake was estimated at about 10 to 14 miles deep, and the secondary focus was figured to be about 8 miles deep. Reports indicate that some interesting events in distant parts of the country coincided with the earthquake, including a reduction in the yield of natural gas wells in Pennsylvania and the reactivation of a geyser in Wyoming's Yellowstone valley after four years of inactivity.

Chichon, El, volcano, Mexico The 1982 eruption of El Chichon generated a controversy over the possible effects of that eruption on global climate and weather. Laser measurements of the cloud from El Chichon indicated it was much more effective at blocking light than the cloud from

Bent reinforcement rods demonstrate the power of a volcanic eruption at El Chiçhon in Mexico in 1982 (*Earthquake Information Bulletin*/USGS)

Mount St. Helens had been two years earlier. The eruption immediately preceded a disturbance in circulation patterns in the atmosphere and in the Pacific Ocean. The change accompanied heavy rainfall in California and drought in Australia. Scientists suspected that the cloud from El Chichon might have played a role in the extraordinary weather of 1982–83, although such disturbances have been observed on other occasions without connection to volcanic events. Apparently the cloud from El Chichon did not cool global climate appreciably, but the cloud lingered in the stratosphere until 1985.

Chijiwa Bay, Japan Located near Unzen volcano, Chijiwa Bay is joined to the volcano by a GRABEN several miles wide, running east to west. Volcanic and seismic activity have occurred in the Chijiwa Bay area frequently over the last several centuries. Earthquakes just before the 1792 eruption of Unzen began in late 1791 and were most noticeable along the western side of the Shimabara Peninsula, which extends south of the caldera. Landslides and sounds of detonations, compared to the noise of artillery, were reported. An especially powerful earthquake occurred on April 21, 1792, and was likened to the sound of naval gunfire. Several hundred earthquakes occurred on April 21 and 22. Thereafter, earthquake activity diminished temporarily, although a powerful earthquake took place on April 29. More than 60 houses and more than 280 stables were destroyed in a landslide in April near the village of Nakakoba. This incident also reportedly damaged Shimabara castle. Strong earthquakes began again in late May and were accompanied by a great avalanche that set off a destructive TSUNAMI. The earthquakes, avalanche and tsunami together are thought to have killed some 14,000 people.

In 1922 a swarm of earthquakes killed more than 20 people. Other earthquake swarms occurred between 1968 and 1974, and in 1984 a swarm of more than 6,000 events was recorded. Active faults in the Chijiwa area include the Chijiwa Fault, to the northeast of the Chijiwa caldera, and the Fugendake, Futsu and Kanahama faults, all to the south of the Chijiwa Fault and east of the caldera.

Chile Running north-south in a narrow band along the western shore of South America, Chile has experienced numerous powerful earthquakes and tsunamis. Seismic activity in Chile is so frequent that the nation is the site of approximately 20% of the earthquakes that are recorded annually. Chile also contains numerous features associated with volcanic activity, such as copper mines and hot springs. Destruction from earthquakes in Chile has not been restricted to the country itself. In 1960, for example, an earthquake in Chile generated tsunamis that reportedly killed 60 people in Hawaii and 438 in Japan and the Philippine Islands; this tsunami reached more than 1,500 feet inland along the Chilean coast.

One of the most destructive earthquakes in Chilean history occurred on January 24, 1939. The earthquake lasted three minutes and shook a section of Chile more than 400 miles long, killing perhaps 50,000 people, most of them children. Damage appears to have been most severe in the cities of Chillán, Coihueco and Concepción. One account says that only three buildings were left standing in Chillán out of more than 140 city blocks, and more than 300 persons died in Chillán at one location, a theater that collapsed on the audience. In Concepción, 70% of the buildings were destroyed, and coal mines collapsed on miners within them. More than a dozen cathedrals were demolished.

A series of earthquakes in central Chile in May of 1960, which generated the aforementioned tsunami, affected more than 90,000 square miles and altered topography greatly in some locations. The earth in one place reportedly sank 1,000 feet over an area of approximately 25 square miles. Lakes vanished; new lakes formed; two small mountains were apparently leveled; and volcanic eruptions were reported in several places. An especially violent eruption occurred at Lake Ranco, where a lava flow entering the lake caused the lake to leave its banks. This earthquake occurred only three months after the catastrophic earthquake at AGADIR, Morocco, and the Chileans were able to benefit from the international relief effort already organized in response to that earthquake. Some grisly stories emerged from this Chilean earthquake and the tsunami associated with it, including a report of the murder of a six-year-old boy whose heart was torn from his body and offered as a sacrifice to the imagined gods of the sea by the Mapuche Indians. More than 100,000 people were left homeless following this earthquake, and in some areas, the army had to disperse crowds rioting for food.

Central Chile experienced still more destructive earthquakes on March 28, 1965. More than 400 people were killed, although damage was restricted to a small area. Much of the loss of life resulted from the collapse of a dam near the village of El Cobre. The resulting flood buried the village in mud to a depth of seven feet. (See also CONCEPCIÓN.)

China China's history of destructive earthquakes is almost as long as history itself. A partial list of seismic disasters in China includes a 1556 earthquake said to have taken more than 800,000 lives, another quake in 1920 that is believed to have killed almost 200,000 and, in 1976, the great TANGSHAN earthquake that is supposed to have killed more than half a million. China's susceptibility to earthquakes has two sources. One is the northward movement of the Indian crustal plate, which collides with China in the vicinity of the Himalaya Mountains in the south. The other is the existence of the so-called RING OF FIRE, the circum-Pacific belt of intense earthquake and volcanic activity, just to the east of China. SEISMOLOGY is believed to have originated in China, and the Chinese have expended considerable time and effort studying the problem of quake prediction. (See also KANSU.)

Chirpoi, caldera, Kuril Islands, Russia Chirpoi has a record of historical activity dating back to the 18th century. The island is believed to have emerged inside a pair of overlapping calderas. Another volcanic island, Brat Chiropev, stands immediately to the south of Chirpoi and is believed to be a piece of a volcano that existed at the site before caldera formation. Two volcanoes occupy Chirpoi: Chernyi volcano, in the middle of the island, and Snow Volcano, a parasitic cone on the south side of Chernyi. Chernyi is named for one Captain Chernyi, a Cossack who visited the island in 1770. Snow Volcano is thought to have arisen only a few years after his visit, prior to a visit by a Captain Golovin in 1811. Snow may have originated around the time of a very strong tectonic earthquake in January of 1780. An eruption on Chirpoi was recorded in 1811, and another eruption may have occurred in 1854. Explosive eruptions are recorded in 1857, 1879 and 1960, and a lava flow was associated with the 1879 eruption.

Cinder Cone, California, United States About one-half mile wide and 600 feet tall, Cinder Cone is a volcanic cone several miles northeast of Lassen Peak. Lava flowed from Cinder Cone as recently as 1851. The lava flow from this eruption extends toward Butte Lake and is approximately three miles long. Cinder Cone's lavas consist of a rare quartz basalt. Ordinarily, quartz is seldom found in basalt. The volcano appears to have formed during a series of eruptive episodes with many years separating them. Cinder cone is also a geological term for a volcanic cone made of cindery pyroclastic material.

Cinque Dente, caldera, Italy Also known as Monastero, Cinque Dente is located on the island of Pantelleria, in the Strait of Sicily between Sicily and Tunisia. The island has two calderas, Cinque Dente and La Vecchia. Cinque Dente experienced strong earthquakes in 1890. Fumarolic activity was observed, and uplift occurred along the northeastern shore of the island. Uplift continued into 1891. A fracture about 600 feet long opened near the uplifted section of coast. More uplift and strong earthquakes occurred in mid-October of 1891. On October 17, an explosive submarine eruption began on the northwest side of the island. A second underwater eruption may have taken place south of the island in December.

Clarion Fracture Zone, Pacific Ocean The Clarion Fracture Zone extends more than 2,000 miles under the Pacific Ocean in an east-west direction and appears to extend into the Mexican mainland as the fracture zone along which the volcanoes Colima and Popocatépetl are located.

clastic rock Clastic rock is sedimentary rock made up of fragments of other rocks. Individual fragments are known as clasts.

climate, volcanoes and Volcanic eruptions can have a dramatic short-term effect on the climate of the earth and possibly a long-term effect as well. A familiar example of the climatic impact of volcanic eruptions is the global cooling that followed the eruption of KRAKATOA in 1883. The explosion of the volcano cast large amounts of finely divided solid material into the upper atmosphere, where the dust intercepted incoming sunlight and thus reduced surface temperatures. A similar phenomenon was observed following the eruption of the volcano TAMBORA in 1815. Much remains to be learned about the physics and chemistry of the interaction between climate change and volcanic eruptions, but it has been suggested that a period of prolonged and voluminous eruptions of volcanoes might suffice to start an ice age, or glacial period. (See also "YEAR WITHOUT A SUMMER.")

Coast Mountains, British Columbia (Canada) and Alaska (United States) The Coast Mountains extend approximately 1,000 miles along the Pacific shore of Canada and Alaska and reach altitudes of more than 10,000 feet in some places. The mountains represent the boundary between the westward-moving crustal plate bearing North America and the oceanic crust beneath the Pacific. Seismic and volcanic events occur frequently along this coast, especially in ALASKA, where earthquakes and eruptions have caused great damage in the 20th century. The potential for generating tsunamis is also great along the Coast Mountains; the 1964 Good Friday earthquake in Alaska, for example, was accompanied by a powerful tsunami that caused extensive damage along the Alaskan shore and was responsible for considerable destruction as far south as CRESCENT CITY, California. Another earthquake, the Yakutat Bay event of 1958, also produced large waves, in this case as a result of a slide. (See also GOOD FRIDAY EARTHQUAKE; LANDSLIDE).

Coatepeque, caldera, El Salvador Located in a chain of volcanoes, Coatepeque caldera is situated just east of the Santa Ana volcano. Two other volcanoes in the area are Izalco and San Marcelino. The caldera contains several lava domes. Santa Ana was very active in the 16th century. Explosive eruptions are thought to have occurred in 1520, 1524, 1570 (this date is not certain) and 1576. The volcano was relatively quiet until the late 19th century, when eruptions were recorded in 1874, 1878, 1879, 1880, 1882 (this date also is not certain) and 1884. Other eruptions occurred in 1904 and 1920. Izalco has erupted every few years from the late 18th to the late 20th century. San Marcelino has been relatively quiet, with two eruptions on record, one in the 17th century and the other in 1722. The 1902 eruption of Izalco coincided, within three days, with the eruptions of Mount PELÉE on Martinique and of La SOUFRIÈRE on the island of St. Vincent.

Cocos Plate The Cocos crustal plate lies beneath the Pacific Ocean off the western shore of Central America. The subduction and destruction of the Cocos Plate are believed to be involved in generating volcanism in Central America. (See also PLATE TECTONICS.)

cohesion Internal bonding between particles in soil or sediment. In plain language, *cohesion* means how well soil or sediment "holds together" in an earthquake.

Colombia One of the most destructive earthquakes in the history of the Americas occurred in Colombia on May 15, 1875. More than 16,000 people are believed to have been killed in this earthquake, which smashed Santiago and several other communities. A volcano is said to have erupted during this earthquake and caused extensive damage in Cúcuta.

Colorado, United States Although Colorado is characterized by dramatic orogenesis (mountain formation), it is not nearly as active from the seismic standpoint as some other states, such as California. It does contain numerous signs of past volcanic activity. The San Juan Mountains are products of volcanoes. Eruptions there deposited hundreds of cubic miles of ash, which contributed to the formation of a great plateau of basalt and tuff that became, after extensive erosion, the San Juan Mountains. This volcanic output buried the Needle Mountains, which were later uncovered by erosion. Thermal baths are located at Ouray. The rich ore deposits of Colorado have played an important part in its history and that of the United States as a whole.

Colorado Plateau, United States The Colorado Plateau occupies 130,000 square miles between the Rocky Mountains and the Basin and Range province in the United States and exhibits abundant evidence of seismic and volcanic activity. The plateau covers portions of Arizona, New Mexico, Colorado and Utah and has numerous structures of volcanic origin, including volcanic necks, cinder cones, mesas topped by lava flows and "dome mountains" formed by intrusive activity.

Volcanic activity on the Colorado Plateau is thought to have occurred as recently as the 11th century. The Datil section of the Colorado Plateau, in Arizona and New Mexico, contains thick flows of lava. Two spectacular volcanic formations are visible near Mount Taylor in this section. At Cabezon Peak, a volcanic neck 2,000 feet wide has been exposed by erosion and towers above benches of sedimentary rock. Several miles away, a plug 500 feet wide in a vent of an ancient volcano has been partly exposed by erosion. Immediately north of the Datil section is the Navajo section, characterized by large numbers of volcanic necks such as those at Shiprock, New Mexico and Monument Valley. The Canyon Lands section is located north of the Navajo section and includes such signs of volcanism as the La Sal Mountains and Henry Mountains, which are believed to have been formed as magma moving underground forced overlying rocks into domes. Some of the igneous rock that produced this effect may be seen at Mount Ellsworth and Mount Hilliers. The High Plateaus section on the northwestern border of the Colorado Plateau has many lava-capped plateaus separated by GRABENS. On the southwestern edge of the Colorado Plateau, the Grand Canyon section contains lava flows from the San Francisco Mountains and from some other volcanoes. Faults divide the western portion of this section into large blocks. Seismic activity along the boundaries of the Colorado Plateau indicate the plateau still may be rising with respect to the lands around it. The epicenters of major earthquakes in the Colorado Plateau are concentrated along its southeastern and western edges. Along the southwestern, northern and eastern boundaries of the plateau, earthquake activity is less pronounced. Earthquakes beneath the Colorado Plateau itself tend to be weak and infrequent.

Columbia Plateau, northwestern United States A broad mass of volcanic rock on which Mount ST. HELENS, Mount HOOD and other famous volcanoes of the CASCADE MOUNTAINS rest, the Columbia Plateau is made up of numerous individual lava flows and extends from northern Washington southward into northern California and Nevada. The plateau covers more than 200,000 square miles. According to one hypothesis, the Columbia Plateau is the terrestrial equivalent of a lunar sea, or mare, formed in the aftermath of a large meteorite impact in prehistoric times that sent molten rock flowing outward over the surface from the point of impact. In this scenario, the meteorite struck close to what is now the border between Oregon and Idaho, blasting out large amounts of rock from the crust at the site. Removal of this crustal rock allowed molten material to rise from below and emerge on the surface as large flows of basaltic lava, some of which extended to the Pacific Ocean. Flow after flow of lava occurred and gradually built up the Columbia Plateau. One lasting result of the postulated meteorite impact was a "hot spot," an enduring area in which heat flows freely to the surface from the earth's interior. As the North American crustal plate moved westward over the hot spot, volcanism continued at that location, resulting in occasional, tremendous releases of ash caused by buildup of water-rich magma. Eruptions over the hot spot, if this scenario is correct, produced a series of calderas across the SNAKE RIVER PLAIN in Idaho. The hot spot is now thought to underlie Yellowstone National Park. This impact and its attendant hot spot also have been implicated in the origin of the basin and range region of the western United States. Studies of one lava flow, the Roza Flow, indicate something of the conditions under which the Columbia Plateau is believed to have formed. The Roza Flow, which apparently occurred in two great releases of magma separated by several hundred years, is believed to have originated from a zone of fissures several miles wide and more than 100 miles long in what is now the eastern half of Oregon and Washington. Lava flowed westward from the fissure zone at a temperature estimated at perhaps 2,000°F. Evidence of the lava's tremendous heat comes from deposits of volcanic glass more than 100 miles from the origin of the flow. It appears that lava coming into contact with lake waters here formed glass so pure that very little cooling can have occurred during the lava's movement overland. This flow covered the land with about 100 feet of basalt across a front perhaps 60 miles wide. Another flow, the Pomona Flow, extended from its origin in Idaho to the Pacific Ocean more than 300 miles distant. (See also PLATE TECTONICS.)

composite volcanoes See VOLCANOES.

Concepción, Chile The city of Concepción has suffered greatly from earthquakes on numerous occasions. An earthquake in the summer of 1757 reportedly killed 5,000 people, injured perhaps 10,000 more and appears to have been accompanied by a tsunami that submerged much of the community, but recorded details of the earthquake and associated events are few.

The earthquake of February 20, 1835, has gone down in history in part because one of the most famous naturalists of the 19th century, Charles Darwin, happened to be visiting Chile at the time as a passenger on HMS *Beagle,* which was under the command of Captain FitzRoy. In his memoir *The Voyage of the Beagle,* Darwin described the earthquake and its results (quoted in Appendix B).

Darwin's speculations about the origins of the South American mountains brought him very close to the formulation that German meteorologist and geologist Alfred Wegener would devise, early in the 20th century, to account for the global distribution of earthquakes and volcanic mountain ranges: a theory that in turn would be modified to provide the modern theory of plate tectonics, which serves as the basis of modern geology.

conduit A channel through which MAGMA rises from an underground reservoir to the surface, where the eruption of magma produces a volcano.

cone, volcanic See VOLCANIC CONE.

Connecticut, United States Connecticut has experienced numerous earthquakes in the 18th, 19th and 20th centuries. Hartford experienced an earthquake on April 12, 1837, that caused bells to ring and startled residents into rushing outdoors. Much of Connecticut was shaken by an earthquake on August 9, 1840. The epicenter is thought to have been a few miles north of New Haven. The Hartford earthquake of November 14, 1925, caused widespread alarm and some damage to buildings and knocked objects off shelves at Windham; noises were reported at East Haddam. An earthquake caused considerable fright at Stamford on March 27, 1953. On November 3, 1968, an earthquake in southern Connecticut affected an area 30 miles in length along the Connecticut River between Glastonbury and Lyme; some damage was reported at Madison, and effects of Mercalli intensity V were noted at Chester, Deep River and Essex.

continent A major mass of dry land. The continents of the earth are North America, South America, Europe, Asia, Africa and Antarctica, with Australia sometimes classified as a distinct continent. Continents occupy about one-third of the earth's surface. Boundaries between continents may be the sites of intense seismic and volcanic activity, and crustal rock under the continents is thought to be generally much thicker than beneath the oceans. (See also PLATE TECTONICS.)

continental drift See PLATE TECTONICS.

continental shelf The portion of a continent that is covered by ocean. A continental shelf exhibits a very gradual slope (only a fraction of 1°) and may be the site of considerable mineral wealth and other valuable natural resources. The continental shelf gives way to the continental slope when the slope increases to several degrees. Strong earthquakes can occur on the continental shelf. The powerful GRAND BANKS EARTHQUAKE, for example, generated a huge, fast-moving turbidity current (that is, an undersea landslide) that destroyed several telegraph cables beneath the North Atlantic Ocean. The width of a continental shelf may vary greatly. Off the eastern shore of the United States, for example, the continental shelf is much broader than off the Pacific coast, where the ongoing collision between the North American crustal plate and the crust underlying the Pacific Ocean has resulted in a relatively narrow shelf. The shelf off California is especially vulnerable to undersea earthquakes because faults involved in seismic activity extend offshore into the seabed.

continuous deformation When earth materials are deformed by flow—that is, gradually and steadily—rather than by sudden failure, the deformation is said to be continuous as distinct from a discontinuous deformation, or fracture.

contour A line on a chart or map that links points of equal value, such as elevation or depth. Simply looking at the contours on a topographic map can reveal a history of intense volcanic and/or seismic activity in a given location. The radial symmetry of volcanoes, for example, is characteristic of them, as is the central crater or caldera. An impact structure, such as Arizona's Meteor Crater, also may show up unmistakably on a contour map.

convection current In general, a convection current is a circulation pattern characterized by heated fluid rising through a comparatively cool medium; in geology, the term usually refers to postulated currents of magma rising through the earth's mantle and providing mechanical energy to move crustal plates on the surface. (See also EARTH, INTERNAL STRUCTURE OF; PLATE TECTONICS.)

Copahue, caldera, Chile Copahue is located on the border between Argentina and Chile. The caldera includes a central cone with more than one crater. The level of the lake in Los Copahues crater reportedly fell more than 100 feet between 1940 and 1945, and temperature and acidity increased.

Corbetti See ASAWA.

Cordillera Nevada, caldera, Chile The Cordillera Nevada caldera has experienced eruptions in 1921–22, 1929, 1934 and 1960. The caldera is similar in some ways to the LONG VALLEY CALDERA in California.

core See EARTH, INTERNAL STRUCTURE OF.

Coseguina, volcano, Nicaragua Located on the Gulf of Fonseca, Coseguina has a history of violent eruptions, notably that of 1835, which reportedly was heard from Colombia to British Honduras, a distance greater than 1,000 miles. This eruption lasted more than four days. The caldera rim of Coseguina rises to an altitude of more than 2,800 feet, and the inner walls of the caldera extend almost 3,000 feet in places above a crater lake.

The 1835 eruption of Coseguina occurred after a long period of apparent quiescence, and it was widely believed before this eruption that the volcano was extinct. Much of the information on this eruption comes from the account of the commandant of the port of La Union in El Salvador, approximately 30 miles from the volcano. A large white cloud emerged from the summit of Coseguina on the morning of January 20, 1835. The cloud then changed color, becoming first gray, then yellow, then red. Evidently, no seismic activity preceded the eruption. Shortly before noon, the volcano's emissions had produced such darkness over the area that lamps had to be lighted in La Union. Fine PUMICE began falling in great quantities that afternoon, and darkness in some places became virtually total. Several inches of ash accumulated by late afternoon at one community in Honduras, some 40 miles north of Coseguina. Ash fell in El Salvador, more than 100 miles northeast of the volcano, by nightfall. Darkness continued the following day, and residents of the area around the volcano felt strong earthquakes and heard noises from underground. Darkness covered the entire nation of Honduras, and chunks of pumice more than an inch in diameter fell some 20 miles from the volcano. The winds changed direction on January 22 so that areas previously free of heavy ashfalls now were covered by ash. On this day, a tremendous noise, almost nonstop in some areas, emanated from the volcano. Listeners compared it to the sound of artillery fire, and in Belize, soldiers prepared to cope with what sounded like naval gunfire offshore. There was a similar response in Guatemala City, where soldiers concluded an attack was imminent and braced for an assault. The noises were loud enough to cause distress at locations as far away as Jamaica. Ash fell in Mexico and along the border of Costa Rica. The eruption started subsiding on January 23. Ashfalls stopped by January 27, but noises continued to be heard until the end of the month. Fumarolic activity may have continued in the crater for several decades after this eruption, which cast out tremendous quantities of solid material. Because much of the material was lightweight pumice, it floated on the waters and covered the sea for about 150 miles, according to one account.

Costa Rica The Central American nation of Costa Rica has a long history of volcanic activity, notably the 1963–65 eruption of IRAZU, which cost the Costa Rican economy an estimated $150 million. Costa Rica also is extremely vulnerable to earthquakes because of its position in seismically active Central America. One of the most destructive earthquakes in the history of Costa Rica occurred on August 27, 1841, killing about 4,000 people out of a population of fewer than 20,000. Another earthquake on May 4, 1910, killed more than 200 people and caused extensive damage.

Cotopaxi, volcano, Ecuador Located near Quito, capital of Ecuador, Cotopaxi stands about 19,300 feet tall and is famed for its 1877 eruption in which an outpouring of lava from the crater is said to have melted ice on the summit and have sent melt water rushing down the volcano's flanks in such quantities that the resulting flood affected lowlands more than 200 miles distant.

coulee This expression refers to several phenomena in geology. One is a flow of viscous lava with a steep front. Another is a long, dry gorge that was carved by meltwater from a sheet of ice. GRAND COULEE in Washington in the United States is a famous example of a coulee. Grand Coulee cuts through lava flows and is the site of a dam.

A view into Halemaumau Crater at Kilauea, February 24, 1961 (D. H. Richter, USGS)

country rock The rock surrounding a mineral deposit (such as gold or diamond) or an intrusion of igneous rock.

crater A roughly circular depression at a volcano's summit or on its flanks, from which LAVA and TEPHRA emerged during eruption.

Crater Lake, Oregon, United States One of the most spectacular calderas in the world, Crater Lake occupies the BASAL WRECK of Mazama, a volcano that evidently collapsed following an eruption many centuries ago, leaving a basin that filled eventually with water. The lake is about 35 miles in circumference and some 1,900 feet deep. A comparatively small volcano, Wizard Island, rises more than 700 feet above the lake's surface and stands as evidence that Mazama's fall did not mean the end of volcanic activity at that site. Credit for "discovering" Crater Lake goes to John Wesley Hillman, who rode his mule up the outer slope of Crater Lake one day in 1853 and suddenly found himself at the rim of the caldera. The lake apparently was known to Native Americans, however, long before settlers of European descent came to Oregon. (See also CASCADE MOUNTAINS.)

Craters of the Moon Monument, Idaho, United States Generated by the same Northwestern volcanism that produced the Columbia Plateau, Craters of the Moon Monument exhibits many different kinds of volcanic formations. If dates based on radiocarbon analysis are accurate, the lava flows at the monument occurred 2,000 years ago. Samples for this analysis were collected in a novel fashion. Students dug tunnels under the lava flow in hopes of finding burnt plant material that had been both produced and then preserved undisturbed by the advancing lava. Some material was indeed preserved under the lava, which had advanced as a smoothly flowing mass and covered the plants in its path without displacing their charred remains. Craters of the Moon Monument includes an area of "tree molds" formed when lava surrounded the trunks of trees. The trees burned away, but the impressions of their trunks remained. Here again, charcoal derived from roots of the original tree was preserved under the lava, and radiocarbon dating yielded an date of approximately 2,200 years ago for the lava flow that overwhelmed the trees and produced the charcoal.

craton Not to be confused with a CRATER, a craton is a stable, generally nondeformed portion of the earth's crust. Much of Canada's interior is an example of a craton.

Creede Caldera, Colorado, United States Located in the San Juan Mountains, Creede Caldera is almost 15 miles in diameter and has a central mountain (Snowshoe Mountain) rising from the caldera floor.

creep The gradual deformation of rock subjected to small but steady stress over a prolonged period.

Crescent City, California, United States Crescent City, in northern California near the Oregon border, experienced heavy damage from the tsunami that followed the 1964 GOOD FRIDAY EARTHQUAKE in Alaska. Eleven of the 13 persons killed by that tsunami died in Crescent City, and damage was estimated at about 11 million dollars. Current velocities were estimated at some 30 miles per hour. More than 50 homes in Crescent City were destroyed, and 37 others were reported damaged. Some 40 businesses were destroyed completely, and more than 100 others experienced major or minor damage. The tsunami lifted houses from their foundations and deposited the houses in nearby streets. The harbor area experienced serious damage, including the capsizing of 15 boats. Buoyancy associated with the tsunami did considerable damage to a dock, which also was damaged by a loaded barge driven against it by the tsunami.

crust The rigid outer layer of the earth in or immediately under which earthquake activity takes place. The crust varies greatly in thickness and composition. Under major mountain ranges, the

Crater Lake, Oregon is the water-filled remnant of what was once a huge volcano (H. R. Cornwall, USGS)

Mold of charred tree in lava at Craters of the Moon National Monument, Idaho, 1926 (H. T. Stearns, USGS)

crust may be tens of miles in thickness. Under the oceans, by contrast, the crust tends to be relatively thin. The crust is divided into several major plates and many smaller plates, interactions among which produce earthquakes, volcanoes and many other geological phenomena. On its surface, the crust exhibits a wide variety of materials and landforms, ranging from broad plains of sediment to rugged ranges of volcanic mountains and from carbonate rock such as limestone to dark igneous rocks like basalt. The crust is widely broken by FAULTS and FISSURES, generated and expanded by a variety of processes related to PLATE TECTONICS. Motion of crustal rock along faults may generate earthquakes. Crustal rock is believed to be produced by solidification of magma along MID-OCEAN RIDGES, from which the newly formed crust moves outward on either side in much the same manner as goods on a conveyer belt. The rate of production of fresh crust at these sites varies from one location to another but is usually only a matter of inches, or fractions of inches, per year, on the average. The boundary between crust and mantle is known as the Mohorovicic discontinuity, or "Moho." It was found that velocities of seismic waves traveling through the earth's interior increased sharply at the depth of the "Moho."

Crustal rock has been classified as follows:

1. *Alpine crust.* Named for Europe's Alps; highly elevated and characterized by rugged mountains and by high seismic instability.
2. *Basin and range crust.* Also highly elevated and seismically very unstable; exhibits volcanic activity; much of the western United States occupies this category.
3. *Deep ocean basin crust.* Consists largely of basalt with thin overlying layers of sediment, as one finds under large areas of the Pacific Ocean. Island arcs exhibit intense earthquake and volcanic activity as well as abundant faulting and folding; Alaska's Aleutian Island chain is a good example.
4. *Mid-continental crust.* Tectonically stable; underlies moderately thick layers of sediment.
5. *Midocean ridges.* The area around Iceland is a familiar example; characterized by volcanic and earthquake activity, basaltic rock (sometimes found in bizarre formations where magma has solidified on contact with cold seawater), high heat flow and little or no sediment.
6. *Shield crust.* Very stable; consists of ancient igneous rock with little or no overlying sediment; a famous example is the Canadian shield, which occupies much of the country's interior.

The Moho dips deeply under thick regions of crust, such as the Alpine, and rises relatively near the surface under thinner portions of crust, such as the deep ocean basins. The behavior of the crust during earthquakes depends on many different factors, including the magnitude of the earthquake, depth of the earthquake focus, degree of faulting and folding, depth of overlying sediment and the distribution of groundwater in that sediment. The extreme variability of the composition and structure of the crust means that two closely situated localities may respond to the disturbance from a single earthquake in greatly different ways. An area underlain by heavily faulted and folded rock may undergo a powerful earthquake but also see damage from the earthquake localized by the shock-absorbing properties of the fractured crust; by contrast, another locality, atop less broken crust, may undergo more serious and widespread damage from a comparatively weak earthquake because the underlying rock provides a more effective medium for transmitting vibrations over long distances. (See also EARTH, INTERNAL STRUCTURE OF; PLATE TECTONICS; SEISMOLOGY.)

crystal A geometrically regular mass of a given mineral in which the internal molecular arrangement of the substance is reflected in its external geometry. Igneous rocks often occur in crystalline form; granite is a familiar example.

Crystal Ice Cave, Idaho, United States The ice formations of Crystal Ice Cave are located in a volcanic fissure in the southern portion of Idaho's Great Rift. Lava flowed from this fissure at one time and covered an area of more than one square mile at an average of more than 20 feet thick. One spectacular feature of Crystal Ice Cave is King's Bowl, a large depression thought to have been formed by a phreatic blast produced when ground water came into contact with extremely hot rock. (See also PHREATIC ERUPTION.)

crystallization The process by which molten rock cools and solidifies, forming crystals of various minerals, such as quartz and feldspar. (See also IGNEOUS ROCKS.)

D

dacite A fine-grained igneous rock light in color, dacite is actually a group of various kinds of rock, including feldspars, biotite, quartz and hornblende. Dacite is commonly found in products of eruptions along the continental side of SUBDUCTION ZONES along the borders of oceanic and continental crustal plates.

Daisetsu-Tokachi, Graben, Japan Now mostly filled, the Daisetsu-Tokachi GRABEN is a large depression on the island of Hokkaido. Several VOLCANIC DOMES (Ushiorasahi-dake, Kuma-dake, Hokuchin-dake, Ryoun-dake, Keigetsu-dake, Kuro-dake, Eboshi-dake and Hakuun-dake) are found in the vicinity and are thought to occupy points along the rim of a buried Daisetsu caldera. The volcano Tokachi-dake occupies the south end of the graben. Tokachi-dake was active in the 1920s, starting with increased emissions of steam and other gases in 1923. Also in 1923, a solfatara near the rim of the crater was replaced by a pool of either boiling water or melted sulfur. A hot spring near the crater also exhibited a dramatic rise in temperature.

Activity at the volcano, including minor explosions, continued through late 1925 and early 1926, and a fairly strong explosion took place on April 5. By early May, minor earthquakes were occurring. An explosion and mudflow were observed on May 24, followed later that day by a more powerful explosion, another mudflow and a landslide. A strong earthquake on September 5 preceded another eruption by three days. Activity then diminished at the volcano, and by 1928 the only signs of continued volcanism were changes from time to time in the output of fumaroles in the crater of Tokachi-dake.

Fumaroles became notably more active in August of 1952 when a new crater, called Showa, formed and continued puffing for the next seven years. A strong earthquake off the coast of Hokkaido in 1962 preceded a large increase in fumarolic activity at the crater. Temperatures of fumaroles rose for about two months afterward. Earthquakes became more frequent in the vicinity of the volcano, and rockfalls occurred along the eastern wall of the crater in late April and in May. Steaming and rumbling increased sharply on May 30, and numerous earthquakes occurred on May 31. Muddy hot water emerged from fumaroles on the northwest side of the central cone on June 29. Seismic and fumarolic activity then diminished until about midnight, when a large explosion took place. By late August, earthquake activity had dropped again to its normal level.

A very strong tectonic earthquake on May 16, 1968, almost 200 miles from Tokachi, appears to have touched off renewed activity at the volcano. Just after the tectonic earthquake, a swarm of earthquakes occurred at Tokachi-dake. This activity diminished for a few months but intensified again in December of 1968 and reached maximum activity in January and March of 1969. The swarm of activity included several thousand recorded volcanic earthquakes. No eruption, however, occurred at this time. Small eruptions occurred in June of 1985.

Dakataua, caldera, Papua New Guinea The Dakataua caldera is situated on the Talasea Peninsula on the northern shore of New Britain Island. The caldera is thought to have formed only about 1,150 years ago. Inside the caldera is a central volcano, Mount Makalia, that contains a crater lake more than 300 feet deep. Eruptions have occurred from ring fractures in the caldera. Lava flowed from the summit in an eruption around the year 1890. Warm springs exist at the base of Mount Makalia, and a cinder cone in the summit crater emits steam.

dams Artificial and natural dams both can present potential hazards during earthquakes and volcanic eruptions. Waters impounded behind a dam can be released with devastating effect if the dam should fail, or landslides or other phenomena could raise the water level above the dam, sending water rushing along the valley downstream.

In the 1971 SAN FERNANDO EARTHQUAKE in southern California, for example, the Upper and Lower Van Norman Dams sustained severe damage. The Lower Van Norman Dam appears to have come close to failing. Most of the damage to the Lower Van Norman Dam consisted of a slope failure that dislodged a huge segment of the earthfill embankment and sent it to the floor of the reservoir. The slide removed the dam's upstream concrete lining and crest, and one of two intake towers was destroyed. Discharge facilities at the Lower Van Norman Dam were opened to lower the water level behind the dam, and the area below the dam was evacuated for several days after the earthquake while water was drained from the reservoir. The Upper Van Norman Dam's crest sagged considerably, and the dam itself shifted about six feet downstream.

Volcanic eruptions also contain a potential for catastrophes involving dams when an eruption occurs nearby. The 1980 eruption of Mount St. Helens, for example, occurred only several miles from Swift Reservoir. Had the tremendous avalanche of debris from the eruption moved south toward the reservoir, instead of north and east, it might have overwhelmed the reservoir and resulted in the flooding of the lower valley of the Lewis River, a tributary of the Columbia River. Following the eruption, two debris dams in the Toutle and Cowlitz river systems were studied for their flood hazards. One of the dams failed the day after the theoretical failure analysis was carried out, and the hypothetical results agreed closely with the actual results.

Death Valley, California, United States Death Valley—site of Death Valley National Monument—is famous for having the lowest point in North America, the Badwater Depression, 282 feet below sea level. Death Valley is part of the BASIN AND RANGE PROVINCE of the western United States and shows abundant evidence of volcanic activity. Death Valley represents a block of crustal rock that is sinking so quickly that erosion from surrounding areas cannot fill it fast enough to keep up with its descent. Some of the most recent volcanic activity in Death Valley is thought to have occurred several thousand years ago in the vicinity of Ubehebe Crater. About half a mile wide, Ubehebe Crater appears to have originated in a PHREATIC ERUPTION in which a giant blast of steam followed contact between groundwater and magma below the surface. Near the Black Mountains, the action of heated water on deposits of rhyolitic ash has created a spectacular display of colors.

Deccan Plateau, India The Deccan Plateau is made up of FLOOD BASALTS and is comparable in many ways to the volcanic COLUMBIA PLATEAU and SNAKE RIVER PLAIN of the northwestern United States. The Deccan Plateau is suspected of being an impact structure, created when a planetoid struck the earth's surface and generated a "HOT SPOT" that produced large flows of EXTRUSIVE ROCK.

Deception Island, caldera, South Shetland Islands, United Kingdom A spectacular caldera in the middle of Deception Island (located near the Antarctic Peninsula) is thought to have been formed by tremendous eruptions of ANDESITE. Many vents are active along ring fractures near the harbor. Records of eruptions at Deception Island before very recent times are probably incomplete because the island has no permanent settlements. When HMS *Chanticleer* stopped at Deception Island for several weeks in 1829, the captain reported hearing noises like that of "mountain torrents" emanating often from underground. The captain added that these noises were so loud on one occasion that there was concern that they might damage instruments being used to conduct experiments at the island. Large numbers of fumaroles were visible in the caldera.

A lava flow may have been responsible for an observation in 1842 that the south side of the island appeared to be on fire. Pyroclastic deposits indicate that eruptions occurred in the first few years of the 20th century. Near a whaling station, the shore of the island subsided in 1923, and strange phenomena were observed in the bay, where water appeared to boil and paint was stripped from the hulls of ships. Several years later, in 1930, the floor of the harbor subsided some 15 feet during an earthquake. Earthquakes preceded an eruption in 1967 that produced a new island, Yelcho Island, at the northern end of the harbor. Several weeks of seismic activity preceded an eruption in 1969, from vents along the east ring fracture in the caldera. Dramatic uplift was observed in 1970 along the northern part of the bay. The shoreline was moved more than 1,500 feet in some places, and Yelcho Island was joined to the mainland. Satellite photos showed a plume from the island in 1987.

deep gas hypothesis See METHANE.

Delaware, United States The small state of Delaware is located in a region of minor seismic risk. Only a few notable earthquakes have occurred in Delaware. An earthquake on October 9, 1871, reportedly caused damage to windows and chimneys in Wilmington and was felt also in Oxford and New Castle, as well as in Pennsylvania and New Jersey. On March 25, 1879, an earthquake along the Delaware River was felt over an area of about 600 square miles, in Dover and below Philadelphia, Pennsylvania. The affected area extended between Chester, Pennsylvania and Salem, New Jersey. The earthquake of May 8, 1906, near Seaford, Delaware, shook buildings and was felt elsewhere in Delaware and in Maryland.

Deriba, caldera, Sudan The Deriba caldera occupies the summit of Jebel Marra Volcano in the western central portion of the Sudan. Although the geologic history of the volcano is uncertain, it has been suggested that olivine basalt built up during early eruptions at the site, then was eroded away after activity ceased. When eruptions resumed in a second phase of activity, the lavas were more acidic. According to this scenario, the third and most recent phase of activity has involved the formation of many secondary cones. Deriba caldera is thought to have originated during this third phase. A central cone and a lake have formed in the caldera. Although it is not known exactly when the caldera formed, it may have originated in historical times. There have been no recorded eruptions at the caldera, but hot springs and fumaroles were reported in the crater in the mid-20th century.

Devil's Tower, Wyoming, United States Devil's Tower is a famous example of a VOLCANIC NECK.

Diamante, caldera, Chile The Diamante caldera is located in southern central Chile and is thought to have been formed by great eruptions of pyroclastic flow deposits that covered several thousand square miles. The stratovolcano Maipo occupies part of the western portion of the caldera. Several eruptions were reported in the 19th and 20th centuries, but the accuracy of these reports is questionable because the "eruptions" may have been nothing but enduring plumes of steam.

diamond The hardest naturally occurring mineral and one of the most precious gemstones, diamond is associated with volcanic formations, particularly the famous "diamond pipes" of South Africa. Diamonds are believed to have been found first in Borneo and India. Brazil was a major producer of diamonds in the 18th and 19th centuries. Diamonds were discovered in South Africa in 1867 on the shore of the Orange River near Hope Town, and several years later, primary deposits of diamonds were found on the plateau between the Modder and Vaal Rivers. These primary deposits were located in volcanic pipes containing a variety of peridotite called kimberlite. These pipes intersected the surface in circular or elliptical areas. Some diamonds were recovered on the surface from a weathered material called "yellow ground," but others had to be extracted from the harder "blue ground" below. Important South African diamond mines include Bultfontein, De Beers, Du Toitspan, Kimberly and Wesselton. For a time, most of the world's annual output of natural diamonds came from the pipe mines, but later, alluvial deposits in Africa surpassed the pipe mines in production.

Diamond production is not confined to Africa; diamonds have been recovered in parts of the world as widely separated as Mexico, Siberia and Canada. Diamonds not used as gemstones have widespread industrial uses. Industrial-quality diamonds occur naturally in forms including ballas (spherical aggregates of diamond crystals), bort (crystals with irregular shape, bad flaws and numerous inclusions; used in drilling and abrasives) and carbonado (a dark variety of diamond). Although a colorless diamond is pure carbon (as demonstrated by the fact that burning such a diamond in an oxygen atmosphere yields only carbon dioxide), other elements may occur in diamonds and affect the color of the stones; blue, green and yellow gem-quality diamonds have been found.

Dieng Plateau, Java, Indonesia The Dieng Plateau is an area of frequent volcanic activity and is thought to occupy an ancient caldera. Explosive and PHREATIC ERUPTIONS characterize the area. An explosive eruption apparently occurred around the year A.D. 1300. An eruption in 1786 was preceded by several months of earthquakes and involved emissions of sulfurous vapor. An eruption in October of 1826 was accompanied by sounds like cannon fire. Several days later, explosions occurred, and earthquakes were felt at a considerable distance. The source of this activity is uncertain. Except for a minor eruption in 1847, the Dieng Plateau was relatively quiet until 1884, when solfataras and fumaroles became more active and mud

Radial dikes are clearly visible in this photograph of Pena de Bernal, a volcanic neck in Queretaro, Mexico (K. Segerstrom, USGS)

eruptions occurred. Powerful earthquakes shook the area in 1924 and 1928. Eruptive activity started just after an especially damaging series of earthquakes in May of 1928. Gas emissions continued for the next nine years. In 1939 earthquakes preceded minor phreatic eruptions. Late in 1944, several phreatic explosions occurred. One of these killed a large number of people; one estimate puts the death toll over 100. A small ashfall took place in 1953, and a minor eruption occurred in 1954. More than 100 people were killed in a large outburst of carbon dioxide gas in 1979. A minor earthquake swarm was noted in 1981, and another such swarm in 1984 caused some damage to property. Another earthquake was felt over a wide area around Dieng in 1986, but thermal activity appears to have been unaffected. A hydrothermal area in the Dieng Plateau has been investigated for geothermal energy.

dike A vertical, or almost vertical, layer of intrusive volcanic rock that has forced its way, while in a molten state, through surrounding layers of already formed rock. Dikes may radiate outward in a star-like pattern from a central volcanic neck or column of rock left standing after erosion has removed the outer layers of the volcano.

dilatancy Defined as the increase in volume resulting from tiny cracks developing in a rock, dilatancy has been an important concept in attempts to predict earthquakes. There has been disagreement about the process or processes involved in the postulated relationship between dilatancy and seismic activity. According to one hypothesis, water moves into the area affected by the small cracks just before an earthquake, weakening the rock; in another hypothesis, most of the cracks close in the affected region just before an earthquake occurs.

Dominica, Lesser Antilles Dominica is situated in the northern portion of the Lesser Antilles island arc and has an unnamed CALDERA with a history of unrest dating back to the late 17th century. The caldera is estimated to be about six miles wide. Several vents are present, including Grande Soufrière Hills, Morne Macaque (Micotrin), Morne Trois Pitons, the Valley of Desolation and Watt Mountain. There is disagreement as to whether these vents belong to a single volcano. Other vents on the island include Foundland and Morne Anglais. Only one eruption has occurred here within historical times, a minor steam explosion.

There has been considerable earthquake activity at Dominica over the past several centuries, however, starting with the report of a moderately strong earthquake in 1673. Emissions of gas and numerous strong earthquakes were reported in 1765, and more strong quakes occurred in 1816 and 1838. In 1843 a very strong tectonic earthquake took place near Dominica and was felt both there and on nearby islands. Lesser earthquakes occurred at intervals of several years through 1849. A moderately strong earthquake in 1879 was accompanied by a minor PHREATIC ERUPTION and a black cloud rising from the area of SOLFATARAS. It is uncertain whether the earthquake was linked with the eruptive activity. Seismic swarms have been recorded at Dominica in 1967, 1971, 1974 and 1976. A moderately strong earthquake occurred approximately 100 miles north of Dominica in 1976 and was viewed as the result of magma movements.

E

Earth, internal structure of Earthquakes and volcanoes are manifestations of processes at work deep within the earth. These processes in turn are evidence of the earth's internal structure, models of which have undergone considerable evolution in the 20th century. On a large scale, the internal features of the earth may be categorized as follows:

1. *Core.* This is the innermost layer of the earth, a dense, approximately spherical mass of very hot rock that accounts for roughly one-third of the planet's mass and one-sixth of its volume. The core is thought to be divided into a very dense inner section and a more fluid outer layer.
2. *Mantle.* The mantle, the thick layer of rock surrounding the core, contains about two-thirds of the earth's mass and more than four-fifths of its volume. The outer layer of the mantle is called the asthenosphere and is involved closely with earthquake and volcanic activity.
3. *Crust.* The crust, or surface layer, of the earth is very thin compared to the mantle and core (only a few miles thick) and accounts for only a tiny fraction of the earth's total mass and volume. The crust floats, so to speak, atop the asthenosphere by a process called isostasy, in which the relatively lightweight rocks of the crust are supported by the denser rocks below. The crust is thought to be divided into a number of individual, rigid plates that interact with one another through various processes, generating earthquakes and volcanic activity in so doing. The crust contains many cracks, or faults, that are characterized by numerous earthquakes.

(See also PLATE TECTONICS.)

earthquake See SEISMOLOGY.

earthquake light This phenomenon consists of a peculiar glow that sometimes is reportedly seen in the sky during earthquakes. What causes earthquake light is uncertain, but it has been suggested that methane escaping from underground during earthquakes is ignited somehow and burns near the surface, giving off light.

East Pacific Rise A MID-OCEAN RIDGE 6,000 miles long, the East Pacific Rise is the site of remarkable geological formations and ecological communities on the seabed. An expedition to the East Pacific Rise in 1980 discovered that superheated (more than 700°F), mineral-laden water flowing from springs, or "hydrothermal vents," on the seabed had generated peculiar "chimneys" formed of minerals deposited when the hot water met the cold ambient waters of the deep ocean. Around the vents lived giant worms and large clams and crabs in an ecosystem supported by bacteria feeding on hydrogen sulfide, oxygen and carbon dioxide in the water from the vents.

economic effects of earthquakes and volcanoes The economic effects of earthquakes and volcanic eruptions can be tremendous. After witnessing the effects of a powerful earthquake in Chile, the 19th-century British naturalist Charles Darwin wrote:

> Earthquakes alone are sufficient to destroy the prosperity of any country. If beneath England the now inert subterranean forces should exert those powers, which most assuredly in former geological ages they have exerted, how completely would the entire condition of the country be changed! What would become of the lofty houses, thickly packed cities, great manufactories, the beautiful public and private edifices? If the new period of disturbance were first to commence by some great earthquake in the dead of the night, how terrific would be the carnage! England would at once be bankrupt; all papers, records, and accounts would from that moment be lost. Government being unable to collect the taxes, and failing to maintain its author-

Earthquake light, photographed here during an earthquake swarm in Japan, is a glow in the sky that sometimes accompanies earthquakes (*Earthquake Information Bulletin*/USGS)

ity, the hand of violence and rapine would remain uncontrolled. In every large town famine would go forth, pestilence and death following in its train.

A single earthquake or volcanic eruption can cause property damage in the hundreds of millions of dollars, and a very powerful shock or eruption occurring in or near a heavily populated area might cause damage into the billions of dollars. Of particular concern in our time is the anticipated earthquake in the Tokyo, Japan metropolitan area. Tokyo has undergone major earthquakes at intervals of approximately 75 years over the last few centuries, and the most recent major earthquake there occurred in 1928. The cost of rebuilding Tokyo after the next "superquake" has been estimated at $1 trillion. In these days of an increasingly integrated global economy, a highly destructive earthquake in Tokyo might have drastic economic effects in many other countries as well, including the United States.

Ecuador Ecuador is located on the western shore of South America, where the westward-moving continental crustal plate encounters oceanic crust and generates earthquakes where the two plates collide. Ecuador has a history of highly destructive seismic activity, notably the great earthquake of August 5, 1949. This earthquake, of magnitude 7.5 on the Richter scale, was centered 25 miles deep and affected an area of 1,500 square miles along the eastern side of the Andes, particularly Ecuador's highland plateau. The earthquake killed more than 6,000, injured some 20,000 and left approximately 100,000 people homeless. More than 50 cities and towns experienced severe damage, and total damage was estimated at more than $60 million.

Emmons Lake, caldera, Aleutian Islands, Alaska, United States The Emmons Lake caldera is located near Pavlov volcano in an area of frequent seismic and volcanic activity.

Emperor Seamounts, northern Pacific Ocean A chain of submerged volcanic mountains in the northern Pacific Ocean, the Emperor Seamounts are part of a string of volcanic seamounts and islands reaching almost from Hawaii to Russia's Kamchatka Peninsula. The Emperor Seamounts and the Hawaiian Islands are both part of this same chain, which is thought to have been formed as the Pacific crustal plate moved over a stationary "HOT SPOT" of volcanic activity. A sharp bend in the chain marks the division between the Emperor Seamounts and the Hawaiian Islands and indicates the Pacific plate changed its direction of motion abruptly, from north to northwest. (See also PLATE TECTONICS.)

epicenter The location on the earth's surface directly above the hypocenter, or focus, of an earthquake. (See also SEISMOLOGY.)

Erebus, Mount, volcano, Antarctica Located on Ross Island, Mount Erebus stands 12,448 feet high and is the most active volcano in Antarctica. The cone on the volcano's summit occupies a caldera. The main crater is approximately 450 feet deep and has a diameter of 1,500 to 1,800 feet. An inner crater is about 600 feet wide and 300 feet deep. The volcano was discovered by the British explorer James Ross on an expedition in 1841. Ross named it Mount Erebus after his flagship, and he named the nearby volcano Mount Terror after another ship on his expedition. Mount Erebus was evidently in eruption when Ross saw it. Flashes of red light were visible from the mountain by night, and by day, black vapor surrounded the summit. On another expedition in 1908, a British team climbed successfully to the summit of Mount Erebus. The volcano is characterized by eruptions of the Strombolian variety, from pools of lava in the crater.

eruption Generally speaking, any volcanic activity that releases lava, tephra or gases. Eruptions may take many forms, from explosive, highly destructive eruptions like that of Mount ST. HELENS in 1980 to comparatively sluggish and harmless eruptions like those characteristic of volcanic activity in the HAWAIIAN ISLANDS. Volcanic eruptions fall into several categories:

- Hawaiian eruptions are gentle by volcanic standards and generate gently sloping volcanoes known as shield volcanoes. Volcanoes of this kind erupt frequently and may form a lake of lava in the crater, from which fountains of lava arise.
- Strombolian eruptions, named after the STROMBOLI volcano and volcanic island off the Italian coast, are more energetic than Hawaiian eruptions but do not cause widespread damage. The volcano casts out some lava but little ash.
- Icelandic eruptions do not involve the classic pattern of ash and lava emerging from a mountain peak. Instead, great quantities of highly fluid lava flow from fissures in the earth. These voluminous lava flows may fill valleys and form dams in river valleys, blocking the river's flow and causing flooding upstream. Icelandic eruptions are much like Hawaiian eruptions, except that in Icelandic eruptions the layers of lava spread out horizontally and form great sheets rather than pile up into dome-like formations.
- Pelean eruptions, named for the devastating eruption of Mount Pelée in 1902, involve such thick, viscous lava that it plugs the throat of the volcano, so to speak, and forms a dome of solid lava in the crater. Meanwhile, pressure builds up inside the volcano until it explodes, sending a NUÉE ARDENTE sweeping down the flanks of the mountain.
- Vulcanian eruptions are named for VULCANO, a mountain near Stromboli, and resemble the classic picture of such events, with a large cloud of ash, shot through with bolts of lightning, arising from the volcano's summit. Eruptions of this kind are infrequent but strong.
- Vesuvian eruptions are much like the Vulcanian kind but also involve release of lava. Exceptionally violent Vesuvian eruptions are known as Plinian eruptions because they resemble the famous eruption of VESUVIUS in A.D. 79, described by the Roman naturalist Pliny the Younger. In a Plinian eruption, the volcano ejects vast quantities of ash and may send tephra to an altitude of 30 miles or higher.

(See also ICELAND; PELÉE, MOUNT.)

Erzincan, Turkey The city of Erzincan in eastern central Turkey was devastated by an earthquake on December 27, 1939. More than 50,000 people were killed, and many more were left injured and homeless by the earthquake, which produced seven shocks. Severe winter weather made the situation of survivors still more serious.

escarpment Also known simply as a scarp, an escarpment is a steep cliff located at the edge of a flat or nearly flat area. An escarpment may form where there is vertical motion along a FAULT.

Etna, volcano, Sicily One of the most famous volcanoes, Etna stands some 10,750 feet tall and has a circumference greater than 90 miles. The volcano has a gentle slope and resembles a shield volcano, such as those of the Hawaiian Islands, more than a steeply sloping volcano, such as Fuji or Vesuvius. Numerous lava flows have occurred from the flanks of the volcano. Charles Morris, in his 1902 survey of volcanoes, *The Volcano's Deadly Work,* described the activity of Etna:

> There is a great similarity in the character of the eruptions of Etna. Earthquakes presage the outburst, loud explosions follow, rifts and bocche del fuoco ["mouths of fire," or small lava vents] open in the sides of the mountain; smoke, sand, ashes and scoriae are discharged, the action localizes itself in one or more craters, cinders are thrown up and accumulate around the crater and cone, ultimately lava rises and frequently breaks down one side of the cone where resistance is least; then the eruption is at an end.

Etna has erupted on numerous occasions from 1226 B.C. to the present. Virgil, Pindar, Seneca,

Pythagoras and Thucydides all mentioned the volcano in their writings. Perhaps the most famous eruption of Etna occurred in 1669, when powerful explosions destroyed part of the volcano's summit and lava from a fissure on the volcano's flank flowed some 10 miles to the sea and damaged the town of Catania.

The 1669 eruption is notable for an attempt to divert a lava flow. About 50 townspeople from Catania reportedly dug a passage that drained away lava and apparently reduced the immediate threat to Catania; but when the flow changed direction and menaced the village of Paterno, inhabitants of that community drove away the Catanians and ended their attempt to divert the lava. This effort, though ultimately unsuccessful, showed that engineering could alter the direction of lava flows.

An eruption in 1755 was accompanied not only by emissions of lava but also by great flows of water, which resulted when lava came in contact with snow and ice on the summit. The water flowed down the flanks of the mountain, carrying along with it great quantities of unconsolidated volcanic material. The volume of water involved in this eruption was estimated at approximately 16 million cubic feet. It formed a channel 2 miles wide and 34 feet deep in places and flowed at the rate of approximately 40 miles per hour. Another great eruption occurred in 1819.

An extremely violent eruption that lasted more than nine months occurred in 1852. This eruption is thought to have produced flows of lava more than 2,000,000,000 cubic feet in volume over an area of some three square miles.

More than three centuries after the attempt to divert a lava flow during the eruption of 1669, volcanologists in Italy asked the United States military to help with another such experiment during the eruption of 1992. The attempt was called "Operation Volcano Buster" and involved using explosives to blast a hole in a lava tunnel 6,000 feet up the flank of Mount Etna. Helicopters then dropped large blocks of concrete into the hole in the hope of stopping the lava flow. The experiment, however, was considered only a partial success at best.

Eurasian crustal plate One of the major crustal plates that are believed to occupy the surface of the earth, the Eurasian plate generally underlies the continents of Europe and Asia, although portions of Asia also are found in the Arabian crustal plate and Indo-Australian crustal plate. The borders of the Eurasian plate are noted for intense earthquake and volcanic activity. On its western edge, the MID-OCEAN RIDGE in the middle of the North Atlantic Ocean is the site of volcanism, notably in areas such as ICELAND. The eastern edge of the Eurasian plate follows part of the trace of the RING OF FIRE, the belt of seismic activity and volcanism that surrounds much of the Pacific Ocean basin. Volcanic activity is especially pronounced along this border of the Eurasian plate in regions such as JAPAN and the KAMCHATKA Peninsula of Russia. The southeastern edge of the Eurasian plate displays abundant evidence of volcanic activity, both ancient and recent. This is the region of the highly destructive volcanoes of the Indonesian archipelago, such as KRAKATOA, the 1883 eruption of which has become virtually a synonym for a natural catastrophe. Along the southern border of the Eurasian plate, earthquakes are frequent and powerful where the Eurasian plate adjoins the Indo-Australian and Arabian plates. The ongoing collision between the Indian and Eurasian plates is believed to have generated the Himalaya Mountains, among other prominent features of southern Eurasia. (See also INDONESIA; IRAN; PHILIPPINE ISLANDS; PLATE TECTONICS.)

extrusive rock Igneous rock that flows, or has flowed, from a fissure or vent on the earth's surface.

F

Falcon Island, Tonga Also known as Fonua Foo, Falcon Island has been emerging and disappearing beneath the ocean's surface for decades. HMS *Falcon* charted the island in 1865 and described it as a shoal. Eruptions in 1885 produced a cone approximately 300 feet tall, but erosion destroyed much of the cone within months after eruptive activity ceased, and nine years later the cone had been reduced to a shoal again. Another eruption in 1896 raised a small cone about 100 feet above the sea, but this too was destroyed in a matter of several years. Eruptions in 1927 and 1928 produced an island more than two miles long; but a quarter-century later, the island had vanished.

Fantale, caldera, Ethiopia The Fantale caldera is located in the main Ethiopian Rift near the point where Africa's Rift Valley joins the Afar Depression. Although Fantale does not appear to have been an especially destructive volcano in terms of fatalities and property damage, it apparently has a notorious reputation in the surrounding area, where parents are said to warn ill-behaved children that they will melt like the volcano. An eruption in the 13th century reportedly destroyed a village and a church. Several lava flows occurred in the vicinity in 1820. Minor fumaroles were observed at numerous places in the caldera in 1930, and highly energetic fumarolic activity occurred in 1960, but these observations are not thought to represent a great departure from normal activity.

Farallon crustal plate A portion of the earth's crust believed to have occupied the space between the Pacific plate and the North American plate off what is now the coast of California, the Farallon plate apparently was subducted and consumed as North America advanced westward. The subduction and destruction of the Farallon plate is thought to have been responsible for widespread volcanic activity in prehistoric California. (See PLATE TECTONICS.)

fault A fracture in the earth's crust along which displacement of rock parallel to the fracture occurs. Perhaps the world's most famous fault is the SAN ANDREAS FAULT, which runs along California's Pacific coast for several hundred miles and is responsible for the earthquake that destroyed much of San Francisco and neighboring communities in 1906. Faults are abundant in mountain belts, although they also may occur in flatlands. Some faults extend for hundreds of miles across the earth's surface and manifest themselves in dramatic ways. Other faults may be comparatively tiny, only several miles in length.

Faults are described as having strike, meaning the angle between true north and the horizontal direction of the fault, and dip, the angle between the horizontal and the plane of the fractured surface underground that constitutes the fault. Motion, or slip, can occur in several directions along an active fault, and the various kinds of fault are named accordingly. A strike-slip fault exhibits horizontal movement, whereas a dip-slip fault displays vertical movement. An oblique-slip fault shows characteristics of both dip-slip and strike-slip faults. A normal fault is characterized by rocks above the fault plane moving downward with respect to the rocks beneath the fault plane. In a reverse fault, rocks above the fault plane move upward relative to rocks beneath.

Some faults are clearly visible on the earth's surface. Evidence of them may take such forms as displaced hills, offsets in a shoreline where a fault runs out to sea, and elongated lakes along fault lines (for example, Lake San Andreas near San Francisco). Other faults are less evident, however, and in many parts of the world—notably the eastern United States, where bedrock lies buried under

Elkhorn Scarp marks the trace of California's San Andreas Fault (R. E. Wallace, USGS)

deep layers of sediment—active faults may go unsuspected until movement along them generates an earthquake. This was apparently the case with the CHARLESTON, South Carolina earthquake of 1883, which destroyed a large portion of the city, and the extremely powerful earthquakes of 1811–1812 in the NEW MADRID FAULT ZONE in the Mississippi Valley.

There are various ways to identify an active fault and estimate its potential for surface rupture in future earthquakes. One approach is to study the history of faulting and creep. Another is to examine ongoing earthquake activity. Measurements of strain by geodetic surveys are also useful. It is not always easy to differentiate between an active and an inactive fault, because a single fault may vary greatly in its behavior over time. Whereas some faults are clearly active because they display measurable creep or generate frequent small earthquakes, not all faults are so consistently active. A fault that has shown no activity for many centuries may reactivate suddenly and slip 30 feet or more in a single event. The determination of a fault as active or inactive may depend also, in part, on how much harm the fault may cause if it should move significantly again. A fault located near a nuclear power station, for example, may be evaluated on a different basis than a fault that poses less of a potential hazard to human activities. The evaluation of faults as active or inactive is made harder by the fact that not all active faults manifest themselves on the surface.

It is important to determine the slip rate along a fault in projecting future activity. The slip rate is a measure of how fast movement is occurring along a fault. It is also important to know how often earthquakes occur along a fault and how powerful those earthquakes are, as well how much slip occurs per earthquake. The slip rate can be derived from a study of recently formed features along a fault. The slip rate, however, represents only an average over time of movement along the fault and is not necessarily steady. Along a given fault, the slip rate may represent steady activity over a certain period, or a few episodes of comparatively great and abrupt movement. In other words, a fault may be very (even catastrophically) active at some times and quiescent at others. As a rule, faults along the boundaries of major plates of the earth's crust exhibit greater slip rates than faults elsewhere. Because the slip rate has several components—the horizontal, vertical and dip slip rates—a given estimate of slip rate may not equal the true slip rate, although in some parts of California the predominantly horizontal character of movement along certain faults means that the horizontal slip rate comes very close to the true slip rate. In southern California, slip rates tend to be greatest along the San Andreas and San Jacinto faults.

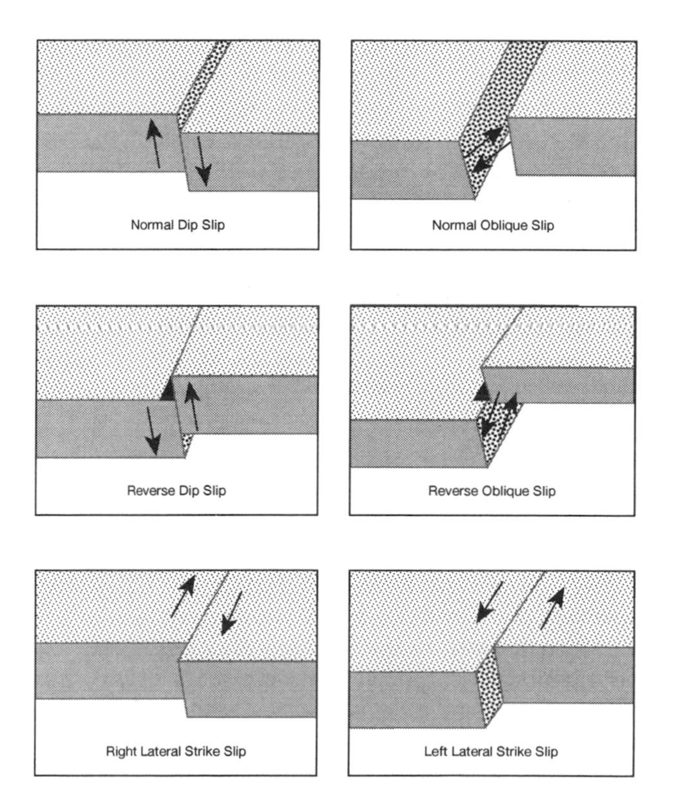

These diagrams illustrate the six kinds of movement along faults (© Facts On File, Inc.)

A fault's potential for generating destructive earthquakes may be judged partly on the basis of earthquake activity recorded by instruments. If such activity is clearly associated with a fault on an ongoing seismological record, then this information can help in associating the HYPOCENTERS of earthquakes with a particular fault at depths of a few miles below the surface. When numerous earthquakes can be linked to a given fault in this way, with their hypocenters lining up (so to speak) along the fault at depth, then one can say with some assurance that the fault is capable of generating more earthquakes in the future. This approach cannot be used reliably to estimate a fault's potential for destructive earthquakes in all cases, however, because a fault can be quiescent but remain seismogenic, or capable of earthquake activity. Moreover, even an active fault is not necessarily dangerous to lives and property. A fault that generates many small earthquakes, for example, may pose little or no danger to residents of its area. Even when a fault exhibits surface rupture, its activity may be so gentle as to pose no particular threat to the surrounding community.

Sometimes risk assessment for a seismogenic fault requires estimating the maximum earthquake that the fault could produce. Here, the historical record of seismic activity is important, because the magnitudes of earlier earthquakes may provide guidance in estimating the possible strength of other seismic events to come. (The strength of the FORT TEJON EARTHQUAKE, for example, has become a benchmark to estimate the magnitude of future major earthquakes in southern California.) Geologists also may find it helpful to compare the history of seismic activity along a particular fault with the history of another fault comparable to it. The behavior of another fault under similar conditions may provides clues to the behavior of the specific fault concerned. The dimensions of a fault may provide some clue to the magnitude of future major earthquakes because big earthquakes are often associated with major faults. In several studies, the length of a fault—especially its surface rupture—has been correlated with magnitude of earthquakes. A fault's dimensions are not a completely reliable guide to its seismogenic potential, however, because a fault may not be traceable for its entire length at the surface and because it is often difficult to foresee how far surface rupture may extend along a fault in an earthquake. In that event, the historical record may provide a better estimate of how powerful an earthquake a given fault might deliver. (In the Los Angeles area, a study of the history of earthquakes and the geology of the region indicates that an earthquake comparable in magnitude to the Fort Tejon earthquake of 1857 was about the most powerful seismic event that Los Angeles may expect.) Because the maximum possible earthquake along a fault may occur only rarely, at intervals of a century or longer on the average, estimates of magnitude and frequency may concentrate instead on the most powerful earthquake that may be expected on a particular fault during a shorter interval, perhaps 50 or 60 years, since this is about the longest period many buildings are expected to last.

One problem with using the historical record to estimate the magnitude and frequency of potential earthquakes is that the historical record does not always extend far enough to make such estimates reliable. This is the case in portions of western North America where records of earthquake activity cover only a few decades. The historical record may be supplemented by studies of the geologic record, which can reveal evidence of prehistoric earthquakes in the form of surface faulting and disturbance to sediments. Such evidence may yield estimates of average slip rate and the average interval between major earthquakes along a specific fault. Studies of the San Andreas Fault in southern California have indicated that major earthquakes there have occurred at an average interval of approximately 150 years. (See also PLATE TECTONICS.)

fault creep Motion that occurs gradually along a fault without constituting an earthquake. (See also SEISMOLOGY.)

fault plane A plane-like surface along which masses of rock have moved along a FAULT. (See also SEISMOLOGY.)

feldspar A category of silicate minerals that make up much of the earth's crust.

Fernandina, caldera, Galapagos Islands, Ecuador The shield volcano Fernandina has a summit caldera whose floor collapsed following eruptions in 1968. Two days after a moderately strong earthquake about 200 miles north of the island, Fernandina began erupting on May 20, 1968. This eruption lasted several days. Further earthquakes on June 11 immediately preceded an explosive eruption that morning, followed by another explosion later in the day. The eruption continued for about one more day, and then the caldera floor started collapsing. The greatest subsidence (more than 1,000 feet) occurred at the southeast end of the caldera. Several hundred moderately powerful

Fire fountains some 10 meters (30 feet) high at Kilauea, 1975 (Earthquake Information Bulletin/USGS)

earthquakes occurred during the collapse, which apparently took place steadily over a period of several days rather than all at once. The volume of the collapse was as much as 10 times greater than the small volume of magma that erupted. Fernandina erupted again in 1978, an hour following an earthquake. Another eruption occurred between late 1980 and early 1982 but was so small that it went unnoticed until later. Activity resumed briefly in 1984, between March 30 and April 11.

fire fountain A prolonged spray of MAGMA above a volcano's vent. (See also HAWAIIAN ISLANDS; LAVA.)

Fisher, caldera, Aleutian Islands, Alaska, United States Fisher caldera is located on Unimak Island. A tremendous eruption believed to have occurred here several thousand years ago resulted in the formation of a caldera approximately 14 miles long and 8 miles wide. In historical times, the record of Fisher's activity is uncertain. An exceptionally noisy eruption in 1850 may have involved Fisher. There is an active field of fumaroles in the caldera.

fissure In general terms, a crack in the earth's crust. Fissures commonly form during strong earthquakes. When lava emerges in large quantities from a fissure, the phenomenon is knows as a fissure eruption. The resulting flow of lava is called a fissure flow.

fissure eruption A volcanic eruption that emanates from a fissure or set of fissures in the earth's crust.

flood basalts Wide, relatively thin sheets of basaltic lava that flowed outward from fissures in much the same manner as floods of water. Flood basalts cover much of the Pacific Northwest of the United States. An individual flow may have a volume of more than a hundred cubic miles. (See also COLUMBIA PLATEAU.)

Florida, United States The state of Florida is generally free from strong earthquakes, but a region of minor seismic risk is found in the northern portion of the state along the border with Georgia. One notable earthquake, of Mercalli intensity VI, occurred in the vicinity of St. Augustine on January 12, 1879, and affected an area of about 25,000 square miles. The earthquake rattled windows and doors in Daytona Beach and caused minor damage in St. Augustine. On October 31, 1900, Jacksonville experienced an earthquake of Mercalli intensity V that included eight separate shocks.

focus The underground point on a fault plane where an earthquake originates. (See also FAULT; SEISMOLOGY.)

Fort Tejon earthquake, California, United States The Fort Tejon earthquake of January 9, 1857, is thought to be the most powerful earthquake to have struck the Los Angeles area since Spanish exploration began. Although no instruments were available to measure the earthquake's magnitude, comparisons with the more accurately measured 1906 SAN FRANCISCO earthquake indicate the Fort Tejon earthquake was considerably more powerful. The epicenter appears to have been located on the San Andreas Fault near Cholame, and there is evidence of surface faulting for more than 360 kilometers (about 216 miles) associated with this earthquake. The earthquake was felt for almost the entire width and length of California, from San Diego to the vicinity of Sacramento, and was perceived as a strong shock in Yuma, Arizona. The earthquake is named for Fort Tejon, a U.S. Army installation built at Tejon Pass in the Tehachapi Mountains. Fort Tejon, a collection of buildings constructed of adobe, was destroyed by the earthquake. Although the earthquake caused no fatalities at the fort, the soldiers stationed there were forced to spend the rest of the winter in their tents because rebuilding was impossible during the cold weather. One fatality was reported at lower altitudes: An elderly woman was killed in a collapsing building. The earthquake is said to have had dramatic effects on local rivers, throwing the Mo-

Fumarole area at Yellowstone National Park, Wyoming that developed later into a geyser (*Earthquake Information Bulletin*/USGS)

kelumne River out of its bed and making the Kern River reverse its flow and spill over its banks.

fractionation The process in which a descending plate of crustal rock along a subduction zone melts on its way into the asthenosphere and the plate's lighter components rise toward the surface to provide magma for volcanic eruptions.

Friuli, Italy The Friuli region in northern Italy was the site of highly destructive earthquakes beginning May 6, 1976. The first earthquake struck Friuli in the evening and was estimated at 6.0 on the Richter scale. More than 900 people were killed, and some 50,000 were left homeless. Another strong earthquake in September shook down reconstructed buildings.

Fuji, Mount, Japan Mount Fuji, known to the Japanese as Fuji-san (not Fujiyama), is an andesitic volcano produced by the convergence of the Pacific, Eurasian and Philippine crustal plates. Although Mount Fuji has been less active in recent centuries than some other volcanoes in Japan, it has erupted at various times in history, notably in 1707, when the heat from the eruption resulted in fractured and shattered rocks and fallout from the eruption covered the present location of Tokyo to a depth of several inches. Eruptions of Mount Fuji have created a naturally partitioned set of lakes at the foot of the volcano. Once the lakes were one single large lake, but lava flows divided the lake into five smaller lakes, named Kawaguchi, Motosu, Sai, Shooji and Yamanaka. (See also ANDESITE.)

fumarole A hole in the earth's surface through which hot water, steam and various hot gases escape from underground. (See also GEYSER; HYDROTHERMAL ACTIVITY; VALLEY OF TEN THOUSAND SMOKES.)

Furnas, caldera, Azores The Furnas caldera is located on San Miguel Island in the Azores and is noted for a violent eruption in 1630. Earthquakes on September 2–3 of that year were followed by the eruption of two fiery clouds. Some 200 people are thought to have died in this eruption, most of them killed by mudflows. Several other eruptions appear to have occurred within the last 1,000 years.

G

gabbro A rock of the GRANITE family, characterized by coarse grains and low silica content, gabbro is made up of plagioclase feldspar and sometimes other minerals such as olivine and hornblende. Gabbro may serve as an ore for precious metals including silver and gold.

Galapagos Islands, Ecuador The Galapagos Islands are located off the western coast of South America and are the summits of volcanoes along the worldwide undersea rift zone. The islands are perhaps best known for their giant tortoises and for remarkable variations in the bodily forms of animal populations from one island to another. The most active of the volcanoes in the Galapagos Islands is Fernandina, a major eruption of which took place in 1968.

Galung Gung, volcano, Java Not quite 7,000 feet high, Galung Gung is adjacent to the Plain of Ten Thousand Hills, which actually number several thousand less than that and are thought to have been formed by a LAHAR that accompanied one of Galung Gung's eruptions. Galung Gung erupted twice in 1822 and destroyed no fewer than 114 villages and killed more than 4,000 people.

Garibaldi, Mount, British Columbia, Canada A volcanic mountain at the northern end of the Cascade range.

Gaua Island, volcano, Vanuatu Located in the northern portion of the Vanuatu archipelago, Gaua Island is a stratovolcano with a CALDERA occupied by a lake, Steaming Hill Lake, and by a postcaldera cone, Mount Gharat. From 1963 to 1982, Mount Gharat underwent frequent eruptions.

Geger Halang, caldera, Java, Indonesia The Geger Halang caldera is located in central Java near Juningan and Telaga and has a history of eruptive activity dating back to the late 17th century. How the caldera originated is uncertain. Possibly the caldera collapsed, but landslide activity has also been put forward to account for the caldera's formation. Before the collapse, a stratovolcano is thought to have occupied the site of the caldera. The stratovolcano Cereme is situated on the caldera's northern rim. An explosive eruption in 1698 reportedly caused many deaths. Other eruptions are recorded in 1772, 1775 and 1805. Earthquakes and subsidence in 1876 at a site several kilometers away from the caldera may have been unrelated to volcanism at Geger Halang. Emissions of sulfur gas increased in 1917 as did fumarolic activity in 1924. A series of eruptions in 1937 and 1938 was accompanied by a large number of strong earthquakes. An eruption in 1951 consisted of a single detonation followed by an emission of thick smoke. Tectonic earthquakes occurred in 1973 but evidently were unrelated to volcanism at the caldera.

geomorphology The study of surface landforms and their evolution.

Georgia, United States Although the state of Georgia is not itself highly susceptible to earthquakes, it has been affected strongly on occasion by powerful earthquakes in neighboring states, such as the New Madrid earthquake and the Charleston, South Carolina earthquake. An earthquake in northern Georgia on Nov. 1, 1875, affected an area of about 150 miles by 200 miles. Portions of the state, especially along its coast, are vulnerable to damage from liquefaction in the event of future strong earthquakes.

geothermal energy Geothermal energy may be defined as energy derived by tapping the earth's internal heat. Sources of geothermal energy may include GEYSERS, hot springs and VOLCANOES. Although facilities for exploiting geothermal energy have been built at numerous locations in several countries, the potential for utilizing this source of energy appears to be more limited than some projections would indicate because only in certain areas can heat from the earth's interior be exploited effectively. In these areas, heat from within the earth is concentrated in formations at or near the surface where the heat is stored and can be tapped conveniently.

The geology of geothermal energy resources is complex, but commercially exploitable geothermal resources tend to be concentrated along areas of crustal spreading, where new crust solidifies from molten rock rising to the surface, and along plate boundaries, where converging plates give rise to conditions favoring increased heat flow to the surface. In either case, magma rising from below brings large amounts of heat toward the surface. Heat from the magma converts underground water into geothermal systems, which may be dominated either by liquid or by vapor. The former tend to be high-pressured, whereas the latter are comparatively low-pressured. Geothermal systems have been used for centuries on a small scale for heating, but application of geothermal energy to other uses began only in the late 18th and early 19th centuries, starting in Italy.

Development of geothermal energy resources in New Zealand began shortly after World War II, and the first such facility there was finished in 1958, at the Wairakei fields. Wairakei derives its energy from a huge volume of hot rock believed to be supplied with water by rainwater seeping down from the surface. Development is thought to have touched only part of New Zealand's geothermal energy resources. Geothermal heating facilities are widespread in ICELAND, which lies along the Mid-Atlantic Ridge and is the site of intense and frequent volcanic activity. In this arrangement, hot water from underground is distributed from central locations to users and achieves good results with less pollution than other heating systems based on combustion of fossil fuels. Geothermal heating in Iceland also is used in greenhouses and in various industrial processes such as drying seaweed and washing wool. The preeminent geothermal power facility in the United States is located at The Geysers in northern California near San Francisco and is operated by Pacific Gas and Electric Company. Other sources of geothermal energy exist at numerous locations in the United States, and some are used for heating, but other factors, such as distance from major population centers, restrict the availability of these heat sources for generation of power or other large-scale commercial applications. Japan and Mexico also have, or have plans to develop, geothermal power facilities.

Generating electricity through geothermal energy must deal with various problems such as materials dissolved in underground hot water; these materials may corrode metals and build up on turbines.

In some locations, earthquakes are associated with high heat flow from the earth's interior. Sometimes the earthquakes are only minor, but some disturbances may be much more powerful. Earthquakes in geothermal areas are thought to be connected to motion on faults along which geothermal fluids flow. Volcanic activity also is associated with areas suitable for geothermal power production because volcanoes themselves constitute areas of high heat flow to the surface from within the earth. Although geothermal energy sometimes has been advocated as a "clean" alternative to burning fossil fuels or relying on nuclear fission to produce electrical power, the scarcity of exploitable sites means that geothermal energy is likely to remain, except in a few especially favored areas, a minor and supplementary source of energy. (See also LARDARELLO; PLATE TECTONICS.)

geyser A jet of water that erupts on an occasional basis from a small opening in the earth's surface. (The hole itself also may be called a geyser.) In a typical geyser, heat from rocks underground vaporizes water at depth. Eventually, pressure from the vapor thus created forces overlying water out through the opening, creating a fountain of very hot water and steam that may reach heights of several hundred feet. The cycle of vapor generation and eruption then starts over. Among the most famous geysers is Old Faithful at Yellowstone National Park in Wyoming, United States. Some geysers derive the vapor for their operation from a different source—release of dissolved carbon dioxide.

glaciation The formation of glaciers, great fields of ice resulting from the accumulation and compression of snowfall over long periods, has helped shape the peaks of numerous volcanoes. A case in point is Oregon's Mount Hood, where glaciers have carved valleys in the flanks of what apparently once was a conical mountain. The water contained in glacial ice can melt and generate destructive mudflows during eruptions. Mudflows pose a par-

An erupting geyser at Yellowstone National Park, Wyoming (D. E. White, USGS)

ticular threat to settled areas near the volcanoes of the Cascade Mountains in the Pacific Northwest of the United States. Avalanches also may occur when eruptive activity melts glacial ice on a volcano's summit and weakens the structure of the ice sheets. At three locations in the United States, Mount Wrangell in Alaska and Mount BAKER and Mount RAINIER in Washington, heat and steam from the volcanoes have created a network of caves and passageways in the glacial ice at the summits of the volcanoes. Floodwaters from a glacial lake in what is now western Montana are thought to have played an important part in shaping the landscape of the COLUMBIA PLATEAU in Washington state, stripping away soil and scouring the land down to bedrock. In some places, the floodwaters apparently removed whole layers of BASALT and thus produced clearly visible MESAS and terraces along the Columbia River. (See also WRANGELL MOUNTAINS.)

gold One of the most economically valuable minerals, gold is highly ductile and malleable, a good conductor of electricity and resistant even to powerful solvents. Gold is found in both native (or pure) form and as ores in areas of volcanic activity, where chemical fractionation and hydrothermal activity combine to concentrate the metal. Gold often occurs as minerals called tellurides and also may be found in native form in veins of quartz. Deposits of gold along the Andes Mountains played an important part in the conquest and colonization of South America's Inca civilization by the Spanish empire in the early 16th century. Drawn to the Andes by reports of abundant gold, Spanish troops under the leadership of Francisco Pizarro conquered the Incas by simply taking their emperor Atahualpa prisoner. The Spaniards demanded a heavy ransom in gold for their imperial captive and, when the ransom was delivered, had Atahualpa killed. The Incas' gold reportedly was mined largely from placer deposits, which occur when bits of gold are eroded away from their original location (the mother lode) and laid down in sediments from which they may be extracted more easily than from solid rock. (See also PLATE TECTONICS.)

Good Friday earthquake, Alaska, United States One of the most powerful and destructive

The 1964 Good Friday earthquake in Alaska caused this collapse along 4th Avenue near C Street in Anchorage (USGS)

earthquakes of the 20th century, the Good Friday earthquake struck the southern coast of Alaska along Prince William Sound on March 27, 1964 and lasted between three and five minutes. The earthquake killed more than 100 people and involved displacements of up to 50 feet along various faults. Estimates of property damage from the earthquake range in the hundreds of millions of dollars. The Good Friday earthquake was notable for the tsunami that accompanied it. The wave destroyed the waterfront at Seward, and at Kodiak, the tsunami wiped out much of the downtown area and destroyed almost half of the local fishing fleet. The tsunami caused extensive damage as far south as CRESCENT CITY, California. More than 2,000 landslides and avalanches were attributed to this earthquake, and at one location at Shattered Peak in the Chugach Mountains, an avalanche was found to have traveled several miles atop a cushion of air across a glacier. Later examination showed that the air cushion had preserved structures on the glacier's surface as the avalanche passed. The earthquake had dramatic effects on lakes in Alaska; movement of lake waters cast chunks of ice onto the shore and reportedly caused damage as high as 30 feet to trees. Salt water invaded certain freshwater lakes along the shore. The Good Friday earthquake coincided with unusual observations in other portions of the United States; the water level at one well in South Dakota, for example, is said to have fluctuated more than 20 feet. Similar, though less dramatic, fluctuations were reported from Puerto Rico and Australia at the time of the Good Friday earthquake.

Gorely Khrebet, caldera, Kamchatka, Russia The collapse that created the Gorely Khrebet caldera is thought to have had a volume of perhaps five cubic miles. The caldera is located on the edge of a large negative gravity anomaly that is evidence of the existence of a huge, buried caldera. The Gorely Khrebet volcano occupies the center of the caldera. Eruptions have been recorded in 1828, 1832, 1855, 1869 (uncertain) and 1929–31. Solfataric activity was reported in 1947. Temperatures of fumaroles increased in 1960–61, and small eruptions of ash followed. Fumarolic activity diminished for several years, then increased again in

The 1964 Good Friday earthquake in Alaska devastated the Turnagain Heights district of Anchorage (Earthquake Information Bulletin/USGS)

the late 1970s as new fumaroles appeared and temperatures in fumaroles rose. Gas plumes rose to heights of several hundred meters in 1979. In June of 1980, PHREATIC ERUPTIONS began, and others took place over the following months. Another eruption, similar to that of 1980–81, occurred in 1984–85.

graben A valley formed by the down-dropping of a fault block. The Rhine River valley is a famous example of a graben. During the 1964 GOOD FRIDAY EARTHQUAKE in Alaska, grabens formed and caused extensive damage to property. In one incident, a building toppled off the edge of a horst, or elevated block, and landed upside down inside a graben.

Grand Banks earthquake This very powerful earthquake occurred on November 18, 1929, and was centered under the Grand Banks of Newfoundland, a rich fishing area off the eastern coast of Canada. The earthquake is thought to have measured 7.2 on the Richter scale. Felt all through the New England region of the United States and parts of Canada located south of the St. Lawrence River and the Strait of Belle Isle, the earthquake stopped clocks and shook objects from shelves on land and subjected ships at sea to a powerful shaking. A tsunami, or seismic sea wave, associated with this earthquake reportedly was responsible for extensive damage and some loss of life at Placentia Bay, Newfoundland; minor waves were reported as far south as the shores of South Carolina. The earthquake broke several submarine telegraph cables that were laid across the area of the earthquake's epicenter. (See also LANDSLIDE; TURBIDITY CURRENT.)

Grand Coulee, Washington, United States A deep canyon cut into lavas by ancient floodwaters, Grand Coulee is about 25 miles long and 800 feet deep in places. The Grand Coulee runs approximately northeast to southwest between the Columbia River and the Quincy Basin and is the site of the Grand Coulee Dam. The waters that carved the Grand Coulee in the lavas are thought to have originated from the melting of a glacier immediately to the north. (See also COULEE.)

granite A crystalline, igneous rock, granite is one of the most widespread rocks and is commonly used as a building material. Granite displays an interlocking pattern of crystals of such minerals as quartz and feldspar and comprises an entire family of rocks. Granite is an intrusive rock, as distinguished from an extrusive rock such as basalt. Coarse-grained granite cooled slowly from the molten to the solid state and thus gave large crystals time to form. Fine-grained granite, on the other hand, cooled more quickly, so that crystals in the rock had less opportunity to grow. Granitic rocks often are found in tremendous masses. (See also INTRUSION.)

granitization Conversion of sediments to GRANITE by hot water and gases moving upward from deep inside the earth. Granitization is believed to be responsible for forming PLUTONS that show no clear boundary with surrounding COUNTRY ROCK.

Great Basin, United States The Great Basin is an area in California, Nevada and western Utah where streams find no outlet but drain instead into lower portions of the basin itself. DEATH VALLEY occupies part of the Great Basin. It is part of the BASIN AND RANGE PROVINCE.

Great Rift, United States An area of parallel fractures in present-day Idaho from which lava flows emerged and covered an area of some 600 square miles. The most recent volcanic activity along the Great Rift is believed to have occurred approximately 2,000 years ago.

Great Rift Valley, Africa A region of east central Africa characterized by rifting and associated volcanic activity. Some of the world's most famous volcanoes and calderas are located along the Great Rift Valley. (See also ASAWA; FANTALE.)

Green's function A mathematical function used in seismology to represent ground motion resulting from instantaneous slip on a given point on a fault. Because slip may occur in any or all of three directions (designated along axes x, y and z), Green's function may be represented as a vector G with three components. Various limitations influence the accuracy of Green's function, including uncertainties about changes in the earth's structure (which may affect the velocity of earthquake waves) and problems involved with computing the function (approximations must be used in many cases). Much computer time and computational effort may be required to calculate a Green's function, but once the function has been obtained for a specific location, it may be stored and used to analyze many different hypothetical earthquakes for that location. Sometimes, small earthquakes that occur along a given fault surface may serve as Green's functions for a larger, hypothetical event along the same surface.

Grimsvötn, volcano, Iceland The 1934 eruption of Grimsvötn, located underneath the Vatnajökull glacier, is a good example of a jokulhlaup, or subglacial volcanic eruption. In late March, an unseasonable increase in the volume of flow in the Skeidara River, which carried away melt water from the glacier, was among the initial signs of the eruption. The river was muddy and smelled of sulfur—further evidence of an eruption. The flow of water from beneath the glacier increased over the next several days, and portions of the glacier broke away and were carried off with the water. So great was the flow that it spurted under high pressure from openings in the glacier. The eruption is believed to have created a reservoir of melt water under the glacier. As this pool of water drained away, the glacial ice overlying the volcano subsided until, soon after the jokulhlaup ended, eruptive activity broke through the ice and the volcano expelled gas and ash into the air, depositing a light ashfall over several thousand square miles.

ground motion In general terms, any shaking of the earth's surface resulting from a seismic disturbance.

groundwater In general, water moving beneath the earth's surface. Groundwater plays an important role in liquefaction, the process responsible for much of the damage caused by earthquakes in localities where structures are built on unconsolidated soil with ground water close to the surface. Groundwater is derived largely from rainwater that has percolated downward through the soil and is confined to within about 3,000 feet of the earth's surface. Groundwater may travel through porous underground conduits called aquifers and emerge at the surface in the form of natural springs.

Guagua Pichincha, volcano, Ecuador This volcano, near the capital city of Quito, has not been highly active in the 20th century but was active from the 16th through the 18th century.

Guatemala Guatemala's location in Central America places it in one of the regions most susceptible to earthquakes and volcanic eruptions. Notable earthquakes in Guatemala's history include that of the night of April 18, 1902. This earthquake lasted perhaps 30 to 40 seconds and caused some 2,000 deaths in Quetzaltenango. The earthquake coincided with a torrential rainstorm, and the city lost electrical power and lighting. The resulting darkness reportedly led to numerous deaths when townspeople, fleeing buildings and running into the street, were unable to see where they were going and perished when walls fell on them. The heavy rainfall also resulted in deaths from drowning. Shocks and rainfall continued for three days, making relief work difficult or impossible.

This earthquake occurred at approximately the same time as the great eruptions of Mount PELÉE and SOUFRIÈRE in the Caribbean.

Gutenberg, Beno (1889–1960) American seismologist. Gutenberg, who relocated from Germany to the United States in the years just before World War II, was a colleague of Charles Richter and used earthquake data to estimate the diameter of the earth's core. The boundary between the core and mantle is called the Gutenberg discontinuity.

guyot An undersea, flat-topped volcanic mountain that does not extend to the ocean surface. Geologists believe that guyots once reached the surface and formed islands but wave action reduced them to subsurface levels. Guyots are often found in chains with older, lower mountains at one end and younger, taller guyots at the other.

H

Hakone, caldera, Japan Hakone caldera is located near Sagami Bay on the central island of Honshu near Tokyo. The caldera is thought to have formed through several explosive eruptions. A large eruption, possibly phreatic, is believed to have occurred perhaps 3,000 years ago, along with a pyroclastic flow or landslide that created a natural dam and thus formed Lake Ashinoko. A lava plug formed afterward. Apparently no eruptions of magma have happened here since that time. Although no extremely powerful eruptions have occurred at Hakone within historical times, this caldera is interesting for its seismic activity. Minor earthquakes occurred in early 1917. Other earthquakes in the 1950s and 1960s helped establish correlations between seismic activity and very high hydrothermal temperatures, as well as between earthquakes and intensified activity of fumaroles. There are various explanations for these correlations. One is that hot water or steam rising from great depths causes earthquakes as it expands on the way to the surface. In some areas, it has been suggested, very hot water underground weakens the rock, so that strain is released as earthquakes in these thermal areas. A worker was killed at Hakone in 1933 in an explosion at a solfatara.

harmonic tremors Also known as volcanic tremors, these are small earthquakes, observed in the vicinity of active volcanoes, that indicate molten rock is flowing beneath the surface. Such a tremor has a frequency of several cycles per second. Scientists are not certain how harmonic tremors are generated, but the vibrations have been linked tentatively to turbulence in magma flowing underground and to the emergence of gas bubbles from the molten rock. Although eruptions do not necessarily follow the occurrence of harmonic tremors, the 1980 eruption of Mount St. Helens in Washington state was preceded by harmonic tremors that told geologists that magma appeared to be forcing its way upward toward the surface. (See also SEISMOLOGY; VOLCANISM.)

Haroharo, caldera, North Island, New Zealand The Haroharo caldera and the Okataina volcanic center are located in the Taupo Volcanic Zone on New Zealand's North Island. The area is noted for abundant volcanic and geothermal activity. Haroharo and Okataina have undergone considerable seismic and volcanic activity in the past two centuries. An eruption of TARAWERA on the southern margin of Haroharo in 1886 involved PHREATIC ERUPTIONS that were generated when molten rock underground heated subterranean water. There were apparently some early signs of an approaching eruption, notably earthquakes, increased geyser activity and rising temperatures at hot springs. The eruption began an hour after a series of strong earthquakes on June 10, 1886. A strong geyser, Waimangu, became active in 1900, and very large hydrothermal explosions occurred at the geyser in 1915 and 1917. This latter blast occurred unexpectedly and hurled out rocks and mud, damaging a government tourist department facility nearby. More earthquakes took place in 1962, and in 1973 a small hydrothermal explosion at Waimangu followed an increase in ground temperatures and in the activity of hot springs and fumaroles. A fairly strong earthquake occurred about 100 miles south-southeast of Waimungu approximately half an hour before the explosion, but the earthquake may have been unrelated to the explosion. Earthquake activity increased at Okataina in 1982 and 1983, and considerable seismic activity continued through the late 1980s.

Hawaiian Islands, United States The Hawaiian Islands constitute an archipelago in the midst of the Pacific crustal plate and are the site of some of the most intensively studied volcanic activity.

Lava fountain in Hawaii, 1924 (H. T. Stearns, USGS)

The relatively gentle character of eruptions of Hawaiian volcanoes, compared to the more explosive and destructive activity of volcanoes such as Bezymianny and Mount St. Helens, makes them easy to observe. Hawaiian volcanoes include MAUNA KEA, the tallest volcano on earth; MAUNA LOA; and KILAUEA. Unlike the magma that supplies volcanoes with histories of explosive eruptions, Hawaiian magma tends to be low in dissolved gases. It is also low in silica and rich in ferrous oxide. Lavas from Hawaiian volcanoes are generally basaltic and extremely fluid. The Hawaiian chain includes the LOIHI SEAMOUNT, a volcano that represents the youngest peak among the islands but that has not yet reached the surface of the ocean.

Hawaii, the major island of the chain, is made up of several volcanic structures: Kohala Mountain, at the northern tip of the island; Mauna Kea, in the north central sector; Mauna Loa, immediately to the south of Mauna Kea; Kilauea, on the southeast shore; and Hualalai, on the western shore near Kealakekua Bay. Mauna Kea is the taller of the two principal volcanoes on the island (the other principal volcano is Mauna Loa) and is thought to be inactive. Mauna Loa is active and is situated atop two rift zones running northeast and southwest from the mountain. These rift zones have emitted lava flows frequently in the last two centuries, and in some cases, the flows from the northeastern rift zone have come very close to the city of HILO, about 50 miles northeast of the volcano's summit. The fast-flowing lava from Mauna Loa's eruptions may reach velocities of more than 30 miles per hour. In some locations where lava from Mauna Loa has reached the sea, it has cooled rapidly into dark volcanic glass that has been fragmented by wave action and formed striking beaches of black sand. A spectacular eruption of Mauna Loa in 1950 lasted two weeks and covered more than 30 square miles of the island with lava. Fountains of molten rock shot hundreds of feet into the air from several miles of fissures along the southwestern rift zone. The emissions of lava caused considerable property damage and flowed across the coastal highway. Kilauea, to the southeast of Mauna Loa, erupts more often than Mauna Loa but is perhaps the safest active volcano on earth to approach because of its predictable and nonexplosive eruptions. On the floor of the Kilauea caldera is a huge lava lake inside Halemaumaua Crater. The level of lava in the crater fluctuates. On several occasions in the 20th century, lava has spilled out of the crater. Most of the time, however, the lava remains at a safe level, and sightseers may approach the rim of the crater in safety. At night, visitors may see a dazzling display on the crater floor as solidified lava on the surface cracks and allows the glowing molten rock underneath to be seen, forming dramatic patterns of light. Two rift zones running east and southwest from Kilauea have been responsible for most eruptive activity at the volcano in the recent past. A curious effect of the lava flows produced "lava trees" along the eastern rift zone where lava flowed around live trees and sheathed their trunks. When the trees themselves died and were carried away, the sheaths, or "lava trees," remained. These columnar formations may be seen at Lava Tree State Monument. In 1790 along the southwestern rift zone, toxic gas from an eruption caused numerous deaths.

Oahu, site of the Japanese attack on the United States naval base at Pearl Harbor in 1941, was formed by eruptions of two large volcanoes. One had its caldera where the Waianae Range on the western side of the island stands today. Mount Kaala, a flat-topped mountain approximately 4,000 feet high at the northwest corner of the island, comprises a portion of the original Waianae volcano. The second major volcano's caldera is thought to have been located on the eastern side of the island, at what is now Kaneohe Bay along the Koolau Range. This latter volcano is known for its great numbers of basalt DIKES that may be seen along roadsides in the eastern part of Oahu. The famous Diamond Head appears to be the product of lesser and later volcanism than Koo-

Ground cracks along crater rim at Kilauea volcano, Hawaii, 1983 (*Earthquake Information Bulletin*/USGS)

Mauna Ulu shield, Hawaii, 1970 (D. A. Swanson, USGS)

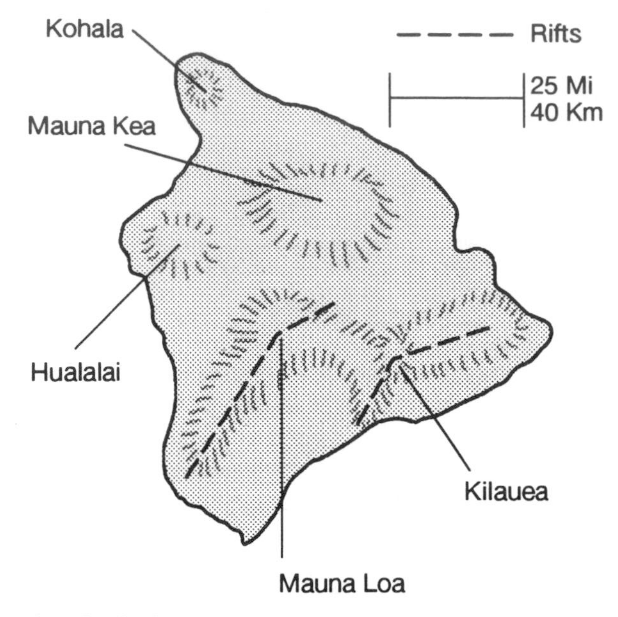

The island of Hawaii showing the location of its volcanoes (© Diagram)

lau's. Diamond Head is part of a chain of volcanic formations that also includes Sugarloaf and Round Top. Another chain of volcanic formations extends along the southeastern shore of Oahu from Koko Head crater near Hanauma Bay to Makapuu Head. Other craters on Oahu include Punchbowl (site of the National Memorial Cemetery of the Pacific), Makalapa and Aliamanu.

The island of Kauai is the location of a single great volcano called Waiualaele, more than 5,000 feet high and surmounted by a caldera some 10 miles wide. Rainfall has eroded the deep Waimea Canyon in the flank of the volcano, exposing the layered evidences of past eruptions. Erosion also has produced spectacular formations along nearby Honopu Valley.

The island of Maui has two volcanoes. On the southeastern side of the island, Haleakala, the larger of the two volcanoes at more than 10,000 feet high, has an exceptionally large crater and still emits steam from time to time. A lava flow that passed through a breach in the crater, the Kaupo Valley gap, reached the sea. The other volcano on Maui is West Maui, on the island's northwestern side. The summit of West Maui, Puu Kukui, is almost 6,000 feet high.

Molokai Island has two volcanoes and is remarkable for its tall cliffs eroded by wave action along its shores. A comparatively recent addition to the island is the Kalaupapa Peninsula, formed by eruption of a small volcano.

Tsunamis are an occasional hazard in the Hawaiian Islands, which stand exposed in mid-Pacific to seismic sea waves accompanying earthquakes along the so-called RING OF FIRE surrounding the Pacific Ocean basin. In 1946 a tsunami from an earthquake in the Aleutian Islands caused extensive damage at Hilo and attained great heights at various points in the Hawaiian chain: 45 feet on the northern shore of Kauai, 55 feet on the northern coast of Hawaii and 35 feet on the northwest coast of Oahu. The islands were battered by another powerful tsunami in 1960, this one approaching from the direction of South America. Waves reached heights of 35 feet on Hawaii, 13 feet on Oahu and 14 feet on Kauai.

Most earthquake activity in Hawaii occurs in the vicinity of the island of Hawaii. The documented history of earthquakes in Hawaii extends back to the early 19th century. An earthquake on February 19, 1834, knocked down stone walls on the island of Hawaii, stopped clocks and made it difficult for standing persons to remain upright. On December 12, 1838, another earthquake similar in its effects to the 1834 earthquake shook Hawaii.

On April 2, 1868, one of the strongest and most destructive earthquakes in the history of Hawaii occurred near the southern coast of the island. The earthquake caused heavy damage to wooden homes and straw houses. Numerous walls were shaken down in Hilo, and landslides occurred as well. Fissures formed in the ground, and mud appeared in brooks. Effects of the earthquake were especially remarkable at Kohala, where ground waves one to two feet in amplitude were reported, and the shock stopped machinery at a sugar mill, including a big 75-horsepower engine that was operating under a full head of steam at the time of the earthquake. On Maui and Lanai, more than 100 miles from the epicenter, rumbling noises were reported, and buildings shook. In Honolulu, the shock was strong enough to stop clocks. The earthquake was felt more than 300 miles distant, on Kauai and Oahu. On Hawaii, a tsunami estimated at 60 feet high or more came ashore along the southern coast and reportedly swept over the tops of palm trees; there were several fatalities, and numerous houses were destroyed. This wave was 10 feet high when it reached Hilo and eight feet high at Kealakekua. A fissure more than two miles long formed at Kohuku, and a volcanic eruption occurred at this fissure on April 7. The earthquake was preceded by foreshocks, one of which, on March 28, was sufficiently powerful to knock down stone walls.

On April 26, 1973, an earthquake of magnitude 6.2, centered near the northeastern shore of the island of Hawaii, affected a large area and was felt as far away as Kauai. The earthquake generated effects of Mercalli intensity VII and caused an estimated $5.6 million in property damage in the area of Hilo. There were 11 injuries, but no one was killed. Two people died in the November 29, 1975 earthquake (magnitude 7.2) that was felt all through the island of Hawaii and also on Oahu, Molokai and Lanai; damage was estimated at $4.1 million, including some $1.5 million in damage from a tsunami that came ashore as a wave some 18 feet high near the epicenter of the earthquake. Numerous foreshocks and aftershocks occurred, and a small eruption was noted at Kilauea volcano less than an hour after the earthquake. (See also PLATE TECTONICS.)

Hayward Fault, California The Hayward Fault runs through the East Bay suburbs near SAN FRANCISCO and was responsible for powerful earthquakes in 1836 and 1868 that destroyed buildings in San Francisco and Oakland. Although less famous than the nearby SAN ANDREAS FAULT, the Hayward Fault has the potential to cause tremendous destruction in a future earthquake because the East Bay is so densely populated. The fault runs through the University of California campus at Berkeley and lies directly under the university's Memorial Stadium. Also, many major highways in the San Francisco Bay area either cross the Hayward Fault or may be affected by damage from any strong earthquakes involving that fault in the future. Among medical facilities in Contra Costa and Alameda Counties, eight acute care hospitals, particularly important in the aftermath of any major earthquake, are at risk because they are located within a mile of the Hayward Fault. These hospitals represent about 30% of such hospitals available to the counties. One projection of the effects of an earthquake of Richter magnitude 7.5 along the Hayward Fault puts the possible number of fatalities as high as 7,000, with the potential for more than 13,000 injuries requiring hospitalization and lesser injuries affecting more than 130,000 individuals. This projection involves an earthquake in early afternoon. Another projection for the hours just after midnight puts fatalities lower, at perhaps 1,500, and injuries around 50,000, with more than 4,000 of those requiring hospitalization. An earthquake of magnitude 7.5 on the Hayward Fault is expected to cause the greatest damage on the western side of the fault, within an area about five miles from the fault. This area encompasses such heavily settled areas as San Leandro and San Lorenzo.

heat flow The flow of heat from the earth's interior to the surface varies greatly from one location on the surface to another. The total heat release at the surface has been expressed as the sum of heat transmitted by conduction (transmission of heat directly through stationary masses of rock) and heat carried by mass transport (that is, the migration of magma from lower levels to the surface). Some areas, such as active volcanoes, represent an unusually vigorous flow of heat to the surface, whereas other parts of the earth's surface are relatively cool. The rate of heat flow to the earth's surface is measured in units called HFUs. One HFU equals one-millionth (that is, 0.000001) calorie per square centimeter per second. Depending on how it is estimated, the mean global rate ranges between about 1.0 and 1.5 HFUs, although in some areas of active volcanism and hydrothermal activity, measurements well above 200 and even 700 HFUs have been made. Various factors are believed to influence the rate of heat flow. For example, radioactivity and the heat released by decay of RADIONUCLIDES may vary from one location to another. Spatial variations in temperature of the mantle also may account for some differences in heat flow. Heat flow shows great variability in North America, with the western half of the nation generally showing higher and more variable HFU values than the east. The dividing line between east and west has been drawn approximately at the eastern boundary of the Colorado Plateau. The higher heat flow in the west is reflected in the relative abundance of hot springs, geysers, and other HYDROTHERMAL ACTIVITY there, as well as the existence of numerous volcanoes, both active and extinct. Eurasia likewise shows great geographical variations in heat flow. "Hot spots" are found, for example, in the Alps, the Carpathian Mountains and the Caucasus Mountains and in Russia's KAMCHATKA Peninsula, with its numerous volcanoes. High heat flow shows up in areas of strong hydrothermal activity, such as LARDERELLO in Italy.

Hebgen Lake, Montana, United States An earthquake in 1959 caused dramatic changes in the shoreline of Hebgen Lake. The lake bed tilted and displaced the lake toward the north, submerging docks and other waterfront property along the northern shore and lifting the southern shore out of the water. The earthquake caused considerable disturbance in the waters of the lake. Near the lake, an escarpment some 10 feet high formed along a fault more than 10 miles in length. This same earthquake caused a huge rockslide, containing almost 40 million cubic feet of rock, along the southern wall of the gorge of the Madison River in the Madison Range. This rockslide dammed the river and created a lake named Earthquake Lake.

Heimaey, volcano, Iceland The 1973 eruption of the volcano Heimaey near the port of Vestmannaeyjar in Iceland's Vestmannaeyjar archipelago is one of the most famous eruptions of modern times and one of a series of more than a dozen eruptions in Iceland since mid-century. The island is about five miles long by three miles wide and is made of basalt only several thousand years old.

The eruption of Heimaey began in the early morning of January 23, 1973, on the eastern side of the island along a fissure running roughly north-south and extending more than a mile in length. Lava fountains emanated from the fissure at first, and volcanic activity occurred underwater at the northern and southern ends of the fissure for the first several days. Later the eruption became concentrated in a relatively small portion of the fissure near the community of Helgafell. In about two days, a cinder cone more than 300 feet high formed. This cone was named Eldfell (fire mountain) and expelled tephra at a rate of more than 100 cubic yards per second. The tephra fell in large amounts on nearby Vestmannaeyjar. Less than a month after the eruption began, the volcano's output of tephra diminished, and lava began to flow from the volcano. A large lava flow encroached on Vestmannaeyjar, and it looked for a time as if the lava might fill the harbor on the northern side of the island. Submarine volcanism continued during this period and broke a power cable that carried electricity from the mainland as well as a fresh water pipeline that served the island.

In late February, Eldfell stood more than 600 feet tall, and lava was flowing gradually along a front ranging from north to east of the volcano. Within several weeks, this flow stood more than 60 feet high at some locations along its front. The average depth of the flow exceeded 100 feet, and at places the lava was more than 300 feet deep. By April the lava flow was about 1,000 yards long and equally wide and moving at a speed of several yards daily. Another lava flow originated from the volcano in late March and began moving northwest. This flow destroyed numerous homes and the town's power facility. Portions of the cone broke away and were carried away with the initial lava flow, including one especially large block known as *Flakkarinn* (the wanderer). Submarine

A house overwhelmed by a lava flow from an eruption of Heimaey in Iceland (USGS)

volcanism continued through late May, and the eruption finally subsided in July.

The eruption produced an estimated 300 million cubic yards of lava and more than 25 million cubic yards of tephra. Among the other products of the eruption were concentrations of poisonous gases—mostly carbon dioxide, with some carbon monoxide and methane—that accumulated in low portions of eastern Vestmannaeyjar and killed one person. The origin of the poisonous gases is uncertain, although it appears likely that the gases migrated from the conduit of the volcano through older volcanic rock and into Vestmannaeyjar. Bulldozers pushed a wall of tephra into place between the vent and the community to halt the gas, and a lengthy trench was dug to carry away steam, but neither of these measures worked with total effectiveness. A rapid evacuation from the island saved almost all of the more than 5,000 residents of the island from harm. Property damage was extensive. Residences near the volcano were destroyed by tephra and lava. Flows of lava from the volcano also destroyed a fish-freezing facility, caused damage to two other such facilities and wrecked a large number of homes in the eastern portion of the community.

The 1973 eruption on Heimaey was notable for efforts to counteract and control the flows of lava. Experiments and observations performed early in the eruption of Heimaey, and also at the earlier eruption of the volcano SURTSEY, indicated that spraying seawater on the lava could cool and harden the molten rock and thus impede its flow. Because the northward- and eastward-moving lava posed the greatest threat to property and activities on the island, including the operation of the harbor, lava-control efforts concentrated on that front. One approach was to spray water on the advancing lava. Another was to build a barrier of lava on the northwest side of the lava flow to block its progress into Vestmannaeyjar. Early in February, about two weeks after the eruption started, a water-spraying operation began. Results were encouraging, and in March a ship capable of directing large amounts of water onto the lava flow was moved into the harbor. Additional pumping equipment was brought in from the United States. Water was pumped straight onto the lava near the harbor and was carried to various portions of the flow through a system of metal and plastic pipes almost 20 miles long. The lava-control effort used a combination of water-spraying and earth-moving activities.

The cooling operation caused a dramatic change in the appearance of the lava flow. When left to cool naturally, the lava solidified into a reddish mass with a surface covered with volcanic BOMBS and varying about three feet in relief. After water cooling, the surface turned gray or black and displayed much greater relief, up to about 15 feet. Cooling the lava with water created peculiar difficulties, such as reduced visibility caused by the large quantities of steam produced by the water contacting the molten rock. Some eight million cubic yards of water were directed onto the lava flows at Heimaey and are thought to have solidified some five million cubic yards of lava. Water cooling appears to have accelerated the solidification of the lava by as much as 100 times. The cooling operation lasted until early July and cost less than $2 million.

One beneficial result of the 1973 eruption of Heimaey was a heating system for Vestmannaeyjar that utilized heat from the still-cooling lava flows. Initial investigations indicated that heat from lava and scoria deposits could be used to supply space heating for the community. Early experiments along these lines met with success, and houses began to be connected with a heating system that exploited the heat from the lava and tephra. Projects were under way by 1979 to exploit heat in several areas of the fresh lava flows. In each area, a set of steam wells extending down into the tephra was connected to a heat exchanger that circulated heated water through the town's central heating system. This system was serving almost all the homes on Heimaey by the early 1980s.

The evacuated population of Heimaey returned gradually to the island after the 1973 eruption. Approximately 80% of the island's residents returned by early 1975. Hundreds of new homes were built to replace those destroyed by the eruption, and tephra deposited by the volcano was used as landfill for the building of many of those homes. Tephra also was used to expand runway facilities on Heimaey's airport, and lava from the 1973 eruption now serves as a breakwater in the harbor.

Hekla, volcano, Iceland Hekla has erupted 20 times since the settling of Iceland in the 10th century. An 1845 eruption of Hekla began on September 2 and was preceded by strong earthquakes. One flow of lava from this eruption was measured at 22 miles long, 1 mile wide at one point, and 40 to 50 feet deep. No humans were reported killed in this eruption, although numerous cattle died.

One remarkably large mass of pumice, its weight estimated at almost half a ton, was carried four to five miles by this eruption. Ice and snow that melted in the eruption flooded rivers. Perhaps the most destructive aspect of the eruption was its effect on pasture land, much of which was covered by volcanic ash and thus made unusable by animals. Even where ash did not cover the land, herbage became toxic and killed cattle that ate it.

Herculaneum, Italy See POMPEII AND HERCULANEUM.

Herdubreid, volcano, Iceland A TABLE MOUNTAIN, Herdubreid is believed to have been formed by eruptions underneath a glacier.

Hilo, Hawaii, United States The city of Hilo has a history of trouble from volcanic eruptions and tsunamis. For example, a lava flow from MAUNA LOA in 1933 posed such imminent danger to Hilo that the U.S. Army Air Force attempted to halt or divert the flow by dropping aerial bombs on selected locations. Whether as a result of the bombardment or from other causes, or possibly

A view of the south side of Mount Hood, showing deposits from pyroclastic flows and mudflows (D. R. Crandall, USGS)

both reasons, the lava flow stopped. Another lava flow in 1881 also threatened Hilo but stopped near the city's edge. The tsunami that struck Hilo on April 1, 1946, did extensive damage in the downtown area and is believed to have killed more than 175 people. Another tsunami, this one from an earthquake in Chile, killed more than 60 people in Hilo and came ashore as a wave estimated at 30 feet high.

Hood, Mount, Oregon, United States One of the most beautiful volcanoes, Mount Hood stands more than 11,000 feet tall and is located near Portland and other major urban areas in the Pacific Northwest of the United States. Mount Hood is believed to have undergone several major eruptions in the last 1,700 years, and further activity is possible. Eruptions in the late 18th and early 19th centuries generated large mudflows, and others about the time of the Civil War cast out PUMICE.

horst An upthrust block of rock or soil between two fault blocks, produced by an earthquake. (See also GRABEN.)

hot dry rock A plan to tap geothermal energy involves pumping water underground into areas of "hot dry rock," where temperature increases quickly with depth. The rocks would be fractured by pumping fluid into them down a well, and water would be circulated through the fractured rock. The heated water then would be returned to the surface through a separate conduit in the form of liquid water or steam and used to generate electricity. This technique is not in widespread use, although results of early tests are said to be promising.

"hot spot" An area of intense, localized volcanic activity. "Hot spots" may occur deep within the boundaries of crustal plates, far from the well-defined belts of volcanic activity that commonly mark plate boundaries. The HAWAIIAN ISLANDS appear to have been generated by the movement of a crustal plate over an underlying hot spot. A familiar example of a hot spot along a MID-OCEAN RIDGE is ICELAND, where volcanic activity has built up an island with large numbers of individual volcanoes. Although the origins of hot spots are not entirely understood, it has been proposed that they are located above plumes of rock rising through the mantle of the earth. (See also PLATE TECTONICS.)

Huascaran, Peru The highest peak in Peru at more than 22,000 feet, Huascaran was the site of a disastrous avalanche following an earthquake on May 31, 1970. The earthquake, centered a few miles off the coast of Peru, dislodged an estimated 10,000 cubic yards of rock and ice from the face of Huascaran and sent an avalanche rolling downslope toward the town of Yungay. The avalanche buried Yungay and approximately 20,000 people in the town. The same earthquake caused extensive damage to coastal communities but relatively little loss of life. At inland locations, however, thousands were killed by collapsing buildings.

hydrothermal activity In general terms, hydrothermal activity is any process involving extremely hot water underground. Hydrothermal activity is involved in the deposition of numerous minerals, including quartz and galena, a widespread ore of lead. Hydrothermal activity differs from geothermal activity in that the latter involves the movement of the waters and need not involve heating of water by contact with a still-hot body of IGNEOUS ROCK. (See also GEOTHERMAL ENERGY; GEYSER.)

hypocenter The underground point of origin of an earthquake. (See also SEISMOLOGY.)

I

Iceland The island nation of Iceland is an exposed segment of the Mid-Atlantic Ridge and is the site of intense volcanic activity. The island is mostly basalt and is, in effect, being torn in two by the spreading action that occurs along the mid-ocean ridge. Iceland's Laki Fissure, a RIFT VALLEY, is evidence of this process. Volcanic eruptions in Iceland generally occur as lava flows from fissures in the ground rather than as outbursts from volcanic mountains. The most famous eruption in Iceland's history took place in June 1783 when strong earthquakes preceded an eruption that cast large quantities of ash into the air, obscuring the sun and harming crops as far away as Norway and Scotland. Two days later, lava erupted from the ground along the Laki Fissure and filled the valley of the nearby Skafta River, forming a natural dam that caused extensive flooding upstream. Flowing southwest toward the sea some 15 miles away, the lava from this eruption covered the land to an average depth of 100 feet. The lava flow was 10 miles wide at maximum. Another eruption on August 8 sent lava flowing southeast and filled the Hverfisfljot river valley, causing more flooding. (See also ASKJA; BARDARBUNGA; GRIMSVÖTN; HEIMAEY; HEKLA; HERDUBREID; JOKULHLAUP; KOTLUGJA; KRAFLA; KVERKFJÖLL; MID-OCEAN RIDGE; ÖRAEFAJÖKULL; SKAPTAR JÖKULL; TORFAJÖKULL.)

Idaho, United States Idaho experienced severe earthquakes in 1916 and 1944, but only minor damage was reported. Volcanic landforms are abundant in Idaho, notably in the SNAKE RIVER PLAIN in the southern portion of the state. (See also CRATERS OF THE MOON MONUMENT.)

Idaho Batholith A tremendous mass of igneous rock some 250 miles long and up to about 100 miles wide, the Idaho Batholith is associated with rich deposits of metal ores at the Coeur d'Alene mining area. (See also BATHOLITH.)

igneous breccia A breccia, a coarse, sedimentary, clastic rock, made up of pieces of igneous rock.

igneous rock Any rock that solidifies directly from a molten or partially molten state is known as an igneous rock. (The word *igneous* is derived from *ignis,* the Latin word for "fire.") Igneous rocks are one of the three principal categories of rock, the others being METAMORPHIC ROCK and SEDIMENTARY ROCK. Basalt, obsidian, granite, pumice and scoria are familiar examples of igneous rocks, many of which have considerable economic importance. Igneous rocks make up some of the most famous and spectacular landforms on earth, including the Cascade Mountains in the United States and Canada, the Hawaiian Islands, the Palisades near New York City and the formations of Monument Valley.

Igneous rocks are classified as felsic or mafic. The term *felsic* derives from feldspar, a dominant silicate mineral in such rocks; mafic rocks have high concentrations of magnesium and iron (ferrum). Felsic minerals and rocks are generally light in color, whereas mafic rocks and minerals tend to be darker.

Grain size differs greatly among igneous rocks. Coarse-grained rocks are called phaneritic (from the Greek *phanero-*, meaning "visible"), while fine-grained rocks are called aphanitic (from the Greek *aphan-*, meaning "invisible"). Granite is a familiar example of phaneritic rock, and basalt is a widespread aphanitic rock. Grain size and degree of crystallinity depend on how quickly the molten rock cools. The finer the grain size, the faster the cooling; large grain size, on the other hand, means slower cooling. In pegmatites, very coarse-grained rocks, one sometimes finds grains several feet wide. At the other extreme are volcanic glasses, such as obsidian, which have an amorphous, grainless structure because rapid cooling gave crystal structure no opportunity to form in the solidifying lava.

Igneous rocks are sometimes classified as acidic or basic, but those terms are holdovers from the 19th century when silicic acid was believed to play a role in depositing silica (silicon dioxide, or SiO_2) in igneous rocks. A rock with high silica content was therefore thought to be acidic, as opposed to basic rock with less silica in it.

Extrusive igneous rocks are those that emerge from the earth and solidify at the surface. Extrusive igneous rocks may occur as LAVA flows, the result of molten rock flowing in sheet form across the surface, or volcanic ejecta, pieces of rock blasted from the throats of volcanoes and cast upward and outward through the air. (See also ASH; MAGMA; PLUTON; TEPHRA.)

ignimbrite A deposit of PYROCLASTIC, pumiceous rock laid down by a PYROCLASTIC FLOW. A given ignimbrite may extend more than 60 miles and be 30 feet thick or deeper. (See also PUMICE.)

Ijen, caldera, Java, Indonesia Located at Java's eastern end, the giant Ijen caldera (about 12 miles wide) is the site of several volcanoes. Kawah Ijen volcano (also known as Ijen Crater) stands inside the caldera, and Merapi and Raung volcanoes have arisen along a ring fracture at the edge of the caldera. A lake occupies the crater at Kawah Ijen. Raung tends to exhibit modest eruptions of ash and occasional flows of lava, preceded by minor earthquakes. PHREATIC ERUPTIONS, many of them occurring through the crater lake, characterize Kawah Ijen. Eruptions of Kawah Ijen are associated with increases in seismic activity and in the temperature of the crater lake.

Ijen appears to be a noisy volcanic site; many reports of eruptions mention loud noises from the volcanoes. The historical record of volcanic activity at Ijen dates back to 1586 and includes dozens of eruptions. In one eruption in 1638, a loud noise that appeared to come from Raung, was comparable to thousands of artillery pieces going off at once. Earthquakes were continuous and so powerful that a person could hardly stand upright. Rivers reportedly dried up for several days, then resumed flowing in a powerful flood. When Raung underwent an ash eruption in 1890, witnesses reported strong vibrations and rumbling noises. A 1902 eruption of Raung was accompanied by a very loud noise like cannon fire. Roaring sounds were heard at Raung during a period of increased activity in 1913–14, and two small earthquakes from Raung were reported in 1915. Water from the lake at Kawah Ijen flowed over a sluice after a large tectonic earthquake near Bali in February 1917; the following month, water in the lake was spouting some 30 feet into the air in a muddy, noisy display. The temperature of the lake rose to scalding intensities in March but dropped back to lukewarm levels by late 1917. The lake heated up again in 1921. Raung erupted in 1921 and again in 1931. A 1952 eruption at Ijen followed immediately after several strong local earthquakes. An eruption cloud appeared, and "boiling" activity was observed in the lake. A minor earthquake preceded an eruption of Raung in 1982, and in 1985 Raung underwent minor ash explosions.

Ijen is especially interesting because of a relationship seen between lake level and eruptive activity. On several occasions when the lake level has been lowered by artificial means, fumarolic activity under the lake has increased, lake temperature has risen, and sometimes hydrothermal or phreatic eruptions have occurred. These phenomena are thought to result from reduced pressure resulting from lowering the lake level.

Iliamna, volcano, Alaska, United States Located in the eastern Aleutian Islands, Iliamna has erupted on several occasions in historical times, notably in 1947. Iliamna is one of many Alaskan volcanoes characterized by explosive eruptions.

Illinois, United States The state of Illinois has undergone numerous strong earthquakes since it was settled by Americans of European descent. One notable earthquake was the Fort Dearborn earthquake of August 24, 1804, felt at Fort Wayne, Indiana some 200 miles away. A series of strong earthquakes occurred in southern Illinois in 1882–83. A severe earthquake at Cairo on August 2, 1887, stopped clocks and was felt over a large area of the Midwest. Illinois is affected by seismic instability in the Mississippi Valley, notably the New Madrid Fault Zone, where a future earthquake comparable to those of 1811–12 could cause tremendous damage in Illinois and other midwestern states.

Ilopango, caldera, El Salvador The Ilopango caldera is about five miles wide and seven miles long and lies between two faults of a GRABEN that runs parallel to the string of volcanoes in El Salvador. The CALDERA is thought to have formed during a huge eruption in the third century A.D. Lake Ilopango occupies the caldera, and a lava dome breaks the surface of the lake at one point to form the Islas Quemadas, or "burned islands." The Islas Quemadas formed during an eruption in January 1880. The lake level rose so dramatically at this time that the Jiboa river valley was flooded on January 9, destroying the town of Atuscatla and killing numerous cattle. This rise in the lake is thought to have been due to the growth of a lava dome underwater. The lake level subsided after this flood, but sulfur gas began to emerge from the middle of the lake, and in late January an eruption of ash and glowing rock occurred. Within three days, the Islas Quemadas formed. Strong earthquakes in 1879 preceded this eruption. Another powerful earthquake occurred in February, was felt all through El Salvador and preceded an intense smell of sulfur in the Ilopango area. Ilopango appears to have been inactive since 1884.

impact structures In recent years, geologists have recognized that the earth's surface bears the marks of numerous planetoid impacts. Major impacts, involving meteorites perhaps a hundred yards in diameter or larger, are by no means infrequent; the impact that created the famous Meteor Crater in Arizona (also known as the "Barringer crater" after D. M. Barringer, a geologist who investigated its origins) is believed to have occurred only a few thousand years ago. Evidence for an impact origin of the Arizona crater began to accumulate in the late 19th century. A mineralogist named A. E. Foote gathered large numbers of iron meteorites, some of them containing minuscule diamonds, at the site of the crater. After Foote completed a report on his findings at the site, G. K. Gilbert of the U.S. Geological Survey examined the crater and found evidence to indicate it was the result of a large object falling from outer space and striking the earth. (The alternative view was that some kind of volcanic explosion had generated the crater.)

A search for commercially workable deposits of nickel and iron led around 1902 to a strong interest in mining the crater. Within the next decade, drilling and excavation of mine shafts at the crater established that there was no workable body of metal buried at the site. Pieces of meteoritic material were found in a BRECCIA extending several hundred feet beneath the surface, but no substantial body of metal appeared. These explorations revealed that the volcanic explanation of the crater's origin was flawed because an unaffected layer of sandstone was discovered under the breccia. Later studies of the crater indicated that a meteorite some 140 feet in diameter blasted out the formation upon striking the earth at a velocity of approximately 45,000 miles per hour. The energy released in the impact is thought to have been equivalent to the explosion of a 15-megaton thermonuclear weapon.

When a planetoid (in effect, a giant meteorite) strikes the surface of the earth, the impact tends to generate a characteristic structure often called an impact crater, although a more general term is *astron* or *astrobleme.* A classic impact character has the following characteristics: round or roughly square shape; depressed central area, with perhaps a central peak where rebound has occurred following impact; a raised rim with overturned strata outside the rim; and clastic material generated by the impact distributed in and around the crater. Arizona's Meteor Crater exhibits these characteristics, except for the central peak. Other indicators of an impact origin include the presence of "shatter cones," peculiar conical structures produced in rock by the tremendous forces released on impact, and minerals, including diamond (found near Meteor Crater) and stishkovite and coesite, two highly dense forms of quartz. A slag-like material associated with impact craters is known as impactite. Pseudotachylite is a rock believed to form from melting at impact sites.

Many well-preserved impact structures are found on the Canadian Shield, including the Manicouagan formation in Quebec, now the site of a ring-shaped reservoir, and Ontario's Brent Crater, some two miles wide and discovered in a study of aerial photos taken by the Royal Canadian Air Force (RCAF). One of the first impact craters identified in Canada was Chubb Crater, named after Frederick Chubb, a prospector who in 1950 noticed a remarkable round lake in an RCAF photo of northern Canada and saw that the lake lay in a formation that looked much like a CALDERA. If a caldera did exist there, Chubb figured, it might be worth investigating for diamonds, which are known to occur in some volcanic formations. Chubb went to see V. B. Meen of the Royal Ontario Museum of Geology and Mineralogy in Toronto and asked Meen if he thought the lake occupied a volcanic formation. Meen thought the lake instead had formed inside an impact crater like Meteor Crater in Arizona. An expedition to the site was formed, funded by a newspaper publishing company, and Chubb and Meen left Toronto in an amphibious plane on July 17, 1950, to investigate the lake. Four other men (the pilot, an engineer, a reporter and a photographer) accompanied Chubb and Meen. On reaching the lake, the men found its surface too icy to allow a safe landing, and the pilot landed the aircraft on another lake in the vicinity. Chubb and Meen made their way from the landing site to the alleged impact crater, a study of which revealed quickly that the formation was not volcanic in origin but rather was formed by meteorite impact.

It has been suggested that much larger features of Canada's landscape are also impact structures, including the Gulf of St. Lawrence and the Nastapoka island arc in Hudson Bay. Other proven or suspected structures of impact origin in Canada are found at Charlevoix, Quebec; Sudbury, Ontario; Lake St. Martin, Manitoba; Elbow, Carswell and Gow Lake, Saskatchewan; Eagle Butte and Steen River, Alberta; Pilot Lake, Mackenzie District; and Nicholson Lake, Keewatin District. One possible feature indicating meteoritic origin is rich metal deposits, including nickel, iron, silver, gold, platinum, lead, copper, cobalt and zinc, such as are found at Sudbury Basin in Ontario. If the Sudbury Basin is indeed the remnant of an impact crater, the event which produced the crater must have been powerful almost beyond imagination. One estimate puts the energy required to blast out such a crater as equivalent to the explosion of 50,000 megatons of TNT. (The origin of the Sudbury Basin has been disputed. Volcanic activity on a tremendous scale, it has been argued, might have produced the basin.)

In the United States, confirmed and possible impact structures (besides Meteor Crater) are found at Serpent Mount, Ohio; Wells Creek, Tennessee; Decaturville, Missouri; Odessa and Sierra Madera, Texas; Kentland, Indiana; Manson, Iowa; and Red Wing Creek, South Dakota, among other sites. One recently identified impact structure in the United States is the Beaverhead astrobleme in Montana. Believed to have been about 45 miles in diameter when formed, the Beaverhead astrobleme is not immediately apparent to observers because various processes have reworked the area extensively since the crater originated. Impressive shatter cones are present in strata extending some 15 miles from north to south at the Beaverhead site, which also exhibits pseudotachylite, as well as suevite, a "microbreccia," characterized by very small fractures in the rock.

Other identified or suspected impact structures are distributed widely around the globe. Germany's Ries Kessel is widely recognized as an impact structure, as is the Vredevoort Ring in South Africa. The Vredevoort Ring, often identified as the largest demonstrated astrobleme on earth, appears to be left over from an impact that created a crater roughly the size of the state of West Virginia. An abundance of shatter cones played an important part in identifying the Vredevoort Ring as an impact structure.

The physics of a planetoid impact are complex but may be summarized in general terms as follows. A planetoid several hundred feet wide or larger is not slowed greatly by passing through the earth's atmosphere and strikes the surface with much of its original velocity undiminished. Depending on the mass and velocity of the planetoid, the impact may release enough energy to vaporize much or all of the planetoid, producing the equivalent of a huge nuclear explosion. The impact may create a crater miles in diameter, as is thought to have happened in the cases of Manicouagan and Vredevoort. Ejecta from the point of impact may fall great distances away from the crater. Secondary meteorites, or sizable bodies of rock cast out by the primary impact, may bring about further destruction, albeit on a smaller scale, when they fall to earth.

The physics of an impact on land are considerably different from those of an impact at sea. On land, a giant meteorite would cause widespread destruction through seismic effects as well as from heat and other radiation from the fireball produced on impact. A land impact also might alter climate by casting large amounts of particulates (finely pulverized rock and smoke from fires started by heat from the fireball) into the upper atmosphere where the material could intercept incoming sunlight and reduce temperatures at the surface.

The damage from an ocean impact, however, is projected to be much greater, because the sea acts in two ways to amplify the destructive effects of the planetoid strike. First, some of the energy released by the impact is transferred to the water as a tsunami, or seismic sea wave, that may travel around the globe many times before eventually dissipating. The comparatively small release of energy from the explosion of the volcanic island Krakatoa produced tsunamis that caused great destruction and loss of life along shores in the vicinity of the island; the tsunami was strong enough to swing ships around at anchor in the harbor at Colombo, Ceylon (now Sri Lanka) and was detectable on the far side of the world. The tsunami from a planetoid impact at sea may be expected to be much more powerful and to cause more extensive destruction, simply because the energies released in a planetoid impact stand to be many orders of magnitude greater than those involved at Krakatoa. In theory, a planetoid strike in mid-Atlantic might generate tsunamis that would come ashore as breakers hundreds of feet high along shores in Europe, Africa and the Americas.

Second, some of the heat released by the impact would go into vaporizing water and creating a convection current in the atmosphere, capable of carrying great amounts of water vapor, pulverized rock and sediment from the seabed into the upper atmosphere. There, the water vapor would condense as fine crystals of ice and produce a high-altitude cloud composed of ice and other particulate material. There might be effects on global climate similar to those projected for a land impact but considerably greater in scope because seawater provides a more effective mechanism for transporting that material into the upper air. Large meteorite impacts and climatic effects associated with them have been suggested as possible mechanisms to account for mass extinctions of prehistoric life on earth.

inclusion A piece of older rock inside igneous rock. The igneous rock may or may not be related otherwise to the inclusion.

India The Indian subcontinent has been the site of numerous strong and destructive earthquakes. India's seismic potential is thought to be largely the result of an ongoing collision between the Indian subcontinent and the Asian landmass immediately to its north. The Himalayas mark the presumed boundary between the Indian and Asian plates. Earthquakes in India tend to take large numbers of lives because of the country's extremely high population density. For example, an earthquake in Calcutta on October 11, 1737 is said to have killed some 300,000 people. Another in Kutch on June 16, 1819 wiped out 1,500 and one in Assam on June 12, 1897 killed about 1,000. On May 31, 1935 in Quetta, an earthquake measuring 7.5 on the Richter scale killed some 60,000 people. An earthquake measuring 6.4 on the Richter scale struck a remote part of the state of Maharashta on September 30, 1993 at 3:56 AM. Approximately 50 villages within a 50-mile radius were damaged and 16 were completely leveled within a few seconds. Many of the people killed were asleep in badly constructed stone houses that collapsed on top of them. Although the exact death toll was not known in October 1993, it was thought to be in the tens of thousands.

Indiana, United States Strong earthquakes affect Indiana occasionally, although the state is not as susceptible to earthquakes as high-risk areas such as California. Strong shocks were reported at Albany on August 6–7, 1827, and another powerful earthquake on September 27, 1909, shook Indianapolis and was felt over much of the Midwest.

Indo-Australian crustal plate A major plate of the crust, the Indo-Australian plate is involved with volcano generation through a SUBDUCTION ZONE in the East Indies. The plate underlies Australia, the Indian subcontinent and part of the Indian Ocean. (See also PLATE TECTONICS.)

Indonesia An archipelago of volcanic islands between Australia and the main Asian landmass, Indonesia has more than 200 active volcanoes by one estimate. Indonesia was the site of the 1883 eruption of KRAKATOA, one of the most famous volcanic events in history. The eruption of TAMBORA in 1815 cast tremendous amounts of finely divided solid material into the atmosphere and was implicated in the unusually cold climatic conditions that affected much of the northern hemisphere over the following year. Earthquakes and tsunami activity also occur frequently in the Indonesian archipelago. (See also BANDA API; SEGARA ANAK; SUKARIA; SUNDA; SUOH DEPRESSION; TENGGER; TOBA; TONDANO.)

insolation In general terms, insolation means the amount of solar radiation reaching the earth's surface. Volcanoes may affect insolation by injecting large quantities of ash into the upper atmosphere during eruptions and thus intercepting sunlight before it can reach the ground. The result is to diminish insolation and reduce temperatures at the earth's surface. This cooling effect has been observed following several major volcanic eruptions over the past several centuries. (See also KRAKATOA; "YEAR WITHOUT A SUMMER.")

intensity In SEISMOLOGY, "intensity" refers to an earthquake's effects at a particular location. Intensity, which may be measured by factors such as the presence (or absence) of cracks in walls, and even of mass panic, is distinguished from magnitude, which refers to the amount of energy released by an earthquake, as expressed in vibrations of the ground. Several scales are used to measure earthquake intensity, including the famous Mercalli scale.

interference Interaction in which two seismic signals arrive at a given point at the same time, and one masks the other.

interstice A void or space inside a soil or rock.

intrusion A body of igneous rock that has entered as molten rock into surrounding COUNTRY ROCK. Intrusion also refers to the process by which such a body of igneous rock, known as intrusive rock, is formed. Intrusion is not restricted to molten rock but also may occur in salt deposits.

intrusive rock See INTRUSION.

Inyo Craters See MONO LAKE.

Io Jupiter's moon Io is remarkable for having active volcanoes. United States space probes have detected and photographed the plumes of volcanoes ejecting sulfur high above the moon's surface. Images returned to Earth by *Voyager* space probes in 1979 and 1980 showed that the whole visible surface of Io appeared to be affected to one degree or another by volcanism. These images showed several plumes of volcanic material, almost 200 miles high and about 600 miles wide, rising from volcanoes. Approximately 200 CALDERAS also were identified on the face of Io. Some of these calderas are believed to contain lakes of lava. Apparent lava flows on the surface of Io are colored black, red and orange. These colors are believed to be the result of temperatures at which the molten material cooled and solidified. Because these colors are those of molten sulfur as it cools, the lava flows of Io are thought to be made at least partly of sulfur.

Volcano in eruption on Jupiter's moon Io (*Earthquake Information Bulletin*/USGS)

Other materials also may be included in the magma and lava because sulfur alone apparently does not have the strength to form some of the steep-sided features on Io's face. It has been suggested that basalt may make up part of the surface rocks of Io. The moon appears to have a very high internal heat flow compared to that of Earth. Io's interior is thought to be kept hot by gravitational influence from Jupiter and from its moon Europa. Evidence of extraterrestrial volcanic activity has been discovered on Venus and Mars as well as on Io.

Iowa, United States Iowa is not characterized by frequent seismic activity but has experienced strong earthquakes on occasion. A severe earthquake at Sioux City on October 9, 1872, was accompanied by a noise like thunder.

Iran Iran is located in an area of intense seismic activity and has been the site of numerous highly destructive earthquakes. These are not always extremely powerful earthquakes; many are moderate but shallow. The southern portion of Iran is the most seismically active. Iran's active faults include, but are not limited to, the Ferdows, Kuhbanan and Nayband faults as well as portions of the Shahrud Fault. The Tehran area reportedly has experienced some earthquakes that are of interest for being evidently unrelated to surface faults. Strong earthquakes have struck Iran repeatedly within historical times, notably in 1611, 1619, 1755, 1879, 1881, 1883, 1911, 1930, 1968, 1969, 1972 and 1977. The March 21, 1977, earthquake (magnitude 7.0), which occurred in southeastern Iran, caused widespread damage and killed about 180 people. A strong aftershock occurred approximately 90 minutes after the main shock, and aftershocks continued for weeks afterward. Another earthquake, in Tabas in September 1978, killed 25,000 and was suspected of having some link to the underground testing of a Soviet nuclear device some hours before near Semipalatinsk in Siberia, approximately 1,500 miles away. Believed to have yielded 10 megatons, the Russian bomb test was unusually powerful and allegedly was set off at a shallow depth of only about one mile. The coincidence between the test and the earthquake was remarkable, as was the unusual set of characteristics observed in the Iranian earthquake. It was very shallow (centered only about 10 miles underground) and reportedly had no aftershocks. Also, a peculiar report from Iran concerned access to the site of the earthquake. As a rule, seismolo-

Damage to a building in Tabas, Iran from an earthquake on September 16, 1978 (*Earthquake Information Bulletin*/USGS)

gists are admitted to the affected area following a large earthquake to assess the damage; but in this case, access was denied, leading some observers to wonder if there was something especially sensitive and deserving of secrecy that had occurred at Tabas.

Irazu, volcano, Costa Rica The stratovolcano Irazu has a double crater and has undergone explosive eruptions on more than a dozen occasions since the early 18th century. Eruptions between 1963 and 1965 generated many ashfalls that caused considerable harm to agriculture and affected the city of San Juan.

Ischia, volcano, Italy The volcanic island Ischia stands close to Vesuvius, in the Bay of Naples, and has been active since earliest historical times. Resurgent uplift at Ischia, similar to that observed at IWO JIMA, has led some observers to speculate that Ischia has a caldera, although this interpretation is not certain. There has been considerable long-term ground deformation at Ischia. One sign of this deformation is the site of a Roman metal foundry on the northeast side of the island, now about 15 to 20 feet underwater. On the island's southern side, a beach has been uplifted almost 100 feet in places above the present water level. At one point on the island, thermal baths built in the late 18th or early 19th century have risen up to approximately 20 feet, at a rate of more than an inch per year, since their construction. In the 20th century, the southern side of the island appears to have started subsiding, thus reversing the earlier trend of uplift. At the same time, Monte Epomeo, the highest point on the island, appears to be undergoing continuing uplift.

Although uplift at Ischia may be volcanic in origin, it also is possible that tectonic activity is involved. Subsidence at Ischia, however, is more difficult to explain in terms of volcanic activity. Major explosive eruptions in prehistoric times appear to have laid down the Tufo Verde, or "Green Tuff," more than 3,000 feet thick. If a caldera does exist at Ischia, it may have formed during the eruption that deposited the Tufo Verde. An eruption around 470 B.C. drove away a Syracusan colony on the island. Residents of the island had to flee yet again in an eruption that occurred between approximately 400 B.C. and 350 B.C. This eruption is said to have followed earthquake activity. A tsunami may have accompanied an eruption, possibly of Monte Epomeo, around 350 B.C. Another eruption may have taken place in 91 B.C., although there is some question whether this eruption involved Ischia, VULSINI or ROCCAMONFINA. The eruptive history of Ischia over the next thousand years is sketchy, but eruptions appear to have occurred around A.D. 80, 180–222 and 284–305. Approximately 700 people were killed in a rock slide in 1228. There was considerable earthquake activity in 1302 as well as an explosive eruption and a flow of lava. In either 1557 or 1559, an earthquake caused a church at Campagnano to collapse. Another church in the same community collapsed during an earthquake in 1762, and a church at Rotaro was demolished in an earthquake in 1767. In the community of Casamicciola (formerly Campagnano), several houses and seven of their occupants were destroyed in a 1796 earthquake. Strong earthquakes shook the island in 1827 and 1828. After one earthquake, residents of the island slept outdoors for more than two weeks. Earthquakes occurred again at Ischia in 1841, 1863 and 1867. Seismic events between 1881 and 1883 may have accompanied an increase in the temperature of wells on the island, but this is not certain. A moderate earthquake occurred in 1961.

island arcs These are arcuate, or crescent-shaped, strings of islands formed at boundaries between crustal plates where one plate is subducted under another. Volcanic islands are formed of material from the subducted plate rising toward the surface and of magma from the asthenosphere. Alaska's Aleutian island chain is a familiar example of an island arc.

isostasy The tendency of lighter crustal rock to "float," so to speak, on the denser rocks below.

Italy The Italian peninsula and its nearby islands and waters are home to a number of famous volcanoes, notably VESUVIUS, which has erupted on numerous occasions within historical times and caused widespread destruction, including the ruin of the cities of POMPEII AND HERCULANEUM. Earthquake activity is also frequently destructive in Italy, which is located in the seismically active Mediterranean basin. GEOTHERMAL ENERGY has been harnessed in Italy. (See also ISCHIA; LARDERELLO; MEDITERRANEAN SEA; PHLEGRAEAN FIELDS; ROCCAMONFINA; VULCANO; VULSINI.

Iwo Jima, caldera, Japan This small island, a famous World War II battleground, occupies a submerged caldera. The name "Iwo Jima" means "Sulfur Island." Much of the island is taken up by Motoyama, a volcanic cone with a high proportion of TUFF to lava. Mount Suribachi, at the southern tip of the island, is made up of tuff and lava. Iwo Jima has a history of PHREATIC ERUPTIONS reaching back to 1889, and the barrenness of certain pumice deposits seen on the island during World War II indicated the deposits were reasonably fresh. Fumaroles are abundant on Iwo Jima and tend to be very hot; one test drilling had to be suspended because the intense heat ruined the drill bits. The island appears to have undergone dramatic uplift over the past several hundred years. Estimates put the rate of uplift at perhaps seven or eight inches per year. It is not clear whether the uplift at Iwo Jima may represent an early indication of another caldera-generating eruption in the future.

Izu-Oshima, caldera, Japan An island located near the point where the Pacific crustal plate meets the Eurasian and Philippine plates, Izu-Oshima is often active, and the historical record of eruptions, mostly explosive but some with lava flows as well, reaches back to the seventh century A.D. The caldera is less than three miles wide and contains a volcano, Mihara-yama, with a summit crater some 450 feet wide.

Izu-Oshima is known for the dramatic and periodic rise and fall of the floor of Mihara-yama's summit crater. The crater floor has been known to rise and fall more than 1,200 feet over 20 years. Periods of rising are correlated with eruptions. Very strong earthquakes may follow these episodes by several years. For example, uplift began in 1908, followed by eruptions in 1912–14 and 1919; a powerful earthquake occurred in 1923. Another episode of uplift began in 1933; eruptions followed in 1950–51 and another major earthquake in 1953. Yet another period of uplift started in 1963; Izu-Oshima erupted in 1974 and again in 1976, and a major earthquake is anticipated. It has been suggested that tectonic forces are responsible for the dramatic up-and-down motion of the crater floor. Tectonic activity, according to this model, may squeeze an underground magma reservoir (somewhat in the manner of a tube of toothpaste) and cause the crater floor to rise; when pressure is relieved, the crater floor then falls. Because Izu-Oshima thus may serve as a natural indicator of tectonic stress, it is studied closely as a possible guide to future strong earthquakes in the Tokyo region. Marked changes in the magnetic field have

Damage from an earthquake in southern Italy on November 23, 1980 (*Earthquake Information Bulletin*/USGS)

ɪeen observed before and after eruptions at Izu-Oshima. These changes may be due to demagnetization of a reservoir of magma about three miles down. (See also PLATE TECTONICS.)

Izu Peninsula, Japan The Izu Peninsula is located a few miles south of Tokyo, at the point of intersection of the Eurasian, Philippine and Pacific crustal plates, and close to Izu-Oshima. Known for its strong seismic and volcanic activity, the Izu Peninsula has numerous volcanic cones and domes. Some of the domes are thought to have appeared only about 3,000 years ago. The KANTO EARTHQUAKE of 1923 occurred only a few miles north of the Izu Peninsula. Instability on the Izu Peninsula in recent times may be due to movement of magma underground.

J

Japan The islands of the Japanese archipelago occupy one of the most concentrated areas of seismic and volcanic activity in the world. The result of a collision between the Pacific crustal plate to the east and the Asian continental plate to the west, Japan's extraordinary history of earthquakes and volcanism has had a profound impact on the history of the nation. Time and again, cities in Japan have been damaged heavily or destroyed entirely by earthquakes and volcanic eruptions. Perhaps the most famous earthquake in Japanese history is the one that struck TOKYO in 1923, killing more than 100,000 and destroying much of the city.

Japan's volcanoes are numerous and spectacular. One of them, Mount FUJI, or Fuji-san, is located within sight of Tokyo and has become a symbol of Japan. An interesting feature of Mount Fuji is a set of lakes—Kawaguchi, Yamanaka, Sai, Shooji and Motosu—at the foot of the mountain; originally these separate lakes were a single large lake, but eruptions produced lava flows that divided the lake into five portions. Other notable volcanoes and CALDERAS in Japan include ASAMAYAMA, ASO, BANDAI-SAN, BAYONNAISE, Oshima, SAKURA-ZIMA, TARUMAI, Unzen and Usu.

Japan is an excellent example of a chain of volcanoes formed along a subduction zone. The volcanoes of the archipelago are located about 120 miles west of the deep ocean trench that marks the actual boundary between the crustal plates. The subduction of crustal rock, and the forces generated and released by that ongoing process, create the intense volcanism and seismicity of Japan. The tallest volcanoes in Japan, as a rule, are located along the eastern side of the belt of volcanism, close to the subduction zone. Hot springs and other geothermal phenomena are commonplace in Japan; indeed, hot springs provide popular attractions for vacationers in many areas. Japan has several major volcanic zones:

1. *Kirishima zone.* This zone is located in the southern portion of the Japanese islands.
2. *Hakusan zone.* Just to the north of the Kirishima zone, the Hakusan zone runs along the southwestern coast of Japan and includes portions of the southern island of Kyushu and the main island of Honshu.
3. *Norikura zone.* This small zone is located southwest of Tokyo and incorporates portions of two mountain ranges, known as the "central Alps" and "northern Alps" of Japan.
4. *Fuji zone.* Located in the vicinity of Tokyo, the Fuji zone includes much of the Kanto Mountains.
5. *Chokai zone.* Running along the northwestern coast of Japan, the Chokai zone takes in the Dewa Mountains.
6. *Nasu zone.* A long, narrow strip of volcanically active land, the Nasu zone runs from the northern island of Hokkaido down the northeast coast of Honshu, Japan's main island, and reaches well into the interior of Honshu, stopping just a few miles from Tokyo. The Oou and Mikuni mountain ranges fall inside this zone.
7. *Chishima zone.* Located in Japan's far north, on the island of Hokkaido, the Chishima zone extends from central Hokkaido outward into the Pacific.

Altogether, these volcanic zones occupy much of the area of Japan and contain some 200 volcanoes.

The Japanese coast is vulnerable to tsunamis, or seismic sea waves, which may be generated along the Japanese coast or elsewhere in the Pacific basin and cause great damage and loss of life when the waves come ashore in Japan. The Japanese shoreline has a long history of devastating tsunamis, one of which, at Sanriku in 1896, reportedly reached the shore as a 100-foot-high wave that killed some 25,000 people and swept away approximately 10,000 buildings. Japan's rise to its current status as an economic superpower has generated cause for concern about the possible economic

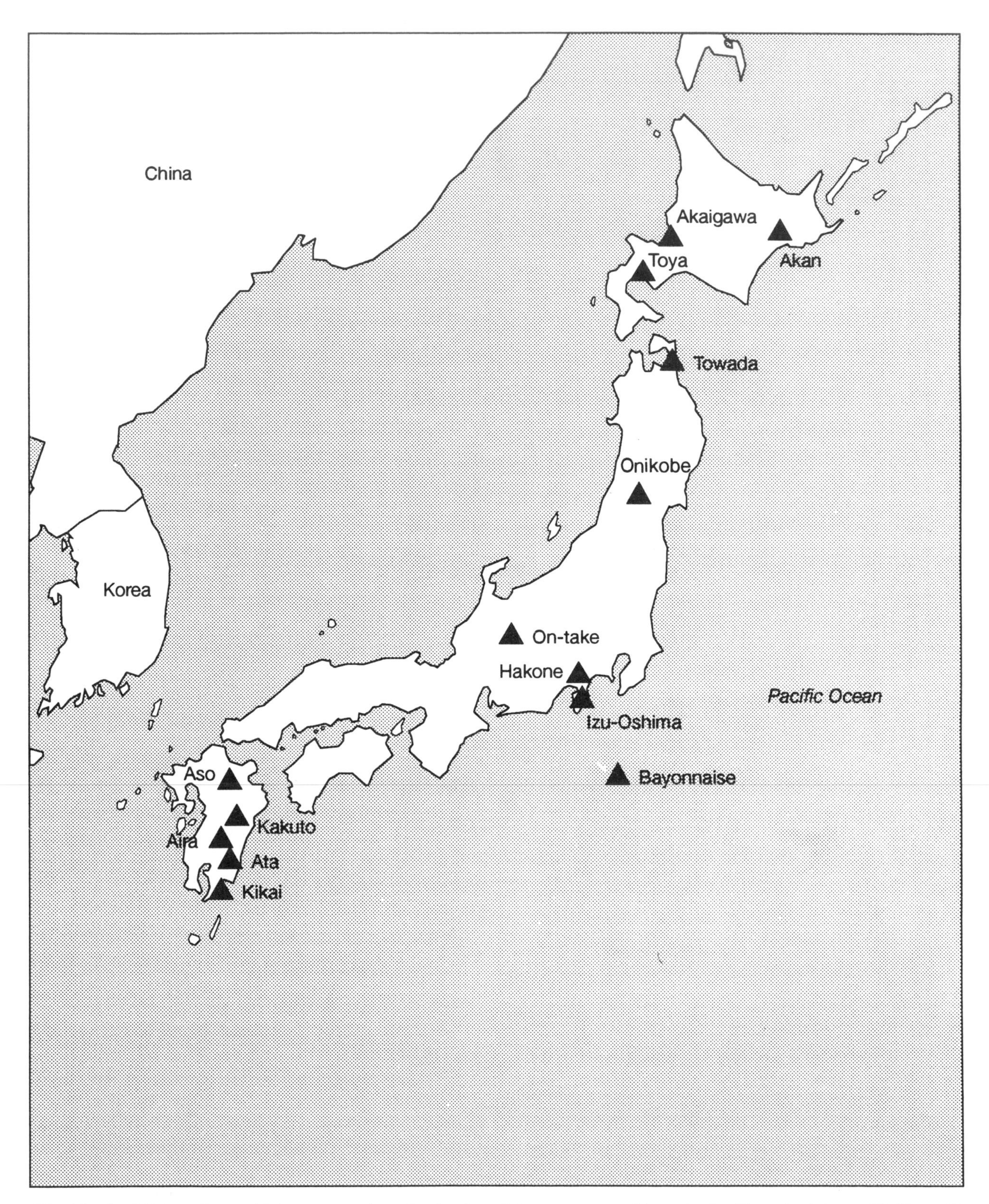

The islands of Japan show a high concentration of volcanic activity, as seen on this map of calderas in Japan (© Facts On File, Inc.)

impact of the next powerful earthquake that is expected to strike the Tokyo area in the near future. Because the economy of Japan is tied now so intimately to the economies of other nations, especially the United States, it has been suggested that a major natural catastrophe such as an earthquake in Japan could have dire consequences for much of the rest of the world. (See also ECONOMIC EFFECTS.)

jokulhlaup Translated from the Icelandic as "glacier burst," jokulhlaup refers to an outpouring of melt water produced when a volcano erupts beneath a covering of glacial ice, as happens from time to time in Iceland. The jokulhlaup may start slowly as volcanic heat melts ice beneath the surface of the glacier and forms a buried pool of melt water. Soon that water makes its way out from under the ice and flows toward the sea. As the waters drain away, the overlying ice settles. (See also GRIMSVÖTN.)

Jorullo, volcano, Mexico Jorullo is located only about 50 miles from PARICUTIN and resembles that volcano in many ways, although Jorullo's birth was studied less intensively than Paricutin's because the area was sparsely populated when Jorullo emerged in 1759. The eruption that gave rise to Jorullo was preceded by earthquakes that began in June, continued through late September, and were strong enough to cause widespread distress among the population and damage to a chapel. On September 29, ash and steam erupted from a ravine. Condensation of steam produced a muddy rain that fell on a nearby hacienda, as the eruption cloud darkened the sky, and a strong odor of sulfur spread over the countryside. The eruption, which soon required the evacuation of the hacienda, continued for two days before lava (or what is believed to have been lava, on the basis of a local plantation administrator's written reports) emerged from the volcano. The eruption destroyed forests in the vicinity and, several days after it began, was reportedly casting out large rocks, some of them the size of cattle. A new vent—one of three that eventually formed—appeared on September 12. Accumulating ash and contaminated water caused great harm to flora and livestock during October and early November. No written records of the eruption appear to have been made from firsthand observation after late 1759, although eruptive activity reportedly continued through early 1760 and recurred intermittently until 1775. Lava extruded from Jorullo covered several square miles.

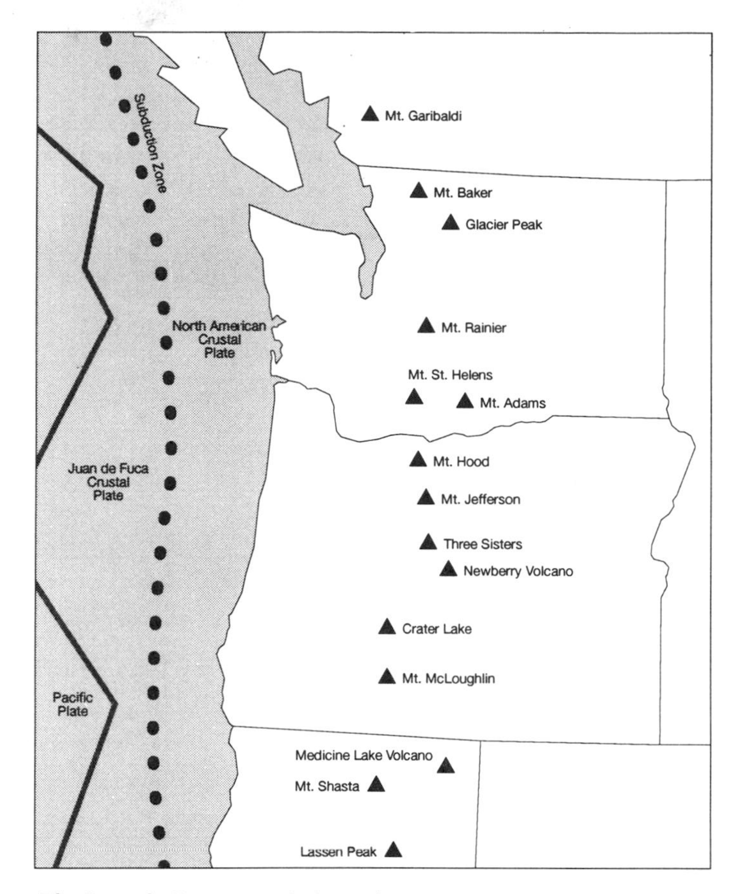

The Juan de Fuca crustal plate adjoins a subduction zone off the Pacific Northwest coast of the United States. Also shown are the volcanoes of the Cascade Range (© Facts On File, Inc.)

Juan de Fuca crustal plate A small plate of the earth's crust compared to the adjacent Pacific and North American plates, the Juan de Fuca plate is nonetheless important to the United States and Canada because its ongoing collision with North America has created the CASCADIA SUBDUCTION ZONE, a source of seismic and volcanic activity affecting the Canadian province of British Columbia and the states of Washington and Oregon in the northwestern United States, as well as parts of northern California. Subduction and fractionation of the descending crustal plate are thought to have generated the magma that formed the volcanoes of the Cascade Mountains, including Mount HOOD, Mount ST. HELENS, Mount RAINIER and MAZAMA. The magma involved in eruptions of these volcanoes has been rich in dissolved gases, presumably as a result of water-rich sediments being subducted along with the Juan de Fuca plate; the water thus drawn into the earth's interior rises back to the surface dissolved in molten rock, generating highly explosive eruptions such as that of Mount St. Helens in 1980.

Although there is abundant evidence of volcanism connected with the Juan de Fuca plate, there is no history of highly destructive earthquakes such as those recorded frequently in California. The reason for this relative absence of strong earthquakes along the Cascadia subduction zone is uncertain, but it has been suggested that the descending Juan de Fuca plate merely "slides" smoothly on its way downward, without such powerful disturbances as threaten portions of California. It is worth noting, however, that the United States government considers the coast of the Pacific Northwest an area of high earthquake potential and has warned that a powerful earthquake remains a distinct possibility for the future.

K

Kagoshima Bay See SAKURA-ZIMA.

Kakuto, caldera, Japan Located close to the Aira and Kikai calderas, Kakuto CALDERA is about six by nine miles wide and is situated northwest of Kirishima volcano. Geothermal fields are situated nearby. Although there have been no eruptions in Kakuto caldera itself within historical times, Kirishima has a record of dozens of explosive and PHREATIC ERUPTIONS dating back to the eighth century A.D.

Activity at Kirishima and seismic activity at Kakuto appear to be interrelated. In 1913–14, several earthquake swarms at Kakuto were followed by minor eruptions at Kirishima at about the same time as a major eruption of nearby SAKURA-ZIMA (although there may have been no connection between eruptions at the two sites). This activity coincided closely with a strong earthquake about 60 miles to the east. Earthquake swarms were noted again in the caldera in 1915, but Kirishima did not erupt this time. Several minor volcanic events in 1923, 1934 and 1936 did not accompany earthquake activity in the caldera. Likewise, a phreatic eruption in 1959 was preceded by only slight earthquake activity, or none at all. Earthquake swarms occurred after this eruption.

In 1961 a strong earthquake about 60 miles east of Kirishima preceded earthquake swarms in the vicinity of the caldera. Earthquake activity, some of it moderately strong, occurred between 1966 and 1969. Data collected between 1967 and 1969 indicate perhaps a couple of feet of uplift to the west, or possibly a comparable amount of subsidence to the east. A minor phreatic eruption without any outstanding seismic precursors occurred at a hot spring at the base of Kirishima in 1971. An earthquake swarm took place in the area later that year. Fumaroles showed increased activity in the early 1970s near Kirishima. Considerable earthquake activity continued from 1974 to 1979. In the winter of 1978–79, a small flow of sulfur, several inches wide and about 150 feet long, is thought to have occurred. Between 1980 and 1983, fumaroles became more active, another small sulfur flow occurred, and a pit inside one crater at Kirishima cast out material resembling tar.

Kamchatka Peninsula, Russia The Kamchatka peninsula, on the eastern shore of Russia, is noted for volcanic activity, notably the eruption of BEZYMIANNY in 1955. (See also KLYUCHEVSKAYA.)

Kansas, United States Kansas does not have a history of frequent and destructive seismic activity but has undergone strong earthquakes from time to time. The Lawrenceville earthquake of April 24, 1867, was characterized by strong shocks and a wave some two feet high reported on the Kansas River. The earthquake caused some damage to buildings and was felt widely through the Midwest.

Kansu, China An earthquake estimated at magnitude 8.6 on the Richter scale struck the Chinese province of Kansu on December 16, 1920, killing approximately 180,000 people. An additional 20,000 deaths were attributed to lack of shelter during bitter winter weather over the following months. Natural dams formed by landslides had to be destroyed to prevent flooding.

Kanto earthquake, Tokyo, Japan The "great Kanto earthquake" on September 1, 1923, destroyed much of Tokyo and the neighboring city of Yokohama. The earthquake, estimated at between magnitudes 7.9 and 8.3 on the Richter scale, struck about one minute before noon and had its epicenter southeast of Tokyo in Sagami Bay. The earthquake and subsequent fire are believed to have killed 140,000 people; tsunamis sweeping shorelines in the vicinity added to destruction. The 1923 earthquake was one in a series of major earthquakes that have struck Tokyo at intervals of approximately 70 to 75 years over the past several centuries: in 1633, 1703, 1782, 1853 and 1923.

Edward Seidensticker, in his history of Tokyo, *Low City, High City,* reports that the initial shocks of the 1923 earthquake knocked out seismographs at the central weather bureau office in Tokyo so that a seismograph at Tokyo Imperial University made the only record of the series of more than 1,700 earthquakes that struck the Tokyo area over the following three days. He adds that the earthquake damaged or destroyed almost 75% of buildings in Tokyo, whereas the fire affected almost 67%. Fire-fighting equipment was destroyed by the earthquake, and so fires burned out of control. The cities were built largely of wood, and oil stored in aboveground tanks added to the conflagration. The fire reportedly generated a curious phenomenon called "fire tornadoes." These were blazing storms with winds strong enough to lift a person off the ground. More than 30,000 people reportedly were killed at a single location, a park on the east bank of the Sumida River, when a storm of fire descended upon refugees who had gathered there.

Destruction was most widespread in the Low City, the eastern and generally lower-class portion of Tokyo built on flat ground consisting largely of unconsolidated and liquefaction-prone sediment and landfill. The comparatively affluent High City built on hilly and more stable ground in the western half of Tokyo came through the earthquake and fire with less damage. The Low City had also suffered extensive damage in another earthquake in 1855, but because of the magnitude of this damage, the 1923 earthquake marked a sharp decline in the influence of Low City culture and hence a change in the social history of Tokyo as well.

Persecution of the unpopular Korean minority in Japan is said to have accounted for several hundred additional deaths following the 1923 earthquake. The traditional Japanese prejudice against Koreans made them a convenient target for public anger. Koreans were accused of arson and of poisoning wells and were sentenced to death, but the government reportedly took a dim view of the persecution and killings because of possible effects on international opinion.

Karkar, volcanic island, near Papua New Guinea Located in the western portion of the Bismarck volcanic arc, the stratovolcano Karkar exhibits a double, nested caldera. It has erupted on approximately 10 occasions since the mid-17th century. An eruption involving an impressive "flame" from the volcano's summit was reported in 1643. In 1885 smoke and ash reportedly issued from Karkar. Numerous earthquakes on June 17, 1895, were followed by an eruption that continued on an intermittent basis through August. A moderately strong earthquake occurred under the northern coast of the island in 1930. An explosive eruption may have occurred in 1962, but this information is uncertain.

No further eruptions were reported until 1974 and 1975, when great amounts of lava emanated from the volcano. Earthquake activity started to intensify in 1976, and in 1978 a periodic variety of earthquake, called a "banded tremor" because of its "banded" appearance on seismograph records, began occurring at Karkar. Also in 1978, ground temperatures increased dramatically and reached the level of incandescence in some areas. Fumaroles also became more active during this time. PHREATIC ERUPTIONS started in 1979 and included an eruption in March of that year that killed two volcanologists. It has been suggested that banded tremors precede, and may provide early warning of, phreatic eruptions, because the tremors may result from processes similar to geysers underground.

Kashmir, India A series of earthquakes in the province of Kashmir in June and July of 1885 reportedly killed more than 3,000 people and injured over 5,000 more.

Katla, volcano, Iceland A subglacial volcano, Katla has a history of eruptions dating back to the 10th century and has erupted on the average every 42 years since the late 16th century. A tremendous fissure eruption was reported in 934. Minor earthquakes preceded another eruption in 1625. Earthquakes occurred again near Katla before eruptions in 1660, 1721, 1755, 1823, 1860 and 1918. The 1918 eruption produced a water flow of about 12 million feet per second for two days. Another eruption has been expected in recent years but has not occurred. It has been suggested that the eruptions of SURTSEY from 1963–67 and HEIMAEY in 1973 may have relieved pressure at great depth and thus delayed the anticipated eruption of Katla.

Katmai, Mount, volcano, Alaska, United States The eruption of Katmai on June 6, 1912, is one of the most famous events in volcanology. In the week before the eruption, earthquakes shook the area around the volcano. Early on the afternoon of June 6, the mountain emitted a thick cloud of ash that cast the land around the volcano into darkness. Great explosions could be heard at a distance of as much as 800 miles from Katmai. Ash began falling and accumulated to such depths on the ocean that rafts of ash were capable of bearing a man's weight. Visibility was reduced to less than 100 feet. On one ship caught in the ashfall, the air was so murky that the helmsman had difficulty seeing the compass in front of him. Ashfalls reached Juneau, 750 miles southeast of the volcano, and the Yukon Valley, some 1,000 miles north of Katmai.

The volcano emitted large quantities of sulfur oxides, which, when dissolved in moisture in the air, produced a solution of sulfuric acid that burned eyes in the vicinity of the volcano and ruined clothes hanging on lines to dry in Vancouver, British Columbia, some 1,500 miles southeast of Katmai. After the eruption, Katmai was found to have collapsed. The summit had lost its inner supports and fallen inward, creating a CALDERA about three miles wide and half a mile deep. Magma had been diverted from Katmai through another volcano called Novarupta. Novarupta put out enough incandescent vapor and ash to cover more than 40 square miles of land to a depth of hundreds of feet. Katmai's eruption created the VALLEY OF TEN THOUSAND SMOKES, a plain of tephra from which hot gases and water vapor arose for years afterwards. The 1912 eruption removed approximately one and a half cubic miles of rock from Katmai. Comparatively minor eruptions of Katmai occured in the 1920s, 1950s and 1968.

Kelud, volcano, Java, Indonesia Also known as Gunung Kelud, this volcano has erupted on numerous occasions over the last two centuries and is known for its highly dangerous lahars (volcanic landslides or mudflows). Kelud is an outstanding case study in using engineering to reduce hazards from volcanos, lahars specifically. Because the volcano's crater lake was involved in generating lahars, a plan was devised for draining the lake through a tunnel. This plan was carried out in the 1920s, and a tunnel drainage system was finished in 1928. This system worked so effectively that an eruption in 1951 generated no lahars. Other attempts to counteract lahars included a wall that was built in 1905 to direct mudflows away from settled areas, but the wall was destroyed in the 1919 eruption.

Kentucky, United States Kentucky is not a very seismically active state but does have a history of occasional strong earthquakes. One such earthquake occurred near Hillman on December 27, 1841, and was accompanied by a rumbling noise and a great disturbance on a river.

Kii Peninsula, Japan The Kii Peninsula is located southwest of the Izu Peninsula and has been the site of several powerful earthquakes in the late 19th and 20th centuries. Very strong earthquakes occurred here in 1899, 1948 and 1952. Other powerful earthquakes have occurred immediately off the coast of the Kii Peninsula, in 1944 and 1946. Swarms of earthquakes were observed along the western side of the Kii Peninsula before the KANTO EARTHQUAKE of 1923. Other earthquake swarms occurred on the Kii Peninsula in the early 1950s, again just before a powerful earthquake, this one some 150 miles southeast of Tokyo in 1953. Earthquakes on the peninsula are believed to be tectonic in origin, although it has been suggested that intrusion of magma at great depth may be involved.

Kikai, caldera, Japan The largely submerged Kikai caldera lies at the southern tip of the island of Kyushu. The caldera occupies part of a large GRABEN that also includes the AIRA, ASO, ATA and KAKUTO calderas. Several small islands and reefs mark the location of the caldera rim. The caldera is thought to have formed in three cycles of eruptions, the most recent of which expelled more than 30 cubic miles of solid material and is considered one of the greatest eruptions in recent geologic history. Kikai has been comparatively quiet in the 20th century, but an eruption that lasted for several months in late 1934 produced a small island. The eruption on September 20, 1934, is said to have been especially dramatic, occurring as it did during the passage of a hurricane. Pressure associated with a surge of water accompanying the tropical storm may have set off this eruption, although this connection has not been proven. Eruptive activity continued through late 1936. Strong earthquakes accompanied the eruptions between 1934 and 1936.

Kilauea, volcano, Hawaii, United States Located on the island of Hawaii, the largest island in the Hawaiian chain, Kilauea is perhaps the most intensively studied volcano in the world. Because eruptions here are gentle compared to the explosive eruptions of many other volcanoes, such as Mount St. Helens, scientists have been able to study Kilauea at close range and at great length. Magma here is low in dissolved gases and flows freely, almost like water, on occasions when the lava spills out of the crater and runs across the

Aa lava flow from Kilauea, 1960 (D. H. Richter, USGS)

surrounding land. Kilauea is one of five volcanoes on the island of Hawaii; the others are MAUNA LOA, MAUNA KEA, Kohala and Hualalai. Kilauea has erupted on dozens of occasions within the past several centuries.

In 1866 the crater of Kilauea surprised observers when the lava lake drained away and disappeared.

Klyuchevskaya, volcano, Kamchatka, Russia The STRATOVOLCANO Klyuchevskaya is one of a group of a dozen closely spaced volcanoes on the Kamchatka Peninsula and has a record of eruptions dating back to the late 17th century. Eruptive activity between 1937 and 1939 involved several locations on the volcano, including the summit and vents along the mountain's flanks. The eruption of BEZYMIANNY in 1955 occurred in this cluster of volcanoes.

K'one, caldera, Ethiopia Also known as Garibaldi, the K'one caldera complex is located east of Addis Ababa, in the Main Ethiopian Rift. Lava flows are thought to have occurred here around 1820, and fumaroles and hot springs exist in the K'one area today.

Kotlugja, volcano, Iceland An 1860 eruption of Kotlugja was reportedly visible 180 miles away at night and audible at a distance of 100 miles. During one eruption, lightning from the cloud is said to have killed two men and 11 horses.

Krafla, volcano, Iceland Krafla underwent a series of eruptions in the 1720s but was dormant thereafter until eruptions began again in 1975. In one eruption in 1977, lava emerged from a steam well at a geothermal facility. Krafla is not a lofty, majestic volcano. Subsidence has been an important factor in its formation. According to one model of activity at Krafla, the flow of magma into the area of the volcano is offset by rifting (that is, the pulling apart of crustal plates on either side of

A tree mold photographed at Hiiaka Crater at Kilauea after draining of lava lake. Note lava-coated boulders in background. (D. B. Jackson, USGS)

the rift that runs through Iceland) so that large volcanoes like those of Japan or America's Pacific Northwest do not have an opportunity to form. The Krafla volcano is "stretched out flat," so to speak, approximately as fast as it is built up by eruptions. Activity at Krafla has been interpreted as the result of magma flowing from great depths into a central reservoir or reservoirs and from there into rifts.

The eruptive history of Krafla has been documented for almost three centuries. In May of 1724, powerful earthquakes in the Krafla caldera were followed by a phreato-magmatic eruption that expelled pumice and scoria. Earthquakes also preceded volcanism at Krafla in January and April of 1725. Strong earthquakes and changes in the level of nearby Lake Myvatn were reported in September of 1875, and these signs are thought to have indicated an intrusion of magma to the south of Krafla. Lava erupted from the caldera in August of 1727, and the flows extended to the north and south. Another eruption of lava, following hours of seismic activity, took place in April of 1728, followed several hours later by yet another eruption a couple of miles to the south. A lava eruption in December of 1728 was followed by a much more powerful eruption of lava in January of 1729. More than one lava flow was reported over the next several months, but it is not known whether the eruption was intermittent or continuous. After the eruptive activity in 1729, Krafla appears to have remained quiet until 1746, when earthquakes and changes in lake level at Myvatn accompanied some kind of volcanism in the caldera.

Krakatoa, volcano, Sunda Strait, Indonesia The catastrophic 1883 eruption of Krakatoa is one of the most famous volcanic event in history. An island in the Sunda Strait between Java and Sumatra, Krakatoa began building up to an eruption in May of 1883. An earthquake on May 20 was followed by a venting of steam and ash from the volcano. As the eruption continued, it became clear that the activity did not center on Rakata, the tallest peak on the island at slightly more than 2,600 feet but rather on comparatively tiny Perboewatan, a 300-foot cone at the island's opposite (northern) end. An exploration party from nearby Batavia arrived at Krakatoa on May 27 and found the eruption had enlarged the CALDERA to a diameter of more than half a mile; the floor was flat, some 500 feet across, and covered by a crust of solidified lava. Steam emerged with a loud noise from a hole about 150 feet wide in the middle of the floor. The eruption continued through July and August, accompanied by loud explosions and mild earthquakes. Lightweight volcanic rock accumulated on the waters around Krakatoa and is mentioned in the logbooks of vessels that passed near the island during the eruption. The captain of the ship *Idomene* noted on August 11 passing through great fields of pumice, and several days later the captain of the bark *West Australia* mentioned in his log that the ship had encountered large amounts of "lava," by which he appears to have meant floating pumice. He added that some pieces of rock were several feet in diameter.

After the island destroyed itself, a postmortem on the eruption reached the conclusion that pressure had built up inside the volcano from gas escaping from magma. The gases inside Krakatoa are believed to have had no adequate outlet because magma came in contact with seawater at the surface, cooled and solidified and thus blocked the gases from escaping. The situation, Professor J. W. Judd of the British Royal Society's committee on the Krakatoa eruption suggested in a report on the event, was comparable to closing the safety valve on a steam boiler while the boiler continued to build up pressure inside. Eventually, the "boiler"—in this case, the volcano—had to give way and explode. The explosion began with a series of powerful blasts on August 26–27.

Loud explosions were heard at Batavia and Butzenbourg, about 100 miles from Krakatoa, early on the afternoon of August 26. A few minutes later, the captain of the ship *Medea,* about 75 miles from the island, saw smoke and vapor rise from the volcano to an estimated altitude of more than 15 miles. Around 5 P.M., some four hours after the final series of explosions began, the noise of Krakatoa's destruction was audible all over the island of Java. Around sunset, the captain of the ship *Charles Bal,* some 10 miles south of Krakatoa, was so intimidated by the eruption that he ordered the ship to retreat eastward. About this time, large chunks of pumice began landing on the decks of the *Charles Bal.* Meanwhile, the captain of the ship *Sir Robert Sale,* also in the vicinity of Krakatoa, witnessed spectacular displays of lightning in the eruption cloud and detected an odor of sulfur in the air. The sulfur odor became intense during the night, as bright flashes of light from the volcano lit up the surrounding area in the manner of a giant strobe light. A thick cloud from Krakatoa covered Batavia soon after dawn on August 27, and a muddy rain began falling, followed by a fall of small clumps of dust held together by moisture. This strange precipitation ended in midafternoon. Although atmospheric effects hid Krakatoa from view during this period, the sound of its final destruction was apparent. Four great explosions starting at 5:30 A.M. on August 27 and ending shortly before 11:00 A.M. sent pieces of rock showering down on ships in the vicinity and were heard as far away as Rodriguez, an island in the Indian Ocean some 3,000 miles away from Krakatoa. The explosions destroyed the northern half of Krakatoa and cut Rakata in two vertically from base to summit.

The explosive destruction of Krakatoa generated tsunamis that caused extensive destruction and loss of life along shores in the region. In some places, the waves are estimated to have come ashore at heights of 120 feet. The men of the ship *Loudon* turned their ship's bow toward the approaching tsunami and rode out the wave's passage, then watched as the tsunami engulfed and destroyed a town. Similar scenes occurred along the coasts of Sumatra and Java near Krakatoa. In one location, a town some 10 miles from the water was submerged by the wave. Some 3,000 residents of Karang Antoe were killed by the tsunami, as were more than 2,000 at Anjer and Batavia and 1,500 at Bantam. The total loss of life from the tsunami that followed Krakatoa's explosion is uncertain, but estimates range in the tens of thousands. The tsunami from Krakatoa was strong enough to twirl ships around at their anchorages at the harbor in Colombo, Ceylon (now Sri Lanka). Some deaths from this eruption were unrelated to the tsunami; at Second Point on the Java coast near Krakatoa, several convicts were reported killed by a lightning bolt.

In 1990, more than a century after the eruption, an expedition to Krakatoa, including a team from the University of Rhode Island's Graduate School of Oceanography, examined the role of PYROCLASTIC FLOWS in generating the tsunamis that originated from Krakatoa during the 1883 eruption. This expedition took samples of clastic material from the ocean floor around Krakatoa. Sediment cores taken in this way showed evidence of a large, poorly sorted main deposit overlain by another deposit of volcanic gravel and sand. Pieces of pumice large enough to impede the action of the sediment-coring mechanism were reported, and the investigators noted, in addition to the chunks of pumice, pieces of rock more than 12 inches across in some cases. These large fragments of rock were contained in a matrix with a silty and sandy character, the same as deposits on land. The similarity between the submarine and terrestrial deposits indicated that the underwater deposits originated as pyroclastic flows. The expedition also determined that submarine mounds near the Steers and Calmeyer island platforms (just north of Krakatoa) were made up of pyroclastics from the eruption in 1883. On the Calmeyer island platform, the investigators noticed signs of TURBIDITY CURRENTS that may have resulted when pyroclastic material at high temperature came in contact with seawater, causing steam explosions that gave rise to ash-filled turbidity currents. At one point on the sea floor west of Krakatoa, the expedition found a pyroclastic-flow deposit some 300 feet thick laid down by the eruption, including fragments of pumice several feet in diameter.

The 1990 expedition found evidence that one or more pyroclastic surges preceded the tsunami from Krakatoa, and two or more additional pyroclastic surges came after the tsunami. Deposits included pieces of coral that appear to have been torn from the sea floor and deposited with the pyroclastic material by the tsunami. The investigators concluded that deposits on land from Krakatoa's 1883 eruption are made up largely of material from pyroclastic flows as are underwater deposits from that eruption. The pattern of these deposits shows they were laid down in all directions around the island, not principally to the north, as had been presumed before. Accumulations are greatest to the west of Krakatoa. Deposition of pyroclastic material on the western side of the volcano is thought to have generated powerful tsunamis.

That the meeting between ocean and pyroclastic flows was not gentle is indicated by the small amount of mixing between seawater and pyroclastics. The pyroclastic materials show poor sorting—evidence that there was little mixing between them and the water. Had mixing occurred on a significant scale, water action would have resulted in more effective sorting of material. This pattern of poor sorting persuaded the scientists on the 1990 expedition that masses of pyroclastic material from the volcano in 1883 displaced water in such amounts as to cause the destructive tsunamis associated with the eruption of Krakatoa. Apparently, five or more units of pyroclastic flow occurred in the eruption. The results of the 1990 expedition indicate that one particular tsunami, at approximately 10 A.M. on August 27, 1883, was linked with the formation of the caldera at Krakatoa during an especially violent portion of the eruption, possibly caused when large amounts of magma came into contact with seawater.

The 1883 eruption of Krakatoa, specifically the fine ash cast into the upper atmosphere by the eruption, has been implicated in the cooler-than-usual winter that followed the volcano's destruction. Mean annual global temperature dropped by one-half to one degree Fahrenheit in 1884, and this cool period is believed to have continued through the 1880s and part of the 1890s too. The high-altitude ash cloud from Krakatoa is suspected of having intercepted enough incoming sunlight to lower temperatures on a worldwide scale.

Atmospheric phenomena attributed to dust from Krakatoa included vivid sunsets and a blue tinge to the sun. Even a "green sun" was reported in India.

The 1883 eruption of Krakatoa is especially notable for the distances at which noises from the eruption were heard. The captain of a ship 500 miles away from the eruption, at Surabaya, heard the explosions clearly. At Singapore, more than 500 miles from Krakatoa, noises from the eruption were interpreted as the sound of a ship firing its guns as a call of distress, and two rescue ships were sent out to see what the problem was. Similar circumstances convinced officials at Macassar and Timor in Indonesia, approximately 1,000 and 1,350 miles respectively from Krakatoa, to send out ships to rescue what they thought was a vessel in distress. Tribespeople in Borneo thought the noise from Krakatoa's eruption was the sound of a Dutch attack on them. On the island of Rodriguez, about 3,000 miles from Krakatoa, the noise of the eruption was clearly audible and resembled the firing of heavy guns. Before the age of electronic communications, this was arguably the only occasion in history when sounds were heard so far from their origin.

An eruption in 1927 generated a new island, called Anak Krakatoa, or "child of Krakatoa," in the midst of the three islands—Rakata, Panjang and Sertung—that occupy what once was the site of the original island of Krakatoa.

Krasheninnikov, caldera, Kamchatka, Russia Several great flows of lava are believed to have emanated from this caldera in the past 3,000 years: the Vodopadny flow (about 3,000 years ago), the Ozerny flow (1,500 years ago), the Yuzhny flow (less than 700 years ago) and the Molodny flow (less than 500 years ago). The area is seismically active, and a fumarole was observed in 1963.

Krozontzky, volcano, Kamchatka, Russia An especially beautiful and symmetrical volcano, Krozontzky erupted in 1922 and 1923.

Kuril Islands, Russia The Kuril Island chain extends more than 700 miles, from northern Japan to Kamchatka. The island arc occupies the boundary between the Sea of Okhotsk and the Pacific Ocean and consists of two chains of islands, the Greater Kurils and the Lesser Kurils. Recently active CALDERAS in the Kurils include CHIRPOI, Golovnin, Ketoi, L'Vinaya Past, MEDVEZHII, MENDELEEV, Nemo Peak, RASSHUA, TAO-RUSYR, Uratman and Zavaritsky.

Kuril Lake, caldera, Kamchatka, Russia The Kuril Lake caldera occupies part of the larger Pauzhetka caldera. The stratovolcano Illiinsky is located on the northeastern rim of the Kuril Lake caldera. Other, nearby volcanoes are Iavinsky, Kambalny and Zheltovsky. Geothermal energy has been exploited in this region.

Kutcharo, caldera, Japan The Kutcharo caldera is located in northern Japan near the southern end of the volcanic Kuril Islands. Kutcharo Lake now occupies much of the CALDERA. The island stratovolcano Nakajima stands in the western portion of the lake. To the southeast of the lake is Atosanupuri, a stratovolcano with an associated set of domes. The volcano Mashu, which has a crater lake similar to the one found at CRATER LAKE in Oregon, is located to the east of Atosanupuri, on the rim of the Kutcharo caldera. Mashu is thought to have erupted most recently some 1,000 years ago. Although no eruptions have occurred in the area of Kutcharo caldera in recent decades, there has been considerable seismic activity, and the caldera has been studied as a possible indicator of major tectonic earthquakes.

Kverkfjöll, volcano and caldera, Iceland The Kverkfjöll volcano has a history of possible subglacial eruptions dating from the 17th century. Confirming these eruptions has been difficult, however, and the principal evidence for them appears to be floods of water. Two CALDERAS at Kverkfjöll are believed to lie under the ice.

L

lahar In general terms, a volcanic landslide or mudflow. A lahar is made up of pyroclastic material and moves down the side of a volcano.

Laki, fissure zone, Iceland A massive eruption of the Laki fissure zone in 1783 was accompanied by the largest release of lava (more than 12 cubic kilometers) in history. The eruption also released gases that contaminated grass in Iceland and caused the death of hundreds of thousands of livestock. Some 10,000 residents of Iceland, or about 20% of the population, died in a famine that followed the eruption. Apparently the particulates released during this eruption remained in suspension over Iceland and portions of northern Europe for months after the eruption, resulting in a peculiar haze that was compared to fog.

Lamington, Mount, volcano, Papua New Guinea The site of one of the most violent eruptions of the 20th century, Mount Lamington is located in the eastern portion of Papua New Guinea, near the communities of Buna and Popondetta, approximately 25 miles from the coast. The mountain rises almost 6,000 feet above sea level and has a deep crater from which a river flows. A large portion of the mountain appears to have been laid down by mudflows and nuées ardentes. The powerful eruption of 1951 at Mount Lamington took many observers by surprise because preliminary activity at the volcano either was not interpreted as eruptive or simply went unobserved. On January 15, 1951, landslides were noted on the crater walls. Vapor emanated from the crater for the next two days, and earthquake swarms affected the surrounding area. The volcano's gas output increased greatly on January 18, and ash came from the mountain as well. Underground noises intensified as the emissions increased. Earthquakes occurred with increasing frequency until they became virtually nonstop. Early on the morning of January 19, a dazzling display of electrical phenomena took place. After daybreak, the top of the cone was seen to be covered with ash. On January 20, a great ash cloud rose to an altitude of five miles or higher. At 10:40 A.M. on January 21, a giant cloud rose from the volcano and reached an altitude of 20,000 feet in less than half an hour. The base of the cloud spread out rapidly over the adjacent land. The NUÉE ARDENTE from this stage of the eruption caused the greatest devastation to the north of the volcano and is thought to have moved at an average speed of about 60 miles per hour, though probably much faster in certain areas. Phenomena similar to tornadoes are believed to have accompanied the nuée ardente and produced dramatic disparities in effects from one place to another. The nuée ardente laid waste some 90 square miles of land and killed almost 3,000 people in the vicinity on January 21. (Total number of deaths attributed to this eruption were 6,000.) Heated dust laden with steam probably accounted for most of the deaths. Many survivors suffered severe burns. The stiff condition of bodies found after the passage of the nuée ardente indicate the heat from the cloud was intense. Two zones of destruction were observed in the area affected by the nuée ardente. Close to the crater, destruction was virtually total because of the combination of intense heat and high velocity; in this zone, some trees and buildings were carried away completely. Close to the crater, the nuée ardente produced a powerful scouring action that effectively ground away everything in its path, down to the level of the abraded soil. A comparatively small outer zone was affected more by heat than by blast. Valleys around the volcano were filled with hot ash that retained its heat for months after the

An earthquake-generated landslide at Hebgen Lake, Montana in 1959 (*Earthquake Information Bulletin*/USGS)

January 21 eruption. Wood buried under this hot ash would catch fire when exposed to the air weeks later.

At the time of the catastrophic eruption of January 21, Mount Lamington, whose noises had been intermittent up to this time, began to give off a steady roar that was audible on New Britain, 200 miles to the north. A few minutes before 9 P.M., another strong eruption occurred. Activity subsided for the following three days, then began again on January 25. Further explosive events occurred in February and March. Less powerful eruptions followed on an intermittent basis. Dome formation began in the crater of Mount Lamington soon after the January 21 eruption. In about a month and a half, a dome more than 1,000 feet high arose, sometimes at an average rate of more than four feet per hour. An eruption on March 5 demolished this dome, but it grew back higher than before by the middle of May. Eventually, the dome reached a height of approximately 1,800 feet above the floor of the crater. The devastated area recovered quickly after the 1951 eruptions, and by the mid-1960s, vegetation reportedly had regrown completely, so that the area affected by the nuée ardente looked identical to adjacent land.

Land of the Giant Craters, Tanzania Located on the edge of Africa's Rift Valley near the volcano NGORONGORO, the Land of the Giant Craters is a plateau made up of ejecta from nearby volcanoes.

landslide Many different kinds of earth movement are described by the comprehensive term *landslide,* which refers in general to a rapid downslope movement of unconsolidated material under the influence of gravity. Landslides may include mudflows, rockfalls and numerous other phenomena. Landslides are commonly associated with earthquakes, especially in mountainous regions, although landslides may occur in relatively flat country when conditions allow mass movement of unconsolidated material down a gentle slope.

In the 1964 GOOD FRIDAY EARTHQUAKE in Alaska, for example, numerous buildings on a plain (that is, flat land) near Anchorage were destroyed by landslides when the earthquake allowed material at and near the surface to slide along underlying wet clay. In some locations, the landslides produced by this earthquake were rotational, meaning the ground surface tilted as blocks of earth rotated while sliding downslope. Elsewhere, nonrotational movement was observed, as the earth broke into an up-and-down pattern of horsts, or elevated blocks, and grabens, or depressed areas. In one widely publicized case, a school building toppled off the edge of a horst and landed upside down in the adjacent graben. Another Alaskan earthquake, in 1958, generated a landslide that caused a spectacular and highly destructive wave to form in LITUYA BAY. This wave reached more than 1,700 feet up the side of the valley where the slide occurred. (The Empire State Building in New York City, by comparison, is only about 1,000 feet high.)

The Columbia river gorge between Oregon and Washington shows evidence of tremendous landslides within recent geologic time, such as the BONNEVILLE SLIDE, thought to have occurred around A.D. 1100 and to have involved almost half a cubic mile of rock, damming the Columbia River temporarily. In the greater Los Angeles area, potential landslides pose a special problem because they stand to block important highways following a major earthquake there in the future, thereby isolating the region.

Landslides may occur either on dry land or on undersea slopes bearing unconsolidated material. Certain submarine landslides are known as TURBIDITY CURRENTS. One famous turbidity current was generated by the Grand Banks earthquake off Newfoundland in 1929 and severed a number of submarine telegraph cables on the bottom of the Atlantic Ocean. Later, the recorded times of cable breaks allowed the maximum velocity of the turbidity current to be calculated at some 55 knots.

lapilli Pieces of PYROCLASTIC material ranging in diameter from 2 millimeters (about 0.08 inch) to 64 millimeters (about 2.5 inches).

Larderello, Italy Larderello, a hydrothermal activity area, is noted for its geothermal power facility, which began to generate electricity for a chemical plant on the site in 1904 and, by World War II, was producing almost 1,000 million kilowatt-hours of electricity each year. The power facilities at Larderello were wrecked by German forces in 1945 but were rebuilt. Larderello also has been the location of a large and profitable chemical industry that began by extracting borax from natural steam. Count Francesco Larderel is credited with the idea of using energy from the natural steam at the site to concentrate boric acid solutions. The hydrothermal activity at Larderello appears to stem from magma underlying an area of several dozen square miles.

Lassen Peak erupts, June 14, 1914 (B. F. Loomis, USGS)

Lassen Peak, volcano, California, United States Located in northern California, just south of Medicine Lake volcano and Mount Shasta, Lassen Peak stands some 14,500 feet high and was considered the most recently active volcano in the 48 contiguous United States until the eruption of Mount ST. HELENS in 1980. Lassen Peak includes three active volcanoes and exhibits hydrothermal phenomena such as steam vents and hot springs. The eruption of Lassen Peak in 1914 began without warning and commenced with minor releases of ash. Continuing eruptive activity over about a year cast blocks of rock from the crater. Explosive eruptions in the spring of 1915 were accompanied by emissions of lava, and one eruption on May 22 deposited tephra in Nevada, some 200 miles east of the volcano. Mudflows descended along the flanks of the mountain, and on May 22 a NUÉE ARDENTE gave rise to a large mudflow that required the prompt evacuation of a valley near the mountain, with no reported deaths. Further eruptions, though less intense, continued through 1917. Volcanic formations include Cinder Cone, which erupted in the 1850s and expelled both ash and lava; and the Sulphur Works, where hydrogen sulfide escapes from hot springs and steam vents.

Lassen Peak differs from the other volcanoes of the Cascade Range in that it is not a stratovolcano but rather a very large dome surrounded on the south and west sides by smaller domes. A few miles south of Lassen Peak one finds the remains of a huge stratovolcano, known as Brokeoff Volcano, that has a diameter up to 15 miles at its base and is believed to have stood some 11,000 feet tall. The collapse of Brokeoff Volcano created a caldera more than two miles wide and left a dramatic fault scarp visible today at Brokeoff Mountain. Lava from Brokeoff Volcano now mostly surrounds Lassen Peak, and these vast outpourings of lava from the ancient volcano may have been partly responsible for its subsequent collapse.

lateral blast An effect sometimes seen in explosive eruptions of volcanoes, lateral blast occurs when a volcano explodes sideways, so to speak, directing the force of the explosion horizontally over the landscape rather than vertically. This phenomenon was observed in the 1980 eruption of Mount St. Helens, when the north flank of the volcano disintegrated and released pressure from inside the mountain, resulting in a blast that sent low-level clouds rushing over the adjacent terrain, destroying large amounts of timber.

lava Lava is MAGMA that flows out onto the earth's surface. The word *lava* appears to be derived from the Italian word *lavare*, "to wash away." The Italian verb is not restricted to the action of lava but also may refer to a flood of water or even to a moving crowd of people. Over the centuries, "lava" has been used to refer to a variety of volcanic phenomena, including mudflows. Now, however, its use is generally confined to fluid, molten rock reaching the earth's surface and to the formations produced by such rock during volcanic eruptions.

Where lava solidifies after flowing for some distance, the resulting deposit is called a lava flow. Lava flows on dry land may form aa, characterized by a broken, sharp-edged surface, or pahoehoe, which has a peculiar ropy texture. Submarine pillow lavas consist of rounded masses about a yard in diameter, formed as molten rock comes into contact with cool seawater. Lava flows show a wide range of grain sizes, depending on their rate of cooling. Open spaces or vesicles may form in lava that has a high concentration of dissolved gases. An example of such vesicular volcanic rock is PUMICE, which is sometimes light enough to float on water. Another widespread, vesicular volcanic rock is SCORIA. Some lava flows include xenoliths, foreign material picked up and carried along by the lava. Lava tubes, hollow tunnels within a lava flow, form as surrounding lava cools and molten rock in the tube drains away. In some cases, lava tubes are large enough to accommodate humans. Spatter cones, small volcanic "mountains" several feet high, may form out of lava spewing from vents in the ground. One impressive phenomenon of some lava flows is columnar jointing, in which solidified lava forms vertical, prismatic columns several feet wide. Giant's Causeway in Ireland is a spectacular example of columnar jointing. Molds are commonplace features of lava flows and are formed when lava surrounds an organism such as a tree and hardens. When the tree decays and disappears, the mold remains in the lava flow. Ashfalls also may produce molds by burying an animal or other organism, of which a mold remains after the organism's body decays. Excavations at Pompeii have revealed detailed molds of animals and humans who were killed in the eruption of Vesuvius in A.D. 79. Plaster was poured into the molds, providing images of the eruption's victims.

Lavas vary in their chemical composition and are categorized according to their silica (silicon dioxide) concentration. Lavas containing 66% or more silica are called acidic lavas because of the acidic action of silica, which combines with other oxides and forms silicate compounds. Basic lavas, at the other end of the spectrum, have a silica content of 52% or less. Intermediate lavas contain between 52% and 66% silica. The behavior of lava during an eruption depends largely on its silica content. As a rule, acidic lavas are "thicker" in character, flow with difficulty and are involved in explosive eruptions. (The explosive character of these eruptions stems from the difficulty dissolved gases have in escaping from the highly viscous, acidic lava. In some situations, gas coming out of solution from magma will accumulate and build up pressure until an explosive event occurs. Well-known examples of such eruptions are those of Mount ST. HELENS in 1980 and BEZYMIANNY in 1952.) Slow-flowing, acidic lavas containing about 74% silica are called rhyolites and are often pink or gray in color. Rhyolite is commonly found in LAVA DOMES. Basic lavas flow more readily than

A lava flow from Kilauea volcano meets the ocean (USGS)

acidic lavas and allow gases to escape easily, thus avoiding explosions. Such eruptions form basalt deposits. Basic lavas are also darker than acidic lavas. The volcanoes of the Hawaiian Islands are familiar examples of volcanoes that emit basic lavas. In between these two extremes of chemical composition are the intermediate lavas, known as ANDESITES. Andesitic rocks occur widely in the western portions of North and South America.

Besides silica, lava may contain many chemical components, including oxides of aluminum, calcium, iron, magnesium, potassium and sodium. Acidic lavas typically are rich in alkalis (potassium oxide and sodium oxide) and poor in calcium, iron and magnesium. These proportions are reversed in basic lavas. Most lavas fall somewhere in between the extremes of acidic and basic composition.

Lava emanating from a volcano may differ in composition from one time to another during an eruption. Also, a single volcano may emit lavas of different compositions from one eruption to another. These differences are thought to involve a process known as magmatic fractionation. Here, fractionation refers to the dividing up of magma, or molten rock still underground, into different components. Initially, the magma is believed to be largely homogeneous. As heavier components of the magma crystallize and settle out of solution, lighter components remain dissolved, and the character of the molten rock changes accordingly. This process alone, however, is not entirely responsible for changes in the composition of magma, geologists argue. Other processes are thought to be involved as well, including the assimilation of rock from the walls of a magma chamber underground.

There is a distinct difference between lavas of the continents and lavas of the seabed. Continental lavas tend to have high concentrations of silicon and aluminum. Therefore, these rocks are commonly known as sial, a name derived from the chemical symbols for silicon (Si) and aluminum (Al). These rocks are also called granitic and are generally the same as the acidic lavas. Lavas on the seabed, by contrast, belong to a category of rock called sima (because it is rich in silicon and magnesium). Sima also underlies the granitic rocks of the continents. When basaltic magma from the sima layer rises toward the surface and passes through sialic rocks, the molten rock takes on some characteristics of the sial. The result is intermediate lava, or andesite.

Lavas may be categorized further into three types: Atlantic, Pacific and Mediterranean. Lavas of the Atlantic type are basaltic and are thought to represent the kind of lava that rises from the sima layer. The Pacific type of lava, exemplified by the volcanoes that run along the so-called andesite line surrounding the Pacific Ocean basin, is more sialic in character. (The andesite line forms the dividing line, so to speak, between lavas of the Atlantic and Pacific types.) The Mediterranean type of lava has assimilated large amounts of limestone, or carbonate rock, on its way to the surface and has become more acidic through this process.

The earth's surface is covered in some places by huge and spectacular outpourings of basaltic lava. These deposits are known as flood basalts and may cover thousands of square miles. In the northwestern United States, flood basalts have covered large areas of the states of Idaho, Oregon and Washington. Great flood basalts are also found in the Deccan plateau of India. Flood basalts may originate as fissure flows, when cracks in the earth's crust disgorge huge quantities of highly fluid, fast-flowing magma. Such floods of molten rock may travel many miles in a single day.

Lava flows can cause great destruction when an eruption occurs in or near a heavily populated area. Such incidents are commonplace in southern Europe and along the Pacific rim, where large communities exist near active volcanoes. Especially dangerous in this respect are volcanoes such as Italy's VESUVIUS and Sicily's Mount ETNA. There have been numerous attempts to divert lava flows before they reach populated areas, but these efforts have met with varying degrees of success.

Lava flows tend to cool gradually because rock is an inefficient conductor of heat. When a relatively cool layer of solidified lava forms on the surface of a lava flow, that layer may provide insulation that allows the lava below to remain in a fluid state for months or even years. Many factors may affect the rate of cooling of a lava flow, however, such as rainfall, cooling through cracks and even flowing into the sea.

lava domes These structures are made up of extremely viscous LAVA that moves upward through the vent of a volcano and forms a domed, craggy surface. Lava dome formation may follow the explosive eruption of a volcano, when gas-rich magma has been expelled and thicker, more viscous lava flows out. A lava dome is commonly surrounded by debris dislodged from the surface of the dome during its growth.

The lava domes formed at Mount St. Helens in Washington state have received extensive study. Out of five episodes of explosive activity at Mount St. Helens in 1980 and 1981, a lava dome formed in three of these episodes after explosive activity diminished. The first dome grew for at least a week after an explosive eruption on June 12, 1980, and eventually grew to be more than 1,000 feet in diameter and some 150 feet high. Part of this dome was destroyed during an explosive eruption in July. A new and smaller dome formed following another explosive eruption on August 7. An eruption in October was followed by the formation of a third dome. Analyses of samples from the dome indicated the magma was very low in sulfur and oxygen, possibly because gases had escaped from the magma before the dome formed. Study of the dome also made possible an estimate of the thermal energy yield of the Mount St. Helens eruptions of May through October of 1980. This estimate puts these eruptions in roughly the same range of energy yield as the eruptions of Hekla in 1970 and Mauna Loa in 1950. Evidence of lava domes on the planet Venus was returned by the U.S. Magellan space probe in 1990. Radar images of a portion of the surface of Venus showed a chain of seven dome-like structures, each about 15 miles wide and up to about 2,200 feet high.

Lipari Islands, Tyrrhenian Sea, Italy Located near the tip of the Italian peninsula, just north of the island of Sicily, the Lipari Islands (also known as the Aeolian Islands) are a cluster of volcanic islands, of which the most famous are Lipari, which contains Campo Bianco; STROMBOLI; and VULCANO. Other islands in the group include Alicudi, Basiluzzo, Filicudi, Panaria and Salina. The volcanoes are seen as part of a line of volcanic mountains extending along the western shore of the Italian peninsula, parallel to the Apennines.

liquefaction One of the principal causes of property damage from earthquakes, liquefaction occurs when earthquake vibrations make soil lose its cohesion and behave temporarily as a liquid. Structures built upon soil undergoing liquefaction may suffer severe damage; for this reason, the character of underlying material is, or at least ought to be, a major safety consideration when planning construction in areas known for frequent and/or energetic earthquake activity. Liquefaction commonly takes place during earthquakes in poorly consolidated soil where ground water exists close to the surface. (These conditions prevail in many areas subject to frequent, strong earthquake activity.) Water in between particles of soil is known as pore water, and the pressure of pore water rises with the passage of waves from an earthquake, thus making the ordinarily solid soil assume a liquid condition temporarily.

Damage may result in several ways. When liquefaction occurs on slopes greater than perhaps 3°, large masses of soil may flow downslope, either in liquefied form or as large blocks of still-cohesive soil sliding on an underlying, liquefied layer of soil. In some cases, vast quantities of material are transported over distances of several miles.

Lateral spread is another source of property damage from liquefaction. In this phenomenon, blocks of soil at the surface slide sideways because of liquefaction of an underlying layer. Lateral spread may not be as dramatic as flow failure and may result in displacements of only a few feet. Even such seemingly small displacements can be highly destructive, however, especially if they sever underground water lines required to fight post-earthquake fires in urban areas. Ground oscillation occurs when liquefaction of subsurface layers of soil allows blocks of soil on the surface to oscillate and display a variety of earthquake-related effects, such as wavelike rippling and fissures. This kind of ground motion may cause severe damage to buildings, which as a rule are not designed and constructed to withstand strong ground oscillation.

Finally, soil simply may lose its bearing strength, or ability to support overlying structures. This occurs when soil at the surface undergoes liquefaction; structures built on it respond in a variety of ways, such as sinking or rolling over. Multistory buildings have been known to capsize, much in the same manner as a ship at sea, during such episodes of liquefaction. Another effect associated with this loss of bearing strength is the rise of buoyant underground structures such as fuel tanks or septic tanks. In evaluating the potential of a given locality to liquefaction, geologists must consider susceptibility to liquefaction (that is, the character of materials near the surface and how they are likely to respond to an earthquake) and the opportunity for liquefaction (meaning how often the locality is struck by earthquakes and how severe those earthquakes are). This determination of liquefaction potential may have to be made on a site-by-site basis rather than a regional basis because conditions may differ greatly from one location to another, sometimes within a distance of feet or yards.

In general, the soils most susceptible to liquefaction have the following characteristics: They were deposited relatively recently; they are made of loose and generally fine sediment (such as sand or silt) that lacks cohesion; and they contain water that rises close to the surface. Particle-size distribution is important here, because liquefaction potential tends to diminish as particle size increases. This means a gravelly soil is less likely to liquefy than a sandy or silty soil is, and deposits of boulders or of mixed cobbles and gravel are unlikely to liquefy at all. At the other extreme of particle size, clay is also resistant to liquefaction. The depth to groundwater is also important because soils in which groundwater approaches to within several feet of the surface are more susceptible to liquefaction than soils where groundwater remains several tens of feet below the surface. Many areas of the United States with histories of strong earthquake activity also show all three of the aforementioned soil characteristics that are associated with high liquefaction potential. The Mississippi river valley; the areas around Charleston, South Carolina and Boston, Massachusetts; and portions of California all exhibit the combination of silty or sandy soil, deposited recently and containing groundwater near the surface.

Two other factors influencing the areal extent and degree of liquefaction are, of course, the magnitude and duration of an earthquake. The more powerful an earthquake, the more intense its associated ground motion is likely to be and therefore the greater the danger from liquefaction of susceptible soils. Proximity to the origin of an earthquake is also a factor; a strong earthquake is likely to cause liquefaction at greater distances than a comparatively mild one.

Lisbon, Portugal One of the most destructive earthquakes in history, the great Lisbon earthquake may have killed some 70,000 people. The shocks began on the morning of November 1, 1755, All Saints' Day, and reportedly killed numerous worshipers in the churches of Lisbon as the buildings collapsed. Three shocks are believed to have struck Lisbon, shaking down most structures in the city. The destruction raised large quantities of dust, which blocked sunlight and cast the city briefly into darkness. The shocks reportedly generated a series of tsunamis that devastated Lisbon's waterfront. Large numbers of refugees made their way to the waterfront, specifically to a large marble quay; although the tsunami appears to have spared the quay itself from destruction, one report indicates that the quay subsided into the waters, leaving not a single survivor behind nor even a trace of the vessels moored to it. Fire following the earthquake added to destruction from the quake itself but appears to have reduced the threat of pestilence by consuming the bodies of numerous victims.

King Joseph was away from Lisbon but was in the vicinity, at Belem, when the earthquake struck. He and his court reportedly were so frightened by the shocks that they spent the night outdoors in their carriages. The next morning, the king returned to Lisbon to take stock of the damage and restore order to the shattered city, which had suffered from looting and murders in the aftermath of the earthquakes. The king's palace had survived with its kitchens intact, and his cooks began preparing food for the hungry populace. The government purchased large amounts of grain from the area around Lisbon and distributed it at modest cost or free of charge to the public. Hundreds of aftershocks during the following year are said to have kept the public's nerves on edge, but eventually the city recovered, and a decade after the earthquake, Lisbon had been rebuilt.

The Lisbon earthquake was felt over much of Europe; it appears to have generated landslides in the mountains of Portugal and to have been felt as a powerful disturbance in Algeria and Morocco. In Morocco, a fissure is said to have opened that sent a village of more than 8,000 residents along the shore tumbling into the water. Tsunamis from the Lisbon earthquake allegedly came ashore as waves more than 20 feet high in the Lesser Antilles, on the opposite side of the Atlantic Ocean. Strong earthquakes also hit North Africa about the time of the Lisbon catastrophe. Water levels in lakes in Scotland and Switzerland rose and fell by several feet immediately after the Lisbon earthquake.

lithology In general terms, the description of rock, especially its texture and mineral components. A low-density, "foamy" volcanic rock such as pumice, for example, has a much different lithology from volcanic glass, or obsidian.

lithosphere The outer, most "rigid" layer of the earth, including the crust and the upper layer of the mantle. (See also EARTH, INTERNAL STRUCTURE OF.)

Little Sitkin, volcanic island, Alaska, United States Little Sitkin is located in the Aleutian Islands and has two calderas. Eruptions were reported in 1776 and between 1828 and 1830, but little is known about these events.

Lituya Bay, Alaska, United States Located on the southeastern coast of Alaska, Lituya Bay is famous for the destruction caused by a wave that resulted from a landslide associated with a powerful earthquake in 1958. A rock slide along the shore of the bay set off a wave that destroyed timber as high as 1,720 feet. At various other locations along the shore of the bay, the wave swept away forest at elevations of 600, 160, 130 and 90 feet. An island in the bay was denuded of timber up to an elevation of 80 feet. Three boats were in the bay at the time of the incident. One was destroyed; the wave carried the second across a spit at the bay's mouth and into the ocean; and the third boat survived the wave in the shelter of the island mentioned earlier.

Loihi Seamount, Hawaiian island chain, United States The Loihi Seamount is located southeast of the island of Hawaii and is some 3,000 feet below sea level. The seamount appears to have undergone recent volcanic activity and has two craters on its summit. Earthquake activity under Loihi in the 1970s may have been linked with eruptions.

Long Island, caldera, Papua New Guinea The Long Island volcanic complex is located on the northern coast of Papua New Guinea and includes two stratovolcanoes, Cerisy Peak and Mount Reamur, on opposite sides of Lake Wisdom, a caldera lake some six miles wide. Motmot Island, a small island within the lake, was built by volcanic activity within historical times. Powerful seismic and volcanic events appear to have occurred at the caldera some 300 years ago, but records were not kept on a scientific basis. Eruptions occurred in 1933, 1938, 1943 and 1953 from a vent or vents under the lake and built up Motmot Island. Discolored water in the lake, a possible indicator of volcanic activity, was observed in 1974, 1975 and 1979.

Long Valley Caldera, California, United States This CALDERA near Mammoth Lakes appears to have undergone a series of eruptions 500 to 600 years ago that produced rhyolite domes in and near the caldera. A much earlier eruption is thought to have been so powerful that it sent a PYROCLASTIC FLOW rushing up the flank of the Sierra Nevada and over the mountains. The output from this eruption is believed to have flowed down Owens Valley for some 40 miles and to have covered almost 600 square miles of land in the present states of California and Nevada with thick ash layers. A famous deposit called the Bishop Tuff appears to date from this eruption. The ashfall from this eruption has been traced into what is now Nebraska. The eruption is thought to have released some 140 cubic miles of magma. Earthquake activity and uplift in the caldera during the 1980s provided evidence that magma was moving underground at that time.

Los Alamos, New Mexico, United States Los Alamos is built on a volcanic plateau made of ash-flow deposits laid down in an eruption that is believed to have resulted in the collapse of the volcano's summit and the formation of a caldera, now called Valle Grande.

Los Angeles, California, United States The Los Angeles area has been the location of numerous earthquakes, some of which in historical times have been powerful and destructive. One of the first earthquakes experienced by European-descended Americans in this part of California occurred during a visit by Captain Gaspar de Portola, governor of Mexico, in July 1769. Another series of earthquakes in 1812 caused such widespread damage to Spanish missions in southern California that 1812 came to be known as *"el ano de los temblores,"* or "the earthquake year." Other notable earthquakes in southern California history include the FORT TEJON EARTHQUAKE of 1857, the San Jacinto earthquake of 1899, the Imperial Valley earthquakes of 1915 and 1979, the San Jacinto earthquake of 1918, the Santa Barbara earthquake of 1925, the Long Beach earthquake of 1933, the Terminal Island earthquakes of 1949, 1951, 1955 and 1961, the Kern County Earthquake of 1952, the SAN FERNANDO EARTHQUAKE of 1971 and the Santa Barbara earthquake of 1978.

It is widely assumed that an earthquake of unprecedented destructive potential is possible in the greater Los Angeles area, although it is difficult to predict when and where it may occur. Projected casualty figures from the Federal Emergency Management Agency, in the case of a postulated earthquake of magnitude 8.3 on the SAN ANDREAS FAULT, put deaths at 3,000 to 12,500 and hospitalized injured at 12,000 to 50,000, with 52,000 long-term homeless. The same agency's projected figures for another postulated earthquake, of magnitude 7.5 on the Newport-Inglewood Fault Zone, are 4,400 to 21,000 killed and 17,600 to 84,000 hospitalized injured, with 192,000 long-term homeless.

The greater Los Angeles area is highly susceptible to earthquakes because of a complex and unstable geology generated by the collision between northward-moving Baja California and the deeply rooted Sierra Nevada range. This collision has produced Southern California's TRANSVERSE RANGES and a network of active faults, some of which have the potential to generate highly destructive earthquakes. Prominent fault zones in the Los Angeles area include the following:

- The Elsinore Fault Zone runs almost 120 miles northwest from the Mexican border to the northern boundary of the Santa Ana Mountains.
- The Newport-Inglewood Fault Zone is actually a collection of small faults running along a line extending past Newport Beach and Santa Monica all the way into Baja California. The powerful Long Beach Earthquake of 1933 occurred along this fault zone.
- The Palos Verdes Hills Fault runs about five miles west of Long Beach
- The San Andreas Fault Zone is the most famous fault affecting the Los Angeles area and runs approximately along a line connecting Palmdale, San Bernardino and Banning and then along the southern boundary of the Little San Bernardino Mountains. The San Andreas Fault Zone was the site of the Fort Tejon earthquake of 1857.
- The San Pedro Basin Fault Zone runs approximately from Point Dune down the San Pedro Channel to the east of Catalina Island.
- The small but nonetheless dangerous San Fernando Fault, about 20 miles south of the San Andreas Fault, was responsible for the destructive San Fernando earthquake of 1971.

Louisiana, United States Destructive earthquakes originating within Louisiana itself appear to be infrequent, although the state occupies the lower end of the seismically active Mississippi Valley. A strong earthquake on October 19, 1930, damaged buildings at Napoleonville, and another earthquake on November 19, 1958, caused alarm in Baton Rouge.

Love waves Named after the British physicist A. E. H. Love (1853–1940), Love waves are earthquake waves that move horizontally with respect to the direction of travel, with no vertical motion involved.

M

Macdonald Seamount, Tubuai Islands and Seamounts, South Pacific Ocean This seamount reaches to within about 100 feet of the surface. Its activity was discovered in 1967 when undersea hydrophones picked up noise from it. Bubbles of gas and fragments of basalt were seen issuing from the seamount in 1987.

magma Molten rock underground is generally known as magma, as distinct from LAVA, which is molten rock that has emerged onto the surface from a volcanic vent. Magma differs greatly in composition and behavior. Some magmas are high in dissolved gases and, when they reach the earth's surface and subterranean pressures on them are removed, form the frothy rocks pumice and scoria as dissolved gases bubble out of solution at the moment the rock hardens. Such gas-rich magma tends to produce explosive eruptions because of its high gas content. Alternatively, magma that is relatively low in dissolved gases may emerge from vents as streams or floods of highly fluid lava that may flow across the land for many miles before solidifying and halting. Rocks produced by such eruptions include basalt and obsidian. Great flows of basalt from eruptions of this kind may be found in the Pacific Northwest of the United States, especially the Columbia Plateau and Snake River Plain. When magma solidifies underground, it may form structures such as dikes, vertical walls of igneous rock created when magma enters a vertical crack in subterranean rock and solidifies there, or sills, horizontal shelves of igneous rock that form in similar fashion but in a horizontal direction. Masses of magma that solidify at depth are known generally as plutons. When magma in the chimney of a volcano solidifies and remains standing after the external layers of the volcano have been eroded away by wind and water, the resulting columnar structure is known as a volcanic neck. Where magma reservoirs near the surface heat groundwater, the result may be an area suitable for GEOTHERMAL ENERGY production. Movement of magma underground is thought to be responsible for minor earthquakes that may signal the imminent eruption of a volcano.

magnitude A measure of the energy released in an earthquake. (See also SEISMOLOGY.)

Maine, United States The northernmost state of New England is known for frequent minor seismic activity but also experiences an occasional strong earthquake originating on or near its borders. Maine is located along the highly earthquake-prone St. Lawrence Valley and is sometimes affected by shocks that originate there.

One of the strongest earthquakes in Maine's history occurred on March 21, 1904, and was felt over much of New England and the Canadian provinces of Nova Scotia and New Brunswick. The earthquake apparently was most powerful near Eastport and Calais in Maine but also was felt strongly near the community of St. Stephen in New Brunswick. The shock was felt in Montreal, 300 miles distant, and at points 380 miles away in Connecticut. This shock was followed by several minor shocks hours later. An earthquake almost as strong, occurred on July 15, 1905, and was felt in central Maine and New Hampshire, a distance of some 100 miles westward from the shore. This earthquake, however, affected a much smaller area altogether than the earthquake of 1904.

Mammoth Hot Springs, Wyoming, United States A terraced natural structure at Yellowstone National Park, Mammoth Hot Springs is a famous example of a formation generated by geothermal activity.

Managua, Nicaragua The Managua earthquake of December 23, 1972, reportedly killed some 11,000 people and injured approximately 20,000 others. About three-fourths of residences in the city were destroyed, and approximately a quarter of a million people were left without homes. Property damage was reported at more than $500 million. This earthquake is believed to have had a minor foreshock approximately two hours before the main shock, with more than 100 aftershocks over the following weeks.

Manam, volcanic island, near Papua New Guinea A small island off the north coast of Papua New Guinea, Manam has undergone frequent small, explosive eruptions since the mid-1970s.

Maninjau, caldera, Indonesia Maninjau caldera is located on the island of Sumatra, along the Semangko Fault, which runs along the southwestern shore of the island parallel to the offshore Java Trench. Volcanoes in the vicinity of the caldera have undergone explosive eruptions on numerous occasions since the late 18th century.

mantle The intermediate layer of the earth's internal structure, between the thin CRUST and the dense CORE, the mantle in recent years has undergone intensive study that has revealed many details of its structure. Once presumed to be homogeneous, the mantle now is known to have considerable heterogeneity. (See also EARTH, INTERNAL STRUCTURE OF.)

mare On the earth's MOON, a plain of igneous rock produced by lava flows following a major meteorite impact is known as a mare (plural, maria), from the Latin word for sea. Viewed from earth, the lunar maria resemble seas, although the effective lack of atmosphere on the moon prevents liquid water from existing there. The lunar maria are thought to be relatively recent additions to the lunar surface because they exhibit few impact craters, a sign that their surface is younger than other, more heavily cratered areas. Comparable structures are thought to have formed on earth when large meteorites struck the surface. Two alleged terrestrial maria are the COLUMBIA PLATEAU in the northwestern United States and the DECCAN PLATEAU of India. In both cases, floods of molten rock have poured out over the surface and created broad, level plains of igneous rock. According to one hypothesis, an impact in what is now the western United States generated a lasting "hot spot," or point of extremely high heat flow to the surface, which produced the volcanic structures of the Columbia Plateau, SNAKE RIVER PLAIN and YELLOWSTONE NATIONAL PARK as the North American crustal plate slid westward over the site. A similar scenario has been put forward to account for the "hot spot" that is believed to underlie the Hawaiian Island chain and produced the islands as the Pacific crustal plate moved over the "hot spot."

Mars Although little is known about the geology of Mars compared to that of Earth, volcanism is known to have played a large part in shaping the surface of the red planet. Unmanned probes to Mars have revealed the existence of large volcanoes on Mars, such as OLYMPUS MONS and the volcanoes along nearby Tharsis Ridge in the planet's northern hemisphere. Although volcanic mountains are evident on its surface, Mars appears to lack the pattern of shifting and colliding crustal plates that characterizes tectonic and volcanic activity on earth.

It has been suggested that material expelled through the volcanoes of Tharsis Ridge came from magma underlying Valles Marinensis, a great valley thought to have been formed by subsidence as molten material underground was removed through eruptions. Volcanism evidently has given the northern hemisphere a geomorphology differing from that of the southern hemisphere in ways besides the mere existence of volcanic mountains. Impact cratering is much more visible in the southern hemisphere, whereas in the northern hemisphere, tephra from eruptions apparently has covered many impact craters and other such features. The Martian volcanoes affect the planet's meteorology on a small but interesting scale; clouds form in the lee of Olympus Mons and are thought to have been observed through telescopes on earth at various times before probes reached Mars, although the true nature of this phenomenon was not known until photographs were returned from Mars by spacecraft. Mars, like the moon, appears to be much quieter than the earth from a seismic viewpoint. Evidence of erosive activity on the Martian surface indicates that Mars once possessed substantial amounts of liquid water and a denser atmosphere than at present. Although the Martian atmosphere is only a fraction as dense as that of the earth, occasional dust storms on Mars are capable of transporting large quantities of sediment aerially. These storms are capable of hiding even the giant Martian volcanoes from view.

marsh gas See METHANE.

Maryland, United States The state of Maryland is not especially noted for earthquake activity, but the state has experienced substantial earthquakes on occasion. The eastern portion of the state has a considerable potential for damage from liquefaction in any future major earthquakes because much of the urbanized area of eastern Maryland is built up atop moist, unconsolidated soil that might be expected to lose coherence in a strong earthquake. Maryland sometimes has felt shocks that originated in adjacent Virginia. On April 24, 1758, Annapolis experienced an earthquake (also felt in Pennsylvania) that lasted about half a minute and followed noises from underground. Shocks in Harford County, Maryland on March 11–12, 1883, made clocks stop. On January 2, 1885, Frederick County was shaken by an earthquake that also was felt in Virginia and knocked small objects off of shelves and furniture; this earthquake was reportedly silent except for rattling noises from windows.

Masaya, volcanic complex, Nicaragua The Masaya complex, a caldera with small stratovolcanoes in its center, has been reported active since the 16th century. Although most Nicaraguan volcanoes have the steep sides characteristic of pyroclastic cones, Masaya has a wide, low configuration, having been built up from basaltic lava flows without a great amount of explosive eruptions. A lava lake like that of Kilauea in Hawaii is believed to have formed in Masaya in the 15th century. Masaya's activity has also included lava flows and explosions.

Massachusetts, United States The state of Massachusetts is located in New England, one of the most seismically active regions of the United States, and has a history of strong earthquakes dating back to colonial times. Some of the most powerful earthquakes in United States history have occurred in Massachusetts. Eastern Massachusetts in particular has a history of strong earthquakes, often accompanied by remarkable acoustical effects, such as rumbling noises. In addition to earthquakes that originate within Massachusetts, the state sometimes undergoes earthquakes originating in other areas of New England and also along the ST. LAWRENCE VALLEY between the United States and Canada. Portions of Massachusetts, especially in the east, appear highly susceptible to damage in any future major earthquake, because parts of the large urban area surrounding Boston, and sections of Boston itself, are built on moist, unconsolidated soil that would be vulnerable to liquefaction as earthquake waves passed through it.

Possibly the greatest earthquake in the history of Massachusetts occurred on November 18, 1755. This earthquake is thought to have had an epicenter east of Cape Ann and a Mercalli intensity of approximately VIII. The earthquake was felt over an area of some 300,000 square miles, from Nova Scotia to Chesapeake Bay, and some 200 miles at sea, where a ship in deep water experienced a shock that made it feel as if the vessel had run aground. (Vessels in harbors along the shore had much the same experience.) Starting with a sound like thunder, the earthquake made it difficult to keep one's footing so that people caught in the earthquake had to hold on to nearby objects to keep from being thrown down. In Boston, chimneys fell, and the ground motion reportedly resembled that of waves at sea. Treetops swayed vigorously, weathervanes fell from buildings, and stone walls were knocked over. In parts of Massachusetts, cracks formed in the earth, and sand moved upward through the cracks to the surface. One curious effect of this earthquake was that it appears to have killed fish in great numbers along the seacoast. Another powerful shock occurred on November 22, 1755, and still more shocks on December 19.

Matsushiro, caldera, Japan The Matsushiro caldera is located on the island of Honshu and has not experienced any eruptions within historical times, but earthquake activity is frequent in the vicinity of the caldera and has received extensive study. There are various opinions on what may be happening under the caldera. One view is that intrusion of magma may be involved. Other processes, including HYDROTHERMAL ACTIVITY, have been suggested to account for unrest at Matsushiro.

Mauna Kea, volcano, Hawaii, United States Mauna Kea is located on the island of Hawaii and is the tallest mountain on earth, more than 30,000 feet from its base on the seafloor to the summit. Unlike MAUNA LOA and KILAUEA, also on Hawaii, Mauna Kea is not active.

Mauna Loa, volcano, Hawaii, United States Mauna Loa, a basaltic shield volcano located on the island of Hawaii, is one of the most active volcanoes on Earth but also one of the least dangerous among these because its lava is low in dissolved gases and therefore not given to producing violent, explosive eruptions. Lava flows from Mauna Loa occur every several years on the average, tend to last only a few days, and have amounted to more than three billion cubic yards in the past century. Lava emerges from the summit crater (called Mokuaweoweo), from rift zones that extend from that crater, and from two pit craters, South Pit and North Bay, adjacent to the main caldera, which was produced by subsidence. One of five shield volcanoes that make up the island of Hawaii, the largest island in the Hawaiian chain, Mauna Loa has erupted on numerous occasions since records of eruptions began to be kept in the early 19th century. An impressive caldera on the summit of Mauna Loa is approximately two to three miles in diameter and formed several hundred years ago when collapse accompanied the withdrawal of magma from an underground chamber. The caldera walls stand about 600 feet high in some places.

In the late 1800s and the first half of the 20th century, Mauna Loa was very active. The volcano fell quiet in 1950 but became active again in 1975. An eruption in 1984 was associated with precursory earthquakes. Relatively little is known about activity at Mauna Loa compared with the nearby KILAUEA volcano because Mauna Loa was quiet from 1950 to 1975 when methods for studying volcanic activity were advancing. Earthquake activity and other evidence usually provide abundant warning that an eruption is forthcoming, and geologists have had considerable success in predicting the volcano's eruptions. Eruptions of Mauna Loa are notable for "curtains of fire," fountain-like eruptions of extremely fluid lava that may last a day or so before subsiding.

On occasion, lava flows from Mauna Loa have threatened the nearby city of HILO. One flow in 1881 stopped only at the edge of town, and another lava flow approaching Hilo in 1935 made history because of an attempt to divert the lava by aerial bombardment. The U.S. Army Air Corps, on the advice of the director of the Hawaiian Volcano Observatory, bombed the lava flow at selected points in hopes of changing its direction. The theory behind this tactic was that bombing might release gas and molten rock beneath the "crust" of the lava flow and thus stop its advance. The bombing appeared to succeed, for the lava flow halted soon after the bombardment. There is disagreement, however, about whether the bombing actually stopped the lava or whether the lava flow was about to cease anyway.

Mayon, volcano, Philippine Islands Mayon has a long history of destructive eruptions. One extremely violent eruption occurred in 1616 and is said to have destroyed dozens of villages near the volcano. In 1766 another eruption, characterized by flows of hot mud down the volcano's flanks, killed some 2,000 people. The 1814 eruption of Mayon caused at least another 2,000 deaths. Three major eruptions took place in 1886, 1888 and 1897. In the 1897 eruption, approximately 400 people are believed to have died. Eruptions in 1914 and 1928 were less costly in terms of lives lost.

Mazama, Oregon, United States Mazama is the name given to the volcano that collapsed and thus created the BASAL WRECK occupied by CRATER LAKE today.

Medicine Lake, volcano, California, United States Located in northern California near Tule Lake, just south of the Oregon border, Medicine Lake volcano has an area of some 900 square miles and rises approximately 4,000 feet above the surrounding land. Although basaltic lava from comparatively quiet eruptions is found on the flanks of the volcano, the mountain also has a history of violent eruption from its summit. One of these violent eruptions, believed to have occurred about 1,000 years ago, deposited pumice on Mount Shasta, more than 30 miles away. Rhyolitic lava flowing from Medicine Lake volcano formed Glass Mountain, and a separate flow on the western rim of the caldera formed Little Glass Mountain. The volcano has been generally quiet since then, although strong seismic activity in 1910 fractured the surface of the ground and was accompanied by a minor eruption of ash. Earthquake activity around the volcano in 1988 indicated possible movement of magma underground. Lava Beds National Monument occupies a portion of the Medicine Lake volcano.

Mediterranean Sea The Mediterranean basin is the location of numerous powerful earthquakes and volcanic eruptions and has a complex geology involving both the northward movement of Africa toward Europe and the interaction of numerous smaller plates. Volcanoes in the Mediterranean include VESUVIUS, ETNA and STROMBOLI. The catastrophic eruption of THERA, comparable in destruction to that resulting from the 1883 eruption of Krakatoa, is believed to have wiped out the Minoan civilization in 1470 B.C. The Italian peninsula especially is known for intense seismic activity as well as volcanism, and the presence of magma near the surface in portions of Italy has enabled that country to satisfy part of its energy needs by drawing on geothermal sources.

Medvezhii, caldera, Kuril Islands, Russia Explosive eruptions at Medvezhii are recorded in 1778 or 1779 and in 1883. The 1883 eruption reportedly involved lava flows. Phreatic activity was recorded in 1946, and activity of undetermined nature was recorded in 1958.

Mendeleev, caldera, Kuril Islands, Russia Mendeleev is thought to have erupted in or around 1880, killing some trees on the island through solfataric activity. Phreatic activity may have occurred in 1900, although this appears uncertain. Some earthquakes in recent years may have been associated with a GEOTHERMAL ENERGY facility on the island.

Merapi, volcano, Java, Indonesia The stratovolcano Merapi has a LAVA DOME on its summit and is believed to have erupted on more than 60 occasions since the beginning of the 11th century A.D. One eruption in 1006 was so destructive that the Hindu rajah of the island moved to nearby Bali and Java fell under Islamic influence.

Mercalli, Giuseppe (1815–1914) Italian scientist who made a study of Italy's earthquake areas. Mercalli is best known for the Mercalli scale, used to rank earthquakes on the basis of their intensity. The Mercalli scale uses Roman numerals, starting with I for the mildest earthquakes and ranging upward to IX or X for the most destructive. The advantage of the Mercalli scale over the more widely known Richter scale of earthquake magnitude is that the Mercalli scale requires no special instrumentation for the user to apply it. The Mercalli scale is based on visual and other noninstrumental observations of the earthquake's effects. (See also SEISMOLOGY.)

mesa A flat-crowned hill slightly broader than a butte.

Mesa de los Hornitos, volcanic formation, Mexico Associated with the volcano PARICUTIN, Mesa de los Hornitos (*hornitos* means "little ovens," or SPATTER CONES) was the site of major lava flows from 1944 to 1947. One flow in 1944 overwhelmed the village of San Juan Parangaricutiro but allowed enough time to evacuate the community.

metamorphic rock Rock that has been altered by recrystallization under the influence of temperature and/or pressure. Changes in content of volatile elements may be involved. (See also IGNEOUS ROCK; SEDIMENTARY ROCK.)

methane A simple hydrocarbon molecule consisting of a carbon atom with four hydrogen atoms attached to it (CH_4), methane is the centerpiece of a recent hypothesis by physicist Thomas Gold (one of the contributors to the "Steady State" model of the universe) that presents release of methane from underground as an explanation for many curious phenomena associated with earthquakes. Gold's theory, known as the "deep gas" hypothesis, suggests that large quantities of methane (also known as "marsh gas"), left over from the formation of the planet, still exist inside the earth and are vented to the surface from time to time in outgassing phenomena. Such phenomena, coinciding with earthquakes, might account, for example, for the fish kills and incidents of "boiling" water that have been seen to accompany earthquakes along the shore. Release of methane into the water would give the sea the appearance of boiling, and a sufficient quantity of the gas could asphyxiate fish, especially if the methane contained other gases such as hydrogen sulfide. (Analysis of gas bubbles rising from the ocean floor off Malibu Point following the SAN FERNANDO EARTHQUAKE of 1971 showed the gas to be 93% methane, with small percentages of nitrogen, carbon dioxide, oxygen and argon.) Methane outbursts accompanying earthquakes also might account for the strange phenomenon of "earthquake light," a glow that has been seen in the night sky during and near the time of earthquakes. Methane gas escaping from the earth and igniting near the surface could burn with enough luminosity to explain earthquake light, although there appears to be no proof that earthquake light originates in this manner. Methane eruptions from the seabed have even been suggested as a possible mechanism for generating tsunamis. A large emission of methane gas from the seabed, approaching the surface, might lift water above it into a dome, which then would collapse and generate a major disturbance in the sea, with a consequent tsunami.

Mexico As part of the westward-moving landmass of North America, Mexico collides with the suboceanic crust beneath the Pacific Ocean to the west of the continent. This collision generates numerous earthquakes along the western coast of Mexico, especially in Baja California, which is part of a northward-moving block of crust that is grinding against the Sierra Nevada range in California and generating numerous earthquakes there. A special set of geological conditions in the Mexico City area has made that metropolis highly vulnerable to damage from earthquakes, even those origi-

Parícutin volcano, Mexico, in eruption around 1944 (*Earthquake Information Bulletin*/USGS)

nating along the coastline far to the west. Mexico City's vulnerability to earthquake damage was demonstrated clearly in the great earthquake of September 19, 1985, when an earthquake originating along Mexico's Pacific shore struck Mexico City during rush hour and caused numerous fatalities and extensive property damage. The capital has been constructed atop unconsolidated sediment that once formed a lake bed. Buildings atop this kind of soil are especially susceptible to damage in earthquakes. Moreover, buildings in Mexico City were designed and built largely without consideration for earthquakes and consequently experienced more damage than newer, quake-resistant buildings such as those required in California. Severe damage and fatalities were reported in four states of Mexico: Colima, Jalisco, Guerrero and Michoacan. Several vessels at sea were reported missing following the earthquake, and the crew of a trawler reportedly witnessed waves some 100 feet high rising from the sea. Mexico also has a long history of volcanic activity, some of it in this century. (See also PARÍCUTIN).

Mexico City earthquake One of the most famous earthquakes of the 20th century, the Mexico City earthquake had an estimated Richter magnitude of 8.1, making it approximately as powerful as the SAN FRANCISCO EARTHQUAKE of 1906. The earthquake occurred early on the morning of September 19, 1985 and originated along the Pacific coast of Mexico. This timing helped reduce the number of deaths from the earthquake, which killed more than 8,000 people and injured some 30,000. Had the earthquake occurred later in the day, the number of deaths might have been much higher because many public buildings vulnerable to damage would have been occupied. Approximately 500 buildings in Mexico City were destroyed or experienced heavy damage. Damage was estimated at about $4 billion, but serious damage was confined to a small total portion of the city. The character of underlying material did much to determine how much damage buildings in the city experienced from the earthquake. Damage was worst in areas underlain by moist, unconsolidated material deposited in the bed of what once was

Lava flow at Church of San Juan Paragaricutiro, Mexico about 1953 (K. Segerstrom, USGS)

Lake Texcoco. The lake was drained by Spanish settlers after the conquest of the indigenous Aztec civilization. Construction began afterward atop the sediments of the lake bed. This kind of soil is especially susceptible to ground motion in a strong earthquake, and any structure built atop such soil is vulnerable to damage from the ground motion. The earthquake's destructive potential was reduced by the distance (some 200 miles) between Mexico City and the earthquake's point of origin. An earlier earthquake in Mexico City, in 1957, produced a pattern of damage similar to that from the 1985 earthquake.

Michigan, United States Although it generally lies within a zone of minor seismic risk, the state of Michigan is subjected from time to time to strong earthquakes originating in the ST. LAWRENCE VALLEY. An earthquake at Calumet on July 26, 1905, is thought to have been induced by conditions associated with mining operations. There was reportedly a great explosion. Chimneys toppled, and windows were broken. The earthquake was felt throughout Michigan's Keweenaw Peninsula. On May 26, 1906, some kind of seismic event at a mine on the Keweenaw Peninsula affected an area up to 40 miles wide and resembled in its effects a powerful earthquake; subsidence was reported, and rails were twisted.

mid-ocean ridges These structures are undersea mountain chains that mark zones where magma rises from below and solidifies, generating new crustal rock that moves laterally outward from the middle of the ridge. Although mid-ocean ridges are part of a worldwide system that is not confined to oceans but may extend into continents as well, the term continues to be used because the first such structure was discovered in the mid-Atlantic, where the ridge includes islands such as the AZORES and ICELAND. Mid-ocean ridges are sites of earthquake and volcanic activity and are of special interest to geologists studying paleomagnetism, or the history of the earth's magnetic field, because igneous rock that solidifies along mid-ocean ridges preserves a record of the orientation of the earth's magnetic field at the time the rock was formed. Exotic chemical and biological environments are found at certain points along mid-ocean ridges. For example, explorers from the Scripps Institute of Oceanography and the Woods Hole Oceanographic Institution examined the mid-ocean ridge near the Galapagos Islands in 1977 and discovered a dense concentration of undersea life around volcanic vents on the seabed at depths of about 9,000 feet. Fauna found at this site included giant clams more than a foot long, and huge polychaete worms that lived in tubes. At another site along the mid-ocean ridge off the western coast of Mexico, mineral-laden hot water rising from vents in the seabed has produced peculiar "chimneys" made of minerals precipitated from the heated water. This process of precipitation may have created extensive deposits of rich metal ores on the seabed along certain portions of the mid-ocean ridges, but such deposits do not appear to be commercially exploitable. (See also PLATE TECTONICS.)

Minnesota, United States Minnesota does not have a long history of strong earthquake activity because the state occupies a zone of minor seismic risk. A strong earthquake reportedly occurred in Minnesota in 1860, but details are unavailable. Another strong earthquake may have occurred in Minnesota between 1860 and 1865, but this is not certain. On September 3, 1917, an earthquake at Staples caused considerable damage to buildings and knocked objects off shelves. Other, minor earthquakes were recorded in 1909, 1925, 1935 and 1968.

Mississippi, United States The state of Mississippi is not known for frequent and strong earthquake activity and lies largely within an area of modest seismic risk. Nonetheless, strong earthquakes have occurred in Mississippi. One of these took place on December 16, 1931, in the northern portion of the state. Walls and foundations were damaged and some chimneys destroyed at Charleston. Damage also was reported at Belzoni, Water Valley and Tillatoba. This earthquake was felt over an area of about 65,000 square miles and was rated at Mercalli intensity VI–VII.

Missoula, Lake, northwestern United States The landscape of the northwestern United States was reshaped extensively by the floods that resulted from the draining of Lake Missoula, which apparently occupied much of western Montana 15,000 to 20,000 years ago. Glacial in origin, Lake Missoula is thought to have begun draining in catastrophic fashion when a dam of ice in the Clark Fork Valley in Idaho gave way and let several hundred cubic miles of water flow westward, carrying away soil and even eroding the underlying bedrock. One result of this assault by water is the Channeled Scabland of Washington, an area of some 2,000 square miles partly denuded by the flood. This same flood is believed to have invaded Oregon's Willamette Valley and reached southward to the location of present-day Eugene. Other floods also originated from Lake Missoula. The floods made the walls of the Columbia River gorge steeper and thus contributed to the conditions that are thought to have resulted in the great landslides along that gorge, notably the BONNEVILLE SLIDE, which may have been set off by a powerful earthquake.

Missouri, United States Missouri and its neighboring states have undergone some of the most powerful earthquakes in the history of the United States. Strong seismic activity in the Mississippi Valley region combines with large areas of unconsolidated, moist soil to produce conditions highly favorable for liquefaction. The most famous earthquakes in Missouri history are thought to have been also the most powerful recorded in North America since European settlement began. These earthquakes occurred between December of 1811 and February of 1812 in the New Madrid area along the Mississippi River and are thought to have affected an area estimated at a minimum of two million square miles. The earthquakes are believed to have altered topography over an area of up to 50,000 square miles. Although the earthquakes have been estimated at Mercalli intensity XI, casualties were few because the area was settled sparsely at the time.

The first earthquake in this series began in the early morning of December 16, 1811, shaking houses and knocking down chimneys. This earthquake had dramatic effects on the landscape; it caused landslides, generated waves in the river and made entire islands vanish. Lesser shocks occurred over the following days. Another shock, comparable in many ways to the initial earthquake, occurred on January 23, 1812. A third great shock, apparently even more powerful than the first two, took place on February 7, 1812, and was followed by occasional aftershocks over the next two years or longer. Considerable uplift and subsidence occurred in these earthquakes. Some areas were uplifted 15 feet or more; one such area, the Tiptonville Dome, is some 15 miles long and as much as eight miles wide. Submergence at Reelfoot Lake in Tennessee occurred to depths of some 20 feet or even greater.

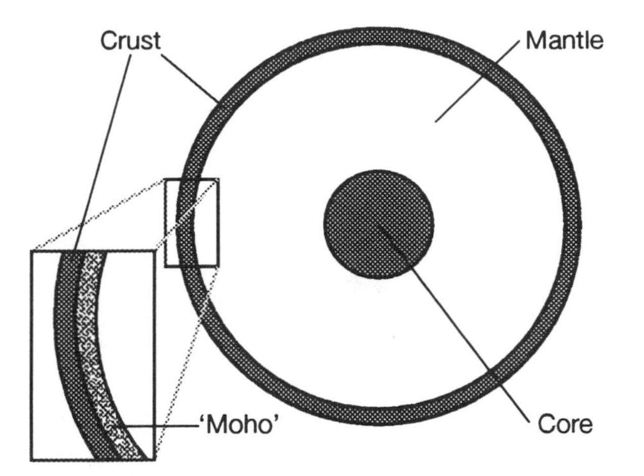

A cross section of Earth. Earthquakes and volcanic eruptions occur in the crust or outermost layer. The Mohorovicic Discontinuity, or "Moho," marks the boundary between the crust and the mantle (© Facts On File, Inc.)

"Moho" See MOHOROVICIC DISCONTINUITY.

Mohorovicic Discontinuity A thin transition zone between the earth's crust and mantle, the "Moho" is believed to represent the uppermost limit at which mantle rock can flow. (Mantle rock should be distinguished here from magma, which is molten rock existing and flowing within the crust.)

Mono Lake, California, United States Located in California near the Nevada border and immediately north of the LONG VALLEY CALDERA, Mono Lake is considered one of the most likely sites for future volcanic activity in the United States and has undergone moderate earthquake activity in recent years. Extending approximately north-south between Mono Lake and Long Valley Cal-

Aerial view of Mono Craters volcanoes in California (C. D. Miller, USGS)

dera is a set of volcanic landforms known as the Mono Craters. Vents in the vicinity of Mono Lake appear to have been active within the past two millennia and to have become active as a series of vents along a zone several miles in length, rather than erupted at a single point, as is the case with many volcanoes. Two major series of eruptions at Mono Lake are thought to have occurred in the 14th century. This activity is believed to have begun when molten rock moved along a fracture to the surface, came in contact with groundwater there and generated steam that caused a powerful explosion. A string of new vents some four miles long formed during this eruption, which expelled large amounts of ash over what now is central California and western Nevada. PYROCLASTIC FLOWS followed this initial eruptive activity but affected only the area immediately around Mono Lake. Next highly viscous lava emerged from the ground and formed domes and COULEES, including Panum Dome. Soon after these eruptive events, another series of explosions commenced at Inyo Craters, immediately to the south of Mono Lake, under circumstances apparently similar to those which led to the first eruptive cycle. (These two outbursts may have been separated by only a few weeks.) This eruption also produced large quantities of ash and steam explosions, followed by formation of domes and flows of obsidian. A series of earthquakes estimated at values up to 6.0 on the Richter scale of magnitude occurred in 1980, and earthquakes of comparable magnitude happened again in 1986 along the nearby White Mountain fault, reminding observers that the area around Mono Lake maintains a considerable potential for seismic and volcanic activity. Depending on the intensity of a future eruption at Mono Lake, as well as other factors including wind direction and velocity, ashfalls might affect areas hundreds of miles downwind.

Montana, United States The state of Montana is moderately active from a seismic standpoint, with much of the earthquake activity concentrated in the southwestern portion of the state. Earthquakes centered in Montana have affected vast areas of the western United States.

One of the greatest earthquakes in Montana's history, estimated at Mercalli intensity VII and Richter magnitude 6.75, occurred on June 27,

The 1959 Hebgen Lake, Montana earthquake warped this fence (J. R. Stacy, USGS)

1925, east of Helena. This earthquake was felt very strongly over an area of some 600 square miles, and the total area affected was more than 300,000 square miles. On an east-west axis, the earthquake was felt from the border of North Dakota to Washington state, and on a north-south axis from the border of Canada to the central portion of Wyoming. Aftershocks continued for months following the earthquake. Damage was concentrated at Logan, Lombard, Manhattan and Three Forks. A large school building at Manhattan was demolished. Chimneys were seen to fall not in one prevailing direction but rather in many different directions; this phenomenon led one observer to speculate that some kind of "twist" was involved in this particular earthquake. Railroad tracks shifted, and rockfalls interfered with service on the Northern Pacific Railroad and the Chicago, Milwaukee and St. Paul. A powerful roar accompanied the earthquake at Bozeman. A church and a schoolhouse were damaged at Three Forks.

The earthquake of August 17, 1959, near Hebgen Lake, Montana was of Mercalli intensity X and was estimated at 7.1 on the Richter scale. This earthquake affected an area of some 600,000 square miles and was felt at locations as distant as Seattle on the west, North Dakota on the east, Utah in the south and Banff, Alberta, Canada to the north. Twenty-eight people were killed, most of them in landslides. One spectacular landslide occurred along the south wall of Madison River Canyon, where a huge mass of earth, rock and trees fell and blocked the Madison River. Within several weeks, this landslide had created a lake with a depth of almost 175 feet. More than 20 feet of vertical displacement was seen near Red Canyon Creek, and prominent fault scarps were noticed near Hebgen Lake. Part of State Highway 287 was lost into Hebgen Lake during the earthquake, and damage to roads and timber alone was estimated at more than $11 million. This same earthquake also shook YELLOWSTONE NATIONAL PARK severely, and fresh geysers started erupting there; a ranger station at Yellowstone reported more than 150 aftershocks during the first day following the main shock, and aftershocks kept occurring for months thereafter. Some of the aftershocks were themselves notable earthquakes.

Montserrat, volcanic island, Caribbean Although there has been no eruption at Montserrat within historical times, the island has exhibited considerable earthquake and solfataric activity. It has been suggested that unrest at Montserrat is related to intrusions of magma under the Soufrière Hills at the southern end of the island.

moon Instruments placed on the moon have reported large numbers of seismic events, but the pattern of lunar seismicity differs from that of the earth in important ways. Although several thousand moonquakes are believed to occur each year, the moon appears to release much less energy in such quakes annually than the earth does; total energy released in terrestrial quakes each year is thought to be more than a billion (1,000,000,000) times greater than on the moon. Some evidence of moonquakes is visible to the unaided eye, as where seismic events have dislodged boulders from positions on slopes and caused the rocks to roll downhill. Lunar quakes also are believed to occur at greater depths than similar events on earth. These differences reflect the fact that the earth is more active geologically than the moon, both on the surface and at depth. The ever-changing pattern of plate tectonic activity, with its characteristic distribution of earthquake foci and volcanic eruptions, is absent from the moon. The character of lunar rock also reflects the difference between processes at work on the moon and on earth. Lunar rock is igneous (BASALT in the maria, or "seas," and GABBRO and other igneous rocks in the highlands), whereas terrestrial rock may be igneous, sedimentary or metamorphic, as a result of erosion, sedimentation and other processes related to the action of wind and water on earth. Impact structures are prominent on the lunar surface. The maria are believed to have formed when large meteorites struck the moon and caused great flows of molten rock. Some of the rills, channel-like features on the lunar landscape, are thought to have formed through flows of lava. (See also MARE.)

mudflow A mass movement of fluid mud.

"mud volcanoes" These are not, strictly speaking, volcanoes at all but rather hot springs whose waters mix with sediment to form mud, which behaves in a manner roughly comparable to that of lava. Mud volcanoes look like SPATTER CONES and tend to be only a few feet high.

Myozin-Syo volcano, Bonin Islands, Pacific Ocean Myozin-Syo volcano, about 300 miles south of Japan, emerged from the Pacific in a 1952 eruption. Two ships arrived at the site to watch the eruption, but one ship was destroyed, with no survivors, when it strayed over the top of a vent.

N

Naples, Italy The area around Naples is noted for earthquake and volcanic activity. Located near Naples is the famous volcano VESUVIUS, whose eruptions have caused repeated and extensive damage within historical times, notably the destruction of POMPEII AND HERCULANEUM in A.D. 79. Major earthquakes have occurred in the vicinity of Naples, notably in A.D. 63 and 1456.

Nazca crustal plate The Nazca crustal plate is one of the smaller major plates of the earth's crust and is located under the Pacific Ocean immediately to the west of South America. The ongoing collision between the South American plate and the Nazca plate is believed to account for much of the belt of strong seismic and volcanic activity running along the ANDES MOUNTAINS of western South America, through BOLIVIA, CHILE, ECUADOR and PERU. The Nazca plate is bounded on the north by the smaller Cocos plate, on the west by a mid-ocean ridge and on the south by the Antarctic plate. (See also PLATE TECTONICS; "RING OF FIRE.")

Nebraska, United States Earthquake activity in Nebraska appears to have been minor since the territory was settled by Americans of European descent. A strong earthquake was reported in eastern Nebraska on November 15, 1877, and was felt over a wide area of Minnesota. Two shocks 45 minutes apart were reported. Some damage occurred to property, and the earthquake was felt over an area of about 300 by 600 miles.

Nevada, United States The state of Nevada is located in a seismically active region of the country and has been the site of several notable earthquakes in the 20th century. The earthquake of October 2, 1915, had an estimated Richter magnitude of 7.75 and generated effects of Mercalli intensity X at its epicenter in Pleasant Valley near Winnemucca. This earthquake was felt over a vast area of the western United States, from San Diego, California to Oregon but did little damage because the affected area was settled so lightly. The earthquake wrecked adobe houses, tossed persons from bed and caused mine tunnels to collapse. A rift with a new vertical scarp 22 miles long and up to 15 feet high appeared along the base of the Sonoma Mountains. Western Nevada experienced another strong earthquake, this one of magnitude 7.3 with effects of Mercalli intensity X, on December 20, 1932; again, there was little damage because the region affected was so underpopulated. Considerable faulting occurred in a valley near Mina between the Pilot and Cedar Mountain ranges; several dozen fissures, with lengths up to four miles, were reported. The shock was felt over much of California, from San Francisco to San Diego. The Dixie Valley earthquake on December 16, 1954, affected an area of some 200,000 square miles, had a Richter magnitude of 7.1 and produced effects of Mercalli intensity X.

Nevado del Ruiz, volcano, Colombia Nevado del Ruiz has been the site of lethal mudflows. In a 1985 eruption, mudflows from the volcano traveled down the Lagunillas River and killed approximately 22,000 people in the town of Armero.

Nevados de Chilian, caldera, Chile Located in southern central Chile, Nevados de Chilian has a history of explosive eruptions dating back to the 17th century. Although the nearby city of Chilian stands some 60 miles from the shore, salt water and mud erupted in large quantities after a strong earthquake in 1835. Numerous new hot springs formed at this time. (It is not certain whether this activity occurred at Chilian itself or at Nevados de Chilian.) There may have been a minor eruption at Nevados de Chilian in 1906, several days before a powerful earthquake that wrecked Valparaiso. A new volcano, Volcan Nuevo, formed in this eruption. Activity, including an incident of dome formation, continued in the 1970s and 1980s.

New Hampshire, United States New Hampshire is located in one of the most seismically active areas of the United States, in northern New England along the ST. LAWRENCE VALLEY. Earthquakes, some of them powerful, have shaken the state frequently within historical times. The earthquake of November 29, 1783, was felt from New Hampshire to Pennsylvania. On November 9, 1810, an earthquake at Exeter, New Hampshire involved a peculiar noise like that of a strong explosion underfoot. The earthquake lasted a minute at Portsmouth, New Hampshire and was strongest between Portsmouth and Haverhill, Massachusetts. A ship in Portsmouth Harbor appeared to hit the bottom, and the earthquake broke windows in Portsmouth. The earthquake was felt also in Maine.

Ancient volcanism in the White Mountains of New Hampshire created a curious set of circular and nearly circular structures that formed when molten rock oozed up around the edges of cylindrical masses of rock overlying volcanic chambers far underground. These circular structures can be found at the Ossipee Mountains, Mount Lafayette, the Pliny Range and the Crescent Range.

New Jersey, United States New Jersey is located in a region of generally minor seismic risk but has experienced strong earthquakes from time to time. On September 1, 1895, an earthquake in Hunterdon County, estimated at Mercalli intensity VI, affected an area of some 35,000 square miles; this earthquake knocked articles off shelves in Hunterdon County and was felt throughout the city of Newark. Strong shocks were reported also at Burlington and Camden, and windows were broken in Philadelphia. A rumbling sound accompanied an earthquake on January 26, 1921, in Moorestown and Riverton. The coast of New Jersey between Toms River and Sandy Hook experienced three shocks on June 1, 1927; between Long Branch and Asbury Park, chimneys and plaster walls were damaged, and objects were knocked off shelves. On December 19, 1933, an earthquake near Trenton was felt sharply and knocked pictures off walls in Lakehurst.

The Salem County earthquake of November 14, 1939, was estimated at Mercalli intensity V and was felt over an area of some 6,000 square miles, from Trenton to Baltimore, Maryland and from the Philadelphia area to Cape May. Western central New Jersey experienced an earthquake on March 23, 1957, that caused considerable damage and alarm in the area of Hamden, Lebanon and Long Valley. An earthquake on December 27, 1961, along the border of New Jersey and Pennsylvania was estimated at Mercalli intensity V and caused much fright in the affected area, which included Bristol and the northeastern Philadelphia area, as well as Langhorne and Levittown. The December 10, 1968, earthquake in southern New Jersey broke windows and caused considerable alarm and was felt also in portions of Pennsylvania and Delaware; this earthquake was estimated at Richter magnitude 2.5, and effects of Mercalli intensity V were reported at some locations.

New Madrid Fault Zone, Missouri, United States The New Madrid (pronounced MAD-rid) Fault Zone along the Mississippi River was the center of what are believed to be the most powerful earthquakes in the history of the United States, in the winter of 1811–12. These earthquakes are believed to have had magnitudes between 8.4 and 8.8 and were felt as far away as Washington, D.C. Pedestrians in Richmond, Virginia allegedly had trouble standing on their feet as the vibrations from the earthquake passed, and in Kentucky the earthquake was felt so strongly as to leave people feeling dizzy and disoriented. The New Madrid area was sparsely settled at the time, and casualties were few, compared with other, similar earthquakes in urban communities; only 11 men, women and children were formally reported as lost in the New Madrid earthquakes, although the total is thought to have been greater.

Detailed records of the earthquakes and their effects have been compiled from personal journals, letters and newspaper accounts. The earthquakes caused the banks of the river to collapse into the water, carrying fishermen with them. Submerged trees were dislodged from the riverbed and floated to the surface, and one report said the current in the river was actually reversed briefly because of elevation of the riverbed. There were numerous observations of strange behavior on the part of animals (both wild and domestic) before and during the earthquake. Naturalist John James Audubon noted that the horse he was riding stopped and began to groan at the time of the earthquake. A naturalist visiting the New Madrid area indicated later that birds behaved abnormally. Waterfowl congregated on boats, and on land small birds settled on houses. In some cases, birds were even seen to alight on humans. Panthers, foxes and wolves were said to have been seen next to wild deer.

Peculiar atmospheric phenomena appear to have accompanied the New Madrid earthquakes. One correspondent for the *Louisiana Gazette* mentioned that fog obscured the sky at the time of one shock and that as another shock occurred observers noticed what looked like frost covering houses and fences. On close examination, however, the "frost" turned out to be a vapor that lacked the chill of frost. Another interesting phenomenon reported during the New Madrid earthquakes was EARTHQUAKE LIGHT, in the form of either bright flashes of light or a dull glow in the sky.

The sociological and psychological effects of the New Madrid earthquakes have been reported in detail, specifically a religious revival that occurred among settlers in the area. Church membership increased dramatically in the weeks and months following the earthquakes but apparently the revival was short-lived.

Another strong earthquake struck the New Madrid area again in January 1843. A report from Little Rock, Arkansas indicates the shaking lasted for about one minute and caused windows and cupboards to rattle. Hundreds of residents of Memphis, Tennessee ran into the streets when they felt the earthquake, and some deaths allegedly occurred at auction houses where people were crushed and trampled by crowds struggling to escape the buildings. The noise of the earthquake was compared to that of hundreds of heavily laden wagons moving down a street. In 1895 another earthquake centered near New Madrid was felt strongly in a zone about 100 miles long and was felt as far away as Georgia and West Virginia. The shocks lasted almost one minute, cracked walls and shook down brick chimneys. As in 1811–12, a strange phenomenon resembling earthquake light was reported as a luminous "streak" that appeared in the sky just prior to the shock. Yet another earthquake, in 1968, startled residents but caused little damage.

If and when another earthquake comparable in magnitude to those of 1811–12 strikes the New Madrid Fault Zone, the results may well be disastrous for metropolitan areas in the Mississippi Valley, notably St. Louis, Missouri and Memphis, Tennessee. Located on the edge of the New Madrid Fault Zone, Memphis may expect severe ground motion, and large areas of the city stand to suffer greatly from liquefaction, which also poses a threat to St. Louis in the event of a major earthquake there. As one U.S. Geological Survey report describes the situation in St. Louis,

> Much of the old part of the city of St. Louis, and particularly the modern highway network, is built on uncontrolled fill. This fill is generally [found] in stream valleys, but there are many rubbish-filled pits in the old portion of the city. All this rubbish and fill is prone to large differential settlements in an earthquake.

The report adds that "parts of the downtown area are underlain by open, underground mines, where clay was mined long ago for making tile. Their locations are generally . . . not known."

Landslides are another danger facing St. Louis, the report points out:

> Landslides [in the St. Louis region] are likely to be commonplace on many natural and highway . . . slopes. The natural slopes most prone to landslides in uplands are thick, silt-rich loess on steep slopes . . . Many highway cuts [will] fail and cause very serious problems to highway traffic. Even today, there are significant landslides in the . . . cuts along I-44, I-244, I-270, and U.S. 67. Some highway fills [material in the roadbed underlying the asphalt] would probably fail, especially on the lake sediments near the airport and on flood plains of the major rivers.

New Mexico, United States Central New Mexico has a long history of earthquakes, mostly localized but strong. Earthquakes in July and November of 1906 affected a area of about 100,000 square miles. New Mexico also has numerous volcanic landforms. (See also BANDELIER TUFF; RIO GRANDE RIFT; VALLES CALDERA.)

Newport-Inglewood Fault Zone, California, United States A fault zone running directly through the LOS ANGELES, California area, the Newport-Inglewood Fault Zone appears to pose a serious threat to the city in the event of any major seismic events along the fault zone in the future.

New York, United States New York state borders on the seismically active ST. LAWRENCE VALLEY between the United States and Canada, and earthquake activity is commonplace in upstate New York, including occasional powerful shocks. Even the New York City area, ordinarily not known for seismic activity, has undergone strong earthquakes over the past three centuries; on December 18, 1737, for example, an earthquake estimated at magnitude VII on the Mercalli scale struck the New York City area. Some other substantial earthquakes to hit the vicinity of New York City within historical times occurred on December 10, 1874 (Westchester and Rockland counties, New York and Bergen County, New Jersey; Mercalli magnitude VI), August 10, 1884 (magnitude VII), March 9, 1893 (magnitude V), June 8, 1916 (magnitude IV–V), May 11, 1926 (New Rochelle; magnitude V) and November 22, 1967 (Westchester County; magnitude V). The 1884 earthquake was especially strong and affected a broad area of the northeastern United States, from Portsmouth, New Hampshire and Burlington, Vermont to points as far south as Baltimore, Maryland and Atlantic City, New Jersey. Maximum damage occurred in Amityville and Jamaica, New York, where large cracks formed in walls. The 1893 earthquake had dramatic effects on New York City itself; the epicenter was located between 10th and 18th streets, and the earthquake distressed animals at the city zoo and interfered with the playing of billiards.

The frequency of small to moderate earthquakes in New York state has made the state a convenient natural laboratory for studying techniques of earthquake prediction, and some success has been reported in predicting seismic activity there.

New Zealand New Zealand has a complex geology and a long history of strong volcanic and earthquake activity. New Zealand and the Macquarie Islands to the southwest are believed to account for more than 1% of the seismic activity on earth. One area of intense seismic activity is Fiordland on South Island.

Notable earthquakes in New Zealand's history have occurred in 1848 at Awatere, 1855 at Wellington, 1881 at Amuri, 1901 at Cheviot, 1929 at Murchison, 1931 at Hawke's Bay, 1932 at Wairoa, 1942 at Wairarapa and 1953 at Bay of Plenty. The Awatere earthquake in 1848 was felt over much of New Zealand and caused serious damage on both islands, although damage reportedly was worst around Cook Strait, notably in Wellington. The 1931 Hawke's Bay earthquake was even more severe and is thought to have had a Richter magnitude of 7.8. This earthquake stands out in the history of seismology because data from it indicated to geologists that the earth had an inner core. North Island exhibits extensive volcanic activity. Its volcanoes include NGAURUHOE, RUAPEHU, TARAWERA and White Island. WHITE TERRACE and Pink Terrace, two beautiful formations deposited by HYDROTHERMAL ACTIVITY, were destroyed by a volcanic eruption in the 19th century. New Zealand has GEOTHERMAL ENERGY facilities developed after World War II on North Island in a thermal area near Ngauruhoe and Ruapehu. This area features many manifestations of geothermal activity, including, but not limited to, geysers, hot springs and fumaroles.

Ngauruhoe, volcano, North Island, New Zealand The stratovolcano Ngauruhoe has erupted frequently since the early 19th century and has produced minor NUÉES ARDENTES on some occasions. Ngauruhoe is adjacent to two other volcanoes, Tongariro and RUAPEHU.

Ngorongoro, volcano, Tanzania Ngorongoro is part of a set of several large volcanoes in the "LAND OF THE GIANT CRATERS," along Africa's RIFT VALLEY. The subsidence crater in Ngorongoro is some 35 miles in circumference and once was home to large herds of wild animals, including wildebeest, hippopotamus, rhinoceros, giraffe, impala and elephant. Hunting reduced the wildlife population of Ngorongoro, but in 1959 a conservation area was established.

Nicaragua Nicaragua has a history of highly destructive earthquakes, notably the one that destroyed much of MANAGUA on December 23, 1972. That earthquake was associated with Nicaragua's Tiscapa Fault.

Nicaragua is also noted for having active volcanoes. The eruption of COSEGUINA in 1835 was one of the most violent of the 19th century. A powerful volcanic eruption occurred in Nicaragua on November 14th, 1867, when a mountain east of the city of Leon erupted for more than two weeks and cast out great amounts of black sand, which was carried as far as the Pacific shore some 50 miles away. Incandescent rocks were reportedly ejected from the crater to a height estimated at 3,000 feet. Seismic and volcanic activity in Nicaragua played an important part in formulating plans to build a canal through Central America to connect the Atlantic and Pacific Oceans. Historian Charles Morris explained in *The Volcano's Deadly Work,* published in 1902, when deliberation on the route of the canal was continuing:

> There are three volcanoes in Lake Nicaragua itself, which body of water it [was] proposed to make the summit level of the projected canal on this line. Indeed, the evidence of geology is that Lake Nicaragua was once an arm of the Pacific, and that the central plateau of Nicaragua was formerly much nearer to the Caribbean coast than at present. The forces which effected so vast a change in the configuration of the land are still active. The eruption in 1835 of Coseguina, which lies but sixty miles from the proposed Nicaragua canal route, was of extraordinary violence. So tre-

mendous was this explosion, and so great was the storm of dust and ashes, that absolute darkness prevailed for thirty-five miles in every direction . . . Coseguina could have filled up ten times in one hour a canal prism which the contractors, with all their . . . labor-saving devices and the employment of tens of thousands of hands, would require eight years to excavate . . .

The danger of such convulsions at Panama is far less. We are told . . . that in Panama there is within a distance of one hundred and eighty miles from the canal no volcano, even extinct. The Isthmus there . . . lies in an "angle of stability," so called by seismographers. Except for rare and not very violent seismic vibrations, originating at distant centers, the Isthmus of Panama has never been affected by volcanic disturbances. One earthquake of some violence, indeed, has occurred there during the historic period, in 1621, when the greater part of Panama City was shaken down. Aside from this the most destructive earthquake known in the history of Panama was that of September 7, 1882. It lasted only a minute, but in that time shook down the court-house and ruined the front of the old cathedral. Yet it may be affirmed that no paroxysmal convulsions have remodeled the geographical features of the Isthmus, as is the case with Nicaragua, and that its hills are nearly if not quite as stable as those of the Appalachian system . . .

The canal route eventually was decided in favor of Panama. (See also COSEGUINA.)

Nisyros, See NYSIROS.

Niuafo'ou, volcanic island, Tonga Niuafo'ou Island is also known as "Tin Can Island" because of the way in which mail used to be delivered to the island. Because the island had no harbor, mail would be sealed in a metal container and dropped over the side of a ship. A swimmer then would carry the can of mail to shore. This method is said to have been discontinued after a shark attacked a swimmer who was retrieving the mail. The island is located in the northern portion of the Tonga archipelago and consists of a lava shield surmounted by a composite volcanic cone. A large lake, called simply Big Lake, occupies much of a caldera on the summit. A second, smaller lake known as Little Lake is also located in the caldera. Eruptions within historical times are recorded as early as 1814, when an explosive eruption apparently was accompanied by phreatic activity and lava flows. This eruption occurred in the caldera itself. Details of a reported eruption in 1840 are unavailable. Eruptions in 1853 and 1867 involved fissures on the flanks of the volcano.

Observers of an eruption in 1886 left detailed accounts of the event, which was heralded by earthquakes and an apparent dramatic uplift in the seabed. The captain of a ship anchored at Niuafo'ou discovered, after an earthquake on August 12, that his ship was floating in only eight fathoms (48 feet) of water, compared to some 20 fathoms the previous day. A very loud noise was heard (compared afterward to the sound of a heavy gun), and some kind of discharge occurred in the lake. Earthquakes stopped at this time. Some subsidence appears to have occurred around the caldera during this eruption.

A curious detail was noted in eruptions in 1919 and 1929: coral was expelled along with lava, even though there appeared to be no recently formed coral on the island. Very little seismic activity was reported before the 1929 eruption. An eruption in 1943 required the evacuation of the island, and the inhabitants did not return for more than a decade. Another eruption in 1946 was preceded by a very strong earthquake just over an hour before the eruption began. Considerable uplift was reported in this eruption. One account says a reef on the northern side of the island was raised 30 to 40 feet out of the water in places. Moderately strong earthquakes occurred in 1985.

North America The North American continent is characterized by extensive earthquake activity and some active volcanism. According to the PLATE TECTONICS model of the earth's crust and its dynamics, much of North America occupies a plate of the earth's crust that is moving slowly westward as new crust is formed along the MID-OCEAN RIDGE running down the middle of the Atlantic Ocean. The ongoing collision between the North American mainland and the oceanic crust underlying the Pacific Ocean has generated a complex geology along the Pacific coast of the continent, marked by major and minor faults with associated seismic activity, although this set of circumstances is not the only source of earthquakes in North America. Much of North America's seismic activity is concentrated in the state of CALIFORNIA, where the SAN ANDREAS FAULT running along the coast has been associated with numerous strong earthquakes, such as that which destroyed much of SAN FRANCISCO in 1906 and the less powerful but also highly destructive SAN FERNANDO EARTHQUAKE of 1971.

In both northern and southern California, but especially in the LOS ANGELES area, urban development has created a tremendous potential for destruction in the event of another major earthquake comparable to that of 1906. Similar dangers exist in other parts of the continent, such as the cities surrounding Puget Sound. Other areas of North America with a history of strong earthquake activity in the continental United States include New England, where earthquakes have caused widespread damage to Boston, Massachusetts within historical times; Louisiana's Gulf coastal plain; upstate New York along the ST. LAWRENCE VALLEY; the Appalachian Mountains, especially the southern portion of the chain, from VIRGINIA through ALABAMA; UTAH, which occupies part of a belt of pronounced earthquake activity extending northward from ARIZONA to the Canadian border; coastal SOUTH CAROLINA, site of the devastating CHARLESTON earthquake of 1886; and the Mississippi river valley in the vicinity of New Madrid, MISSOURI, where the most powerful earthquakes in the history of the United States are believed to have occurred during the winter of 1811–12. Virtually every part of the continental United States has experienced strong seismic activity at one time or another since settlement by Europeans began, although certain portions of North America, such as the stable rocks of the Canadian Shield, are all but free of substantial earthquakes.

The earthquake potential of western North America is generally better understood than that of the eastern portion of the continent, partly because earthquakes occur more often in the west and partly because active faults in the east tend to lie under thick layers of sediment that have hindered efforts to study the seismology of eastern North America. Earthquake activity is also pronounced in ALASKA, the site of one of the greatest earthquakes of modern times, the GOOD FRIDAY EARTHQUAKE of 1964, which was accompanied by a powerful tsunami that caused great damage as far away as Crescent City, California. Earthquake activity in Alaska concentrates along an arc that follows the ALEUTIAN ISLANDS and extends into the central portion of the state.

MEXICO, too, has a long history of strong seismic activity. Earthquakes have caused widespread destruction in Mexico over the centuries, notably in the MEXICO CITY EARTHQUAKE of 1985. Western Mexico shares with the western United States and Canada a potential for devastating earthquakes, as smaller blocks of crust along the coast are squeezed between the westward-moving continent and the oceanic crust to the west. Especially susceptible to earthquakes are the northwestern corner of Mexico, which is subjected to the same set of geological conditions that make southern California susceptible to major earthquakes, and the extreme southern Pacific shores of Mexico, where earthquake epicenters on maps are clustered almost as densely as in California. The depth of earthquake foci differs greatly from one portion of North America to another. Earthquakes in Alaska, for example, may extend much more deeply than in southern California and NEVADA.

Volcanism has done much to shape the landforms of North America. The western states of the United States exhibit numerous signs of past and current volcanic activity. Great flows of basalt in the Northwestern states of the United States testify to past volcanism there. A subduction zone off the coast of the Pacific Northwest of the United States and the Canadian province of British Columbia has given rise to the volcanoes of the CASCADE MOUNTAINS, including Mount HOOD, Mount BAKER, Mount RAINIER, LASSEN PEAK and the recently active Mount ST. HELENS, as well as the photogenic caldera of CRATER LAKE in Oregon. Volcanism in Mexico has a long and colorful history; numerous volcanic mountains stand near Mexico City, and the volcano PARÍCUTIN formed in a spectacular eruption in a Mexican field earlier in this century. Alaskan volcanoes have erupted on many occasions in this century, sometimes with great violence, as in the Mount KATMAI-Novarupta eruption of 1912, which laid down a plain of fumaroles, the VALLEY OF TEN THOUSAND SMOKES. Alaskan volcanism is tied to a subduction zone running along the Aleutian Island arc.

North American volcanism sometimes is extended to cover that of the HAWAIIAN ISLANDS because of their inclusion in the United States. Hawaiian volcanoes include KILAUEA and MAUNA LOA, known for their frequent but generally nonviolent and harmless eruptions. Tsunamis pose a particular threat to the Hawaiian Islands because of their position in mid-Pacific; the tsunami that struck HILO in 1946, for example, caused extensive damage and loss of life.

The presence of large amounts of still-hot magma near the surface in western North America has given rise to abundant geothermal activity and HYDROTHERMAL ACTIVITY in certain locations, such as YELLOWSTONE NATIONAL PARK, where geysers have become tourist attractions. GEOTHERMAL ENERGY has been exploited to generate electricity on a large scale in Northern California.

North American crustal plate The North American crustal plate generally underlies the continent of North America and includes Canada and the 48 continguous United States in addition to Alaska and Mexico. The North American plate is thought to be moving westward relative to the PACIFIC CRUSTAL PLATE, with which it shares a boundary along the western shore of North America. This boundary between the North American and Pacific plate has been the location of some of the most famous and destructive earthquakes and volcanic eruptions in history, although the Pacific and North American plates are not the only ones involved in generating seismic and volcanic activity here; the JUAN DE FUCA CRUSTAL PLATE, for example, is responsible for earthquakes and volcanism along the CASCADE MOUNTAINS of British Columbia and the northwestern United States, including portions of the states of CALIFORNIA, OREGON and WASHINGTON. On the east, the North American plate adjoins the MID-OCEAN RIDGE that runs down the middle of the Atlantic Ocean. Here, geologists believe, new crust for the North American plate is created as molten rock rises from beneath, encounters the cold waters of the deep ocean and solidifies. To the southeast, the North American crustal plate adjoins the Caribbean plate, along the borders of which have occurred some of the most destructive eruptions and earthquakes in the recent history of the western hemisphere. The small COCOS PLATE, off the Pacific coast of Central America, also shares a boundary with the North American crustal plate. (See also PLATE TECTONICS.)

Mayon volcano in eruption, with nuée ardente moving down flank at left (*Earthquake Information Bulletin*/USGS)

North Carolina, United States North Carolina is located in a region of moderate seismic activity in the southeastern United States. Although severe earthquakes have been rare in North Carolina's history, the state has a record of mild seismic activity extending back into colonial times. An earthquake on March 9, 1828, probably centered in Virginia, was felt as a severe shock at Raleigh, North Carolina and was associated with noises like thunder at Hillsborough. Another earthquake on April 29, 1852, again probably centered in Virginia and was felt in Raleigh, Greensboro, Hillsborough and Milton, North Carolina.

A notable series of earthquakes occurred between February 10 and April 17, 1874, in McDowell County. Strong shocks were followed by a curious rumbling sound. The earthquakes shook buildings vigorously. Between 50 and 75 shocks are thought to have occurred altogether, each one associated with rumbling. On some occasions, a noise like artillery fire was also reported. The earthquakes affected an area approximately 25 miles wide.

North Dakota, United States Although characterized by low seismic activity, North Dakota is affected from time to time by earthquakes in nearby states. Notable effects were reported in North Dakota, for example, from the MONTANA earthquake of 1959.

North Pagan, caldera, Mariana Islands Located on Pagan Island, North Pagan has erupted on several occasions since the late 17th century. Details of eruptions are few, but a large eruption appears to have occurred in 1872, and earthquakes reportedly accompanied other eruptions in 1917 and 1923. The caldera appears to have been quiet until 1984, when a small eruption followed an earthquake. Another minor eruption occurred in 1987, and there is evidence of further eruptive activity in 1988.

nuée ardente A French expression meaning "fiery cloud," nuée ardente refers to a violent and potentially highly destructive phenomenon that occurs when a VOLCANIC DOME collapses. A PYROCLASTIC FLOW emanates from the volcano along with an overlying cloud of ash. The resulting cloud of ash, rock and superheated gas flows along the ground at high velocity and incinerates or melts virtually anything in its path. A nuée ardente can devastate the area around a volcano almost as effectively as a nuclear explosion. Perhaps the most dramatic example of a nuée ardente in the 20th century is the one that accompanied the eruption of Mount PELÉE in the Caribbean.

Nyamuragira, volcano, Zaire Nyamuragira is a shield volcano and has erupted frequently in the 20th century. A lava lake was active in the 1920s and 1930s.

Nyiragongo, volcano, Zaire A stratovolcano, Nyiragongo has erupted often in the late 19th and the 20th centuries. Some 300 people were killed by lava flows from the volcano in 1977. In 1958 an expedition to the volcano succeeded in measuring surface temperatures within the lava lake of Nyiragongo. Measurements at the surface indicated temperatures between 1,000° and 1,200° C.

Nyos, Cameroon The village of Lower Nyos was overwhelmed on August 21, 1986, by a cloud of toxic gas that erupted from the bed of Lake Nyos and asphyxiated some 1,200 people in that community, plus some 500 others in the vicinity. Some 3,000 domestic animals also died. The cloud consisted largely of carbon dioxide that had accumulated at or near the bottom of the lake. This phenomenon is thought to be common in volcanic lakes, although most of those lakes undergo sufficient mixing (that is, overturn of warmer upper waters with colder, gas-laden bottom waters) that prevents excessive buildup of toxic gas in deep waters. It is not known exactly what caused the gases to rise from the bottom of Lake Nyos. According to one explanation, the lake had gone so long without overturning that toxic gas had built up to tremendous concentrations in bottom waters; then, when something finally initiated mixing, the asphyxiating gases rose to the surface and were released. An earthquake has been suggested as the possible agent of this overturn; and indeed, a rumbling noise reportedly was heard from the vicinity of Lake Nyos just before the catastrophe. It also seems possible that a minor volcanic eruption expelled the gas from the lake bed, but this hypothesis has not been confirmed.

Nysiros, caldera, Greece The volcanic island Nysiros lies in the Aegean Sea near Turkey. The caldera of Nysiros is about 2.5 miles wide and occupies what was the summit of an andesitic stratovolcano. The date of collapse is uncertain. Several rhyolitic domes have formed inside the caldera. Unrest has been reported at the caldera on several occasions since the early 15th century. In 1422 there was a report of an eruption. Detonations and roaring noises, along with emissions of vapor and hot water, reportedly occurred at the caldera in 1830. In 1871 earthquakes were reported, accompanied by "flames" and detonations; ash and LAPILLI from this eruption are said to have wiped out gardens of fruit on the floor of the caldera. A crater some 20 feet wide formed in 1873 following an earthquake; again on this occasion, lapilli and ash were cast out from the volcano. A lake formed in the caldera from hot, salty water, and eruptions of dark mud lasted for several days. Also in 1873, the caldera ejected lapilli on many occasions, along with salt water and mud. Eruptive activity had formed a crater approximately 75 feet wide by the autumn of 1888. Solfataric activity was reported in 1956. (See also SOLFATARA.)

O

obsidian A dark, glassy volcanic rock, obsidian was used as material for arrowheads by Native Americans. Obsidian cools quickly from the molten state and is often found in large flows on the surface.

Ohio, United States Ohio undergoes significant earthquakes on occasion. Several examples are listed here. The June 18, 1875, earthquake was felt strongly in Chicago, Columbus and Cincinnati and was strongest at Urbana and Sidney. On September 19, 1884, an earthquake near Columbus caused slight damage; this earthquake was felt at the top of the then-unfinished Washington Monument in Washington, D.C. A September 20, 1931, earthquake at Anna caused minor damage but was felt at great distances, in Alabama, Arkansas and Tennessee. In March of 1937, a series of strong earthquakes was reported. An earthquake on March 2 near Anna and Sidney caused alarm and slight damage. On March 3, minor damage was reported from an earthquake that was felt at Anna, Botkins, Jackson Center and Sidney. The March 8 earthquake in western Ohio caused minor damage and was felt on upper stories of high buildings in Chicago, Milwaukee and Toronto.

Oklahoma, United States Oklahoma is not one of the more seismically active states of the United States, but strong earthquakes do affect the state on occasion. A strong shock on April 9, 1952, at El Reno, for example, was felt over an area of 140,000 square miles.

Okmok, caldera, Aleutian Islands, Alaska, United States The Okmok caldera is located on Umnak Island and has two summit calderas overlapping each other. The Tulik volcano is located several miles from the caldera. Eruptions have been reported here since the early 19th century. A major explosive eruption in 1817 deposited several inches of ash over the surrounding area, and large rocks were cast out, some landing more than 30 miles away. Earthquake activity and a tsunami were reported in 1878, although it is not certain how this activity may have been related to the Okmok caldera. Tulik emitted black smoke in 1931, and an eruption took place on the island several weeks later. The island was relatively quiet until 1944, when activity resumed. In 1945 an earthquake was felt at the army base on the island, and several days later pilots reported seeing an ash column. Later investigations showed that ash and a lava flow had emanated from a cinder cone in the caldera. Airborne observations revealed another lava flow in 1958. From 1960 to 1961, explosive eruptions occurred, and a minor ash eruption was observed in 1983. Several small eruptions of ash took place between 1986 and 1988.

"Old Faithful," geyser, Yellowstone National Park, United States The geyser Old Faithful has erupted approximately once every 40 to 80 minutes since the 19th century. Water from the geyser reaches a height up to 150 feet. Each eruption is thought to release some 100,000 pounds of water.

Olympus Mons, volcano, Mars The largest known shield volcano in the solar system, Mars's Olympus Mons stands some 75,000 feet high and was revealed in photographs taken by the U.S. *Mariner 9* probe in 1971 and 1972. Olympus Mons is one of several large volcanoes in the Tharsis region of the planet's northern hemisphere. The Martian volcanoes appear to have played a major role in shaping the topography of that hemisphere in relatively recent times, although the existence of older volcanoes elsewhere on Mars indicates that volcanism has been active on Mars for much of the planet's history. The major Martian volcanoes are larger by far than any on Earth. The

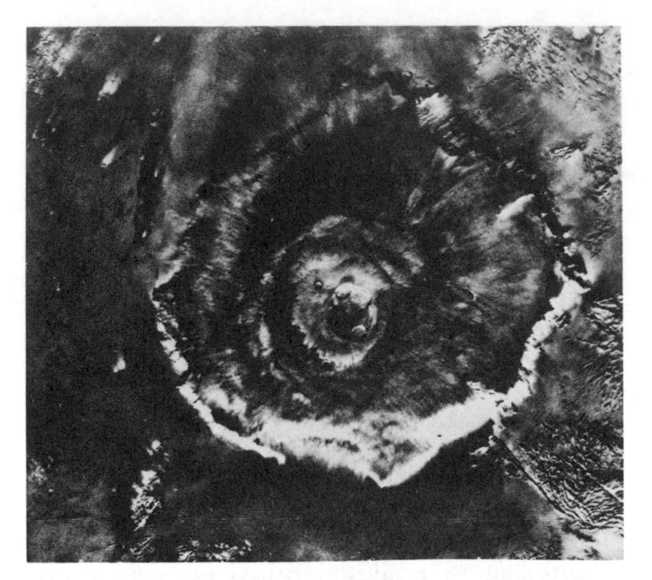

Olympus Mons, a volcano on Mars, the largest known volcano in our solar system (M. U. Carr, USGS)

evident absence of plate tectonics on Mars may account for the impressive size of its volcanoes. Unlike Earth, where horizontal motion of crustal plates can prevent a volcano from growing past a certain height, Mars appears to have a fixed, comparatively immobile crust, which gives a volcano time and opportunity to reach tremendous size by terrestrial standards. Because the processes that create volcanoes along crustal plate margins on Earth are apparently lacking on Mars, it is believed that radionuclides underground may supply the heat required for volcanism there.

Onikobe, caldera, Japan The Onikobe volcanic center is located in the northern portion of the island of Honshu. The volcano Kurikoma stands several miles northeast of the rim of the CALDERA. The Narugo caldera and a group of volcanic domes are situated just southeast of the Onikobe caldera. Although the Onikobe caldera itself appears to have been quiet within historical times, eruptions of Kurikoma were recorded in 1946 and 1950. These eruptions do not appear to have released large amounts of material, although a mudflow was associated with the 1950 eruption. Some unrest was reported at the volcano in 1957, but there is no indication of an actual eruption.

On-Take, volcano, Japan The On-Take volcano is located on the island of Honshu, some 50 miles northeast of Nagoya. The volcano rests within a CALDERA and has a summit caldera more than a mile wide. The volcano's first eruption within historical times occurred in 1979 following several hours of seismic activity in the vicinity. A strong earthquake in 1984 caused extensive landslides. Fumaroles in the area of the 1979 vent appeared unaffected by the earthquake, but changes were noted in a fumarole on the summit of the volcano. Although it is not one of the more active volcanoes in Japan, On-Take has received particular attention because its eruption in 1979 occurred after many hundreds of years of quiescence and because that eruption is thought to be associated somehow with earthquake swarms in the area between 1976 and 1984 and with the strong earthquake near the volcano in 1984.

Opala, volcano, Kamchatka, Russia The Opala volcano has been generally quiet over the last few centuries, although an explosive eruption is reported for the year 1776. About 1,500 years ago, explosions are believed to have formed a crater about one mile wide on the average within the Opala caldera, on the northern rim of which the Opala volcano is located. Intense solfataric activity was reported in 1827, 1854 and 1984 but no actual eruptions.

Öraefajökull, volcano, Iceland Although not one of Iceland's more active volcanoes, Öraefajökull has erupted on two notable occasions within historical times. The first was in 1362, when tephra and JOKULHLAUPS caused considerable damage in the vicinity of the volcano. Powerful earthquakes are thought to have occurred just before this eruption. The next eruption occurred in 1727 and was apparently similar to the eruption of 1362.

Oregon, United States Located in the seismically active Pacific Northwest of the United States, Oregon nonetheless has not had a major earthquake since the territory was settled by Americans of European descent in the 19th century. This does not mean that a powerful earthquake in Oregon is impossible, however; parts of Oregon would stand in risk of severe damage in any such event, especially where numerous buildings are constructed of unreinforced masonry, a kind of construction that is extremely vulnerable to damage from seismic events. Some notable earthquakes in Oregon history are listed below.

On February 3, 1892, a strong shock caused buildings to sway in Portland, and frightened residents of the city ran into the streets. The earthquake was also felt strongly at other locations, including Salem and Astoria. An earthquake at McMinnville on April 2, 1896, involved three shocks and awakened sleepers; the principal shock was felt also in Portland and in Salem. The earthquake of January 10, 1923, was felt in Lakeview and Klamath Falls and probably was very strong in an underpopulated area; minor damage occurred in California.

Oregon shows abundant evidence of volcanic activity in the recent geologic past, notably the peaks of several volcanoes in the CASCADE MOUNTAINS, which run north-south through the western and central portions of the state. Volcanism in Oregon is thought to be the result of the gradual and ongoing destruction of the JUAN DE FUCA CRUSTAL PLATE in the CASCADIA SUBDUCTION ZONE beneath the Pacific Ocean immediately to the west of Oregon. As molten rock from the fractionation of the descending crustal plate rises back to the surface along the Cascade range, the rising magma carries with it a heavy load of dissolved gases thought to be derived from ocean water carried downward into the earth's mantle with sediments from the seabed. As the magma approaches the surface and pressure on the molten rock is relieved, the gases are believed to pass out of solution and into gaseous form, thus giving the Cascade volcanoes the potential for highly explosive eruptions similar to that of Mount ST. HELENS in Washington in 1980. Volcanoes in Oregon include Mount HOOD, Three Sisters and CRATER LAKE, the water-filled CALDERA left from the collapse of Mount Mazama in prehistoric times.

Owens Valley earthquake, California, United States The Owens Valley earthquake on March 26, 1872, is believed to have been about 8.0 on the Richter scale and to have killed from 50 to 60 people. The Owens Valley Fault along the eastern face of the SIERRA NEVADA exhibited surface rupture for more than 100 miles. Vertical displacement in this earthquake was dramatic, especially near Independence and Lone Pine; total relative displacement along the fault exceeded 20 feet. The community of Lone Pine was virtually leveled, and 23 of its inhabitants were killed. The initial shock of the earthquake is said to have been the most powerful. It is said to have created a large wave in Owens Lake and dried up the Owens River for hours. Observers at one bridge saw fish thrown out of the water and onto the riverbank by the earthquake. Earthquakes continued for eight weeks. Witnesses to the earthquake included naturalist John Muir, who was asleep in his cabin near Sentinel Rock when the earthquake occurred. He ran outside to see what was happening and was forced to take shelter behind a tree as boulders dislodged by the shock rolled downhill to the floor of the valley. Muir improvised a primitive earthquake detector (a bucket of water set on a table) to indicate when shocks were occurring. The Owens Valley earthquake provided dramatic proof of how poorly adobe structures withstood earthquakes, notably in the devastated community of Lone Pine.

P

Pacific crustal plate The biggest plate of the earth's crust in terms of area, the Pacific plate underlies much of the Pacific Ocean and is colliding with the Americas on the east and with the Asian landmass on the west. The boundaries of the Pacific crustal plate include much of the so-called "RING OF FIRE," a belt of intense earthquake and volcanic activity surrounding the Pacific basin. Numerous SUBDUCTION ZONES, associated with volcanic activity, are found along the boundaries of the Pacific crustal plate, such as the subduction zone found along the ALEUTIAN ISLANDS arc in Alaska. (See also PLATE TECTONICS.)

Pacific Ocean The Pacific Ocean basin is surrounded by a ring of intense earthquake and volcanic activity known as the "RING OF FIRE," produced by ongoing collisions between the Pacific crustal plate and adjacent plates of the earth's crust. Other, smaller plates of crustal rock are also involved. "HOT SPOTS" of volcanic activity within the Pacific plate have produced some large volcanic islands, of which the Hawaiian Islands are familiar examples. (See also PLATE TECTONICS.)

Pagan, volcano, Mariana Islands The stratovolcano Pagan has erupted on several occasions in the 20th century. Four minor eruptions occurred between 1909 and 1925. A powerful explosive eruption in 1981 produced a cloud of ash that rose some 12 miles above the volcano. Lesser eruptions have taken place since then.

pahoehoe Lava having a "ropy" surface, as distinct from the broken, blocky surface of AA lava.

Palisades, New Jersey, United States The Palisades, cliffs along the Hudson River near New York City, famous for their great height and prominent columnar structure, represent the edge of a huge SILL of BASALT.

Pahoehoe lava flow, Kilauea, Hawaii, 1950 (H. R. Joesting, USGS)

Palmdale, California, United States Located approximately where the Garlock Fault and the SAN ANDREAS FAULT meet, Palmdale was the site of an uplift called the "Palmdale bulge" in the 1970s. The Palmdale bulge drew considerable attention as a potential generator of destructive earthquakes.

Papandajan, volcano, Java, Indonesia Also known as Papandayang, and located in western Java, Papandajan is a stratovolcano. A 1772 eruption killed some 3,000 people and destroyed several dozen villages through explosions and landslides. An eruption in 1883 accompanied the destruction of nearby KRAKATOA. Another eruption occurred in 1925.

Papua New Guinea Some of the most famous volcanic activity in history has occurred in Papua New Guinea, notably eruptions of Mount LAMINGTON in 1951 and RABAUL in 1937. KARKAR and Manam have also been active in historical times.

parasitic cone A conical buildup of material around a vent erupting near the base of a volcano.

A spectacular nighttime eruption of Parícutin volcano in Mexico (W. F. Foshag, USGS)

A parasitic cone may grow to become a large volcano itself.

Parícutin, volcano, Mexico In one of the most spectacular eruptions of the 20th century, the volcano Parícutin arose beginning in February 1943 from a field in Michoacan, Mexico in the Sierra Occidental on the western edge of the Central Plateau. The volcano attained a height of about 1,200–1,300 feet before the eruptions ceased in 1952. There was a particularly violent eruption on June 10, 1944.

Pavlov, volcano, Alaska, United States The stratovolcano Pavlov is located on the Alaska Peninsula, next to the volcano Pavlov's Sister. Pavlov has undergone eruptions frequently since the late 18th century. There is evidence that one of the two volcanoes (probably Pavlov's Sister) was active between 1762–86. A very strong earthquake occurred at about the time of an eruption in 1844, although it is not known what precise relationship the earthquake may have had to the eruption. A fissure on the side of Pavlov erupted for a single day in 1846. An unusually strong earthquake struck the area again in 1866, several months after an eruption in March of 1866. Pavlov's biggest eruption in historical times occurred in December of 1911. Earthquakes accompanied this eruption. Lava flowed from a fissure on the north flank, and large blocks were cast out of the volcano. Earthquakes attributed to explosions became more numerous at Pavlov in July of 1983. An eruption preceded by several weeks of earthquakes may have occurred on November 14, 1983, when an aircraft pilot reported sighting a plume, and a glow was seen. Earthquake activity declined after this event and remained at a generally low level. In 1986 Pavlov entered a new period of eruption.

Lava flows have occurred in 1846, 1958, 1960–63, 1966 and during the 1970s and 1980s. Lava may also have flowed from the volcano between 1936 and 1948.

Pelée, Mount, volcano, Martinique The volcano Pelée's eruption on May 8, 1902, was one of the most destructive in the history of the Caribbean. The eruption wiped out the city of St. Pierre and is thought to have killed more than 30,000 people; only six individuals from the city survived. The first signs of eruption occurred late in April. A cloud of smoke began rising from the volcano on April 23, accompanied by occasional emissions of ash. On May 5, the eruption intensified and sent lava and mud spilling over the rim of the crater and down the valley of the River Blanche; a sugar mill was destroyed, and more than 20 people were killed.

Many firsthand accounts of the destruction of St. Pierre and vicinity were recorded, despite the virtually complete loss of life on shore. Some of those reports are worth quoting at length, for they convey perhaps more clearly than any other documentation of volcanic eruptions the power and destructive effect of such events. One of the most comprehensive and vivid accounts of the catastrophe comes from Comte de Fitz-James, a French traveler, who with his companion Baron Fontenilliat witnessed the destruction of St. Pierre from the relative safety of a boat in the harbor. He is quoted in Charles Morris's 1902 book, *The Volcano's Deadly Work:*

> From the depths of the earth came rumblings, an awful music which cannot be described. I called my companion's name, and my voice echoed back at me from a score of angles. All the air was filled with the acrid vapors that had belched from the mouth of the volcano . . .
>
> From a boat in the roadstead . . . I witnessed the cataclysm that came upon the city. We saw the shipping destroyed by a breath of fire. We saw the cable ship *Grappler* keel over under the whirlwind, and sink as through drawn down into the waters of the harbor by some force from below. The *Roraima* was overcome and burned at anchor. The *Roddam* . . . was able to escape like a stricken moth which crawls from a flame that has burned its wings . . .
>
> When we got ashore we called aloud, and only the echo of our voices answered us. Our fear was great, but we did not know which way to turn, and had it been our one thought to escape we would

The destroyed city of St. Pierre, with Pelée in distance (I. S. Russell, USGS)

not have known how to do so. It was about one o'clock in the afternoon when we reached shore. Our weariness was beyond description. Sleep was the one thing that I wanted, but I overcame the desire and, with Baron de Fontenilliat, set off to make our way to St. Pierre, hoping that we might still render some assistance to the injured . . .

We saw great stones that seemed to be marvels of strength, but when touched with the toe of a boot they crumbled into impalpable dust. I picked up a bar of iron. It was about an inch and a half thick and three feet long. It had been manufactured square and then twisted so as to give it greater strength. The fire that came down from Mont Pelée had taken from the iron all of its strength and had left it so that when I twisted it, it fell into filaments, like so much broom straw . . .

I know that the explosion of Mont Pelée was not accompanied by anything like an earthquake, for . . . when we entered St. Pierre we found the fountains all flowing, just as though nothing had happened. They continued to flow, and are flowing still, unless destroyed by the later explosions.

There was no flow of lava. It was all ashes, dust, gas and mud . . .

The sole survivor of the catastrophe at St. Pierre was a murderer named Joseph Surtout. He survived because he was locked in a cell so far underground that it protected him from the worst effects of the explosion. He was trapped in his cell for four days after the eruption without food and water, though somehow he got enough air to survive. Since his cell was windowless, he had no way of knowing exactly what had happened. He called for help and eventually was rescued.

Workers disposing of bodies found many curious sights among the devastation left by the catastrophe. The charred body of a woman was found holding a silk handkerchief—unburned—to her mouth. Carbonized bodies were found with their shoes undamaged. Severely burned bodies lay adjacent to others that the fire appeared to have touched only slightly. Articles such as purses were discovered virtually intact.

The May eruption of Pelée destroyed approximately eight square miles. The area of destruction was enlarged slightly by another eruption on August 30. Yet another eruption occurred in December.

In October, a lava dome, or great mass of solidified lava, arose in the caldera and grew to a height of some 800 feet by the first of December. This formation was called the Tower of Pelée and stood more than 1,000 feet tall at its maximum height. Later activity at the volcano shook down the Tower of Pelée, however, and only a small portion of the dome remained by the end of 1903. Mount Pelée became active again in 1929; this eruption was less intense than that of 1902 but lasted more than three years. On this occasion, St. Pierre and other communities on the island were evacuated.

Pennsylvania, United States Pennsylvania is located in a region of minor seismic risk but has undergone strong earthquakes on occasion. Very strong shocks were reported at Philadelphia on March 17 and November 29 of 1800 and again on November 11 and 14, 1840; this latter earthquake was accompanied by a large swell on the Delaware River.

Peru Located on the Pacific shore of South America, Peru suffers from earthquakes as a result of the ongoing collision of the South American crustal plate with the NAZCA CRUSTAL PLATE to its west. Earthquakes in Peru and Ecuador on August 13–15, 1863, are believed to have killed some 25,000 people. Another major earthquake in Peru in 1892 killed an estimated 25,000. The Lima earthquake on May 25, 1940, is believed to have killed 200 to 300 and caused some 5,000 injuries in the port of Callao. The Ancash earthquake of November 10–13, 1946, caused some 700 fatalities. More than 100 were reported killed in the Cuzco earthquake on May 21, 1950. Strong earthquakes in Peru, Bolivia and Chile on January 13, 1960, killed more than 2,000 people and caused hundreds of millions of dollars in damage. An earthquake on May 30, 1970, at Yungay killed more than 60,000 and left more than half a million homeless after the inundation of a resort.

An unusual case of a landslide apparently causing an earthquake, rather than vice versa, was observed on April 25, 1974, along the Mantaro River in Peru. A great landslide there produced seismic effects comparable to those of an earthquake of about 4.5 on the Richter scale of magnitude.

Tsunami damage at Lebak, Mindanao, Philippine Islands from earthquake on August 16, 1976, in Moro Gulf, Mindanao (*Earthquake Information Bulletin*/USGS)

The stratovolcano El Misti is located in Peru near the city of Arequipa. The volcano is famed for its beauty more than for its eruptions; in historical times, El Misti has been characterized by mere steaming and occasional minor explosive eruptions.

Philippine Islands Volcanic activity has been frequent and destructive in the Philippine Islands within historical times. The stratovolcano MAYON has undergone explosive eruptions on more than 40 occasions since 1616 and has generated both lava flows and NUÉES ARDENTES. The volcano TAAL, located in the central Philippines, has erupted on more than 30 occasions since 1572, and its eruptions are sometimes accompanied by tsunamis in the lake from which the volcano rises. The island of Luzon has many volcanic mountains. The volcano Malaspina is located on the island of Negros, and the volcano Camaguin erupted on an island approximately 90 miles southeast of Negros in 1876. Volcanoes on the island of Mindanao include Apo and Cottobato. Hot springs and sulfur deposits in the Philippines are further evidence of volcanic and hydrothermal activity there.

The Philippine Islands are also vulnerable to earthquakes, of which perhaps the most famous occurred on July 3, 1863, in Manila. The earthquake destroyed many government buildings, hospitals and churches and is thought to have killed some 1,000 persons. Damage to a single industry—

tobacco—was estimated at more than $2 million. Another earthquake in Manila in 1880 caused widespread destruction but reportedly no loss of life. A 1991 eruption of Mount Pinatubo was one of the most powerful eruptions of the 20th century and forced the United States to abandon an air base near the volcano.

Phlegraean Fields, Italy This volcanic area, several miles west of Naples and Vesuvius, has a high concentration of fumaroles and solfataras. Indeed, Solfatara, one of the volcanoes located in the Phlegraean Fields, provided the name for that phenomenon. Nineteen craters occur within an area of about 25 square miles here. Classical literature is full of references to the Phlegraean Fields, including a mention in Virgil's *Aeneid,* and they are said to have provided imagery for the *Inferno* of Dante.

The field is made up of explosion craters from PHREATIC ERUPTIONS, and cinder cones inside a collapsed caldera some eight miles wide. The caldera is thought to have formed in prehistoric times either in or following the eruption of vast volumes of TUFF. A large portion of the volcanism at the Phlegraean Fields is thought to have occurred between 3,000 and 5,000 years ago. The migration of vents at the volcanic field has followed a pattern: Vents have tended to migrate from the rim of the caldera toward the center and to diminish in the volumes of their eruptions. Another interesting pattern is that early activity at the Phlegraean Fields appears to have taken place underwater, whereas later activity moved onto the land. The patterns of activity at the Phlegraean Fields indicate that a reservoir of magma near the surface is crystallizing slowly and causing activity at the surface to move inward toward the center of the caldera as crystallization proceeds underground. The magma reservoir is believed to be within two to four miles of the surface. The boundaries of the volcanic field, however, are indistinct. The Phlegraean Fields are noted for their history of dramatic uplift and subsidence. The motions may be slow by everyday human standards, but on the geological time scale they are swift.

A case in point is the marketplace at Pozzuoli, which once was under water, as shown by the borings of mollusks in the ancient columns. The marketplace sank by perhaps 30 feet or more by the year A.D. 1000, then rose again by about six feet between 1000 and 1198 when the volcano Solfatara erupted. (Outstanding examples of uplift and subsidence are also recorded at other points in the vicinity. One pier built in the second century A.D. reportedly dropped almost 20 feet by the 18th century.) Following the eruption in 1198, uplift continued. The uplift was accompanied by earthquakes, including a strong tectonic earthquake under the Campanian Apennines that occurred in 1456 and apparently caused severe damage to Pozzuoli. Other powerful local earthquakes shook the area around Pozzuoli in 1488. The rate of uplift increased around 1500. A considerable amount of new shoreline had emerged from the waters by 1503 and was taken over by a local school. Between 1000 and 1503, the rate of uplift was perhaps an inch per year on the average, for a total of approximately 36 feet. Most of that uplift took place after 1500, as the rate of uplift increased to about seven inches per year.

Earthquakes occurred frequently in 1534 and between 1536 and 1538 and apparently originated along the southern and eastern edge of the caldera. (A colorful story is associated with one of these earthquakes, in 1534. The earthquake occurred during a worship service just before Easter, as the gospel account was read of the earthquake that took place at the resurrection of Christ. This coincidence made a tremendous impression on the churchgoers.) Fumaroles became more active between 1536 and 1538 in the vicinity of Sudatoio, slightly to the west of Pozzuoli.

In this area, a new volcano, Monte Nuovo, was about to appear. Numerous earthquakes preceded this eruption, as did a spectacular display of uplift. On September 26–27, 1538, the ground at Pozzuoli rose perhaps 15 feet, and the shoreline retreated some 1,200 feet. Water started gushing from fissures on September 28, and subsidence lowered ground level by approximately 12 feet the following day. On September 29, uplift resumed near what would be the site of the Monte Nuovo eruption. The eruption began in the evening with great noise, and great quantities of ash and pumice mixed with water emerged from the ground. Evidently uplift had caused a considerable retreat of the shoreline, because one account of the eruption mentions townspeople carrying fish that they had picked up along the shore. New springs of water, one hot and the other cold, also reportedly appeared at this time. Fire was reported from this eruption and is thought to have been burning gas from fumaroles. An explosive eruption near the community of Trepergule cast out pumice. Eruptive activity lasted five days. On October 6, 24 persons who had climbed the cone of the volcano were killed in an explosion and in pyroclastic surges. This outburst knocked down trees three miles from the volcano. After the 1538 eruption, subsidence proceeded at an average rate of less than one inch per year, with only a short period of uplift between 1951 and 1952.

Several earthquakes in the late 16th century took place around Pozzuoli. Another strong earthquake occurred in 1832 and appears to have been centered near Monte Nuovo. Later, though weaker, earthquakes occurred between 1887 and 1892. Several earthquakes in the early 20th century are thought to have occurred around the borders of the caldera. A mud volcano at Solfatara erupted with great energy after an earthquake about 30 miles east of the Phlegraean Fields in 1930. In 1969, uplift resumed in the caldera, with accompanying mild earthquake activity. Seismic activity and uplift diminished in 1972. A strong earthquake about 60 miles southeast of the Phlegraean Fields in 1980 preceded a renewal of uplift. Earthquake activity started increasing in 1983. A measuring device at Pozzuoli showed uplift of approximately four feet between mid-1982 and late 1984 and approximately 10 feet between 1969 and 1985. Earthquakes released much more energy during this later period of uplift than in 1969 to 1972. During the 1982–84 episode, much of the earthquake activity was focused around the Solfatara volcano and around a location immediately north of Pozzuoli. Uplift had ceased by the end of 1984, and a very gradual deflation began in January of 1985.

Uplift, volcanism and earthquake activity at the Phlegraean Fields are thought to reflect the presence of a small, shallow magma chamber a couple of miles below the surface. The on-and-off pattern of uplift at the Phlegraean Fields, along with the history of phreatic activity there, also indicates that groundwater heated by a subterranean body of magma plays an important role there.

phreatic eruption Generally speaking, a nonincandescent volcanic eruption. A phreatic eruption involves mud, steam or other nonincandescent material and results from heating of groundwater by igneous material underground.

pillow lava A distinctive form of lava flow in which the solidified rock forms rounded masses with a glassy exterior. Pillow lava is found where lava bearing little or no dissolved gas flows into the ocean from the land and cools or where lava flows onto the ocean floor at depths so great that water pressure prevents gas dissolved in the lava from bubbling out of solution. Pillow lava also is found in table mountains, curious volcanic formations that originated in Iceland where eruptions occurred beneath thick layers of glacial ice. Herdubreid is an example of such a mountain.

Pinatubo, Mount, volcano, Philippine Islands The June 1991 eruption of Mount Pinatubo, after more than 600 years of dormancy, killed hundreds of people and forced the abandonment of a United States air base in the Philippines. The eruption is believed to have cast between 15 million and 20 million tons of sulfur dioxide into the atmosphere, where the sulfur dioxide combined with water to form an estimated 30 million to 40 million tons of sulfuric acid particles. The climatic effects of the Mount Pinatubo eruption, combined with those of a lesser eruption of Chile's Mount Hudson volcano, are believed to have contributed to global cooling the following year. In the lower and middle atmosphere, temperature measurements reportedly showed a change of almost 1°F between 1991 and 1992. The cooling effect might have been greater but for the warming effect of El Niño, a periodic increase in temperature in the Pacific Ocean that influences weather.

Piton de la Fournaise, volcano, Réunion Island An active shield volcano, Piton de la Fournaise has a summit caldera surrounded by a set of curving faults. As a rule, Piton de la Fournaise does not exhibit dramatic precursory phenomena such as strong earthquakes before eruptions. Nonetheless, eruptions do tend to follow minor but notable changes, such as small earthquakes and the formation of fissures in the caldera. Eruptions were recorded in 1708 and 1800 and every several years thereafter up to the late 1980s. Most eruptions have been explosive with associated lava flows. Piton de la Fournaise is thought to be fed by basaltic magma. Geologists believe that the magma is stored in a central reservoir near the surface before erupting through the summit and through rifts on the volcano's flanks. Apparently, inflation occurs at the summit for some weeks or months before an eruption. Reinflation, however, does not appear to start soon after each new eruption. Piton de la Fournaise is presumed to be a "hot-spot" volcano like Hawaii. A volcanic observatory has been set up there.

plate tectonics One of the foundations of modern geology, the theory of plate tectonics describes the interactions among several large, and a number of smaller, rigid plates of the earth's crust. These interactions are complex, but for the purpose of understanding earthquakes and volcanic eruptions, the plate tectonic theory may be de-

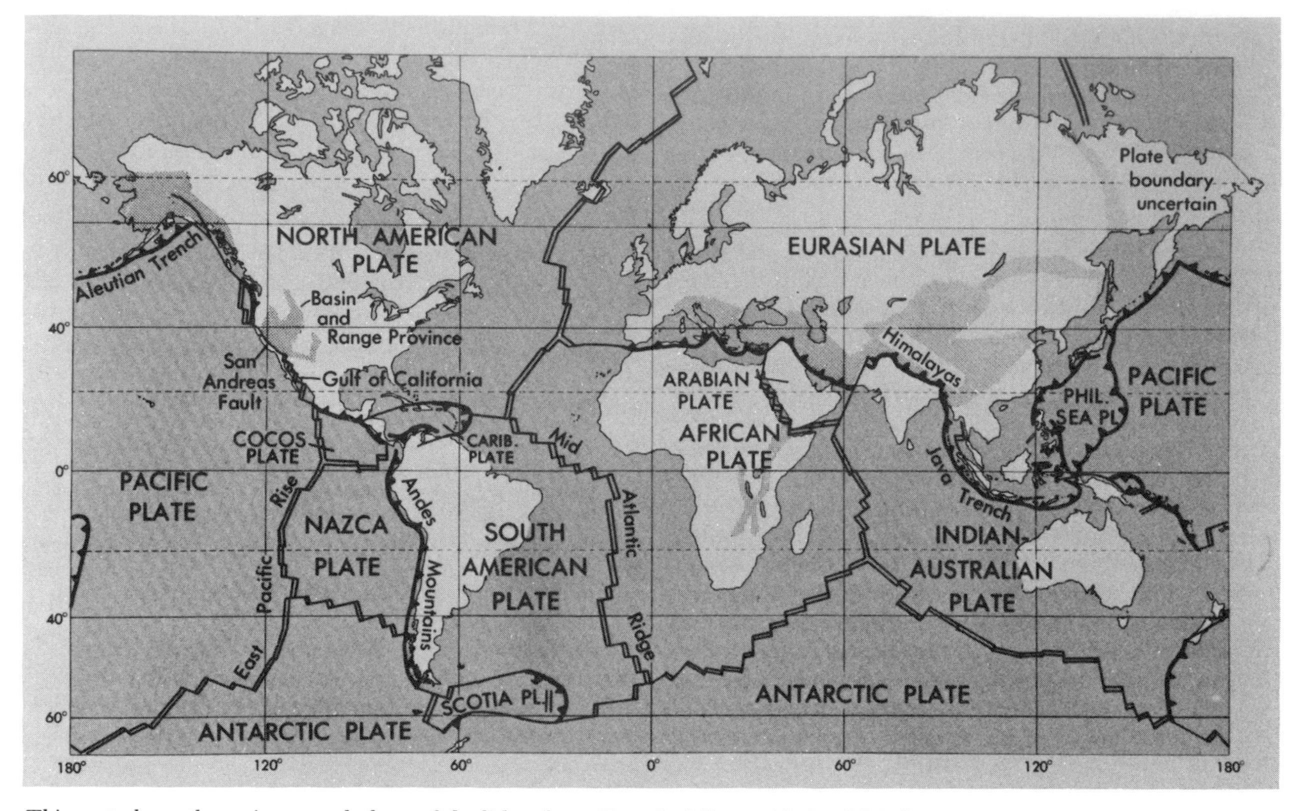

This map shows the major crustal plates of the lithosphere. Note dark lines with "teeth" indicating subduction zones (USGS)

scribed as follows. Crustal plates float, so to speak, atop the denser rocks of the mantle by a process known as isostasy. The plates are formed along mid-ocean ridges, where molten rock rising from the asthenosphere solidifies and forms new crustal rock. The newly formed rock moves outward to either side of the ridge in a pattern much like that of objects on a moving conveyer belt. Volcanic activity is therefore commonplace along the ridges and may build up whole islands, of which ICELAND is a spectacular example. Crustal plates are destroyed along subduction zones, where the plates collide with one another and one plate is subducted, or warped downward, beneath another. As the subducted plate descends into the asthenosphere, the plate melts, and its lighter components rise back toward the surface as molten rock, or magma. This magma finds its way to the surface through openings in the crust and produces volcanoes. Arcs of volcanoes and/or volcanic islands are associated with subduction zones. Examples in North America are the volcanoes of the Pacific Northwest of the United States, produced by the CASCADIA SUBDUCTION ZONE off the coast of Oregon, Washington and the Canadian province of British Columbia, and the ALEUTIAN ISLANDS chain, adjacent to a subduction zone just to the south of the island arc. (Not all volcanic activity is linked to subduction zones; "hot spots," or points where ascending plumes of molten rock are thought to approach the earth's surface, may be responsible for volcanic activity, as in the case of the Hawaiian Islands.)

Earthquakes also occur along subduction zones and along the boundaries of crustal plates generally. In the United States, the famous SAN ANDREAS FAULT in California runs along the boundary between the Pacific crustal plate and the North American crustal plate. This fault and its affiliated faults have been responsible for some of the most destructive earthquakes in the nation's history. The source of energy for the movement of crustal plates has been a subject of debate since plate tectonics was developed from the earlier theory of continental drift. In recent years, it has been widely believed that much of the energy behind the motions of crustal plates is derived from convection cells, or circulation patterns produced within the mantle by convective activity. (See also CONVECTION CURRENT; EARTH, INTERNAL STRUCTURE OF; WEGENER, ALFRED.)

pluton An intrusion of igneous rock. The term *pluton* is sometimes restricted to such intrusions occurring deep underground and resembling granite in texture.

plutonism The formation of plutons.

Poas, volcano, Costa Rica Located northeast of the communities of Naranjo, Grecia and San Pedro, Poas volcano has arisen within a pair of nested calderas and is suspected of having a lake rich in sulfur, underlain by a layer of molten sulfur, in its main crater. Molten sulfur has sometimes been expelled in eruptions of Poas. Most of its eruptions, which have been recorded on numerous occasions in the 19th and 20th centuries, have been small and either magmatic or phreatic in character. Notable eruptions in 1828, 1834 and 1880 deposited ash on nearby communities. Other significant eruptions occurred in 1910 and 1953. The 1910 event began with the eruption of a geyser, and a plume rose to an altitude of more than four miles. Mud covered the whole top portion of the volcano. Numerous strong earthquakes occurred later that year, including one especially powerful earthquake that caused widespread damage in nearby communities. A 1953 eruption eliminated a crater lake and dropped ash 30 miles or more away from the volcano. An eruption column rose about two miles over the volcano in 1976, and geysers erupted in 1979 and 1980. Another eruption was reported in 1988.

Pompeii and Herculaneum, Italy Two of the most famous cities in the history of volcanology, the Roman communities of Pompeii and Herculaneum were destroyed in the eruption of VESUVIUS in A.D. 79. The destruction of the two cities has provided a popular theme for authors, notably Edward Bulwer-Lytton in his novel *The Last Days of Pompeii.* Bulwer-Lytton was apparently influenced by the historian Dion Cassius, who reported that the fallout from Vesuvius buried the people of Pompeii as they sat in the theater. This melodramatic scenario appears to have little basis in fact, and it is thought that Dion Cassius was merely drawing on traditional (and exaggerated) accounts of the catastrophe. Other of Dion Cassius's statements sound improbable, to say the least, especially his report that a great number of giants appeared in the vicinity of the volcano during the eruption. An eyewitness account of the event was made by Pliny the Younger (see Appendix B).

Excavations at Pompeii have revealed much about life in that coastal city some 2,000 years ago. One visitor, quoted in Charles Morris's study of volcanoes, *The Volcano's Deadly Work,* commented on "the weirdness of the scene" at Pompeii and the remarkable preservation of details of life there:

> We have before us the narrow lanes, paved with tufa, in which Roman wagon wheels have worn deep ruts. We cross streets on stepping-stones which sandaled feet long ago polished. We see the wine shops with empty jars, counters stained with liquor, stone mills where the wheat was ground, and the very ovens in which bread was baked more than eighteen centuries ago. "Welcome" is offered us at one silent, broken doorway; at another we are warned to "Beware of the dog!" The painted figures,—some of them so artistic and rich in colors that pictures of them are disbelieved,—the mosaic pavements, the empty fountains . . . the marble pillars and the small gardens are there just as the owners left them. Some of the walls are scribbled over by the small boys of Pompeii in strange characters which mock modern erudition. In places we read the advertisements of gladiatorial shows, never to come off, the names of candidates for legislative office who were never to sit. There is nothing like this elsewhere . . .
>
> The streets of Pompeii must have had a charm unapproached by those of any city now in existence. The stores, indeed, were wretched little dens. Two or three of them commonly occupied the front of a house on either side of the entrance, the ostium; but when the door lay open, as was usually the case, a passerby could look into the atrium, prettily decorated and hung with rich stuffs. The sunshine entered through an aperture in the roof . . .
>
> As the life of the Pompeiians was all outdoors, their pretty homes stood open always. There was indeed a curtain betwixt the atrium and the peristyle, but it was drawn only when the master gave a banquet. Thus a wayfarer in the street could see, beyond the hall described and its busy servants, the white columns of the peristyle, with creepers trained around them, flowers all around, and jets of water playing through pipes which are still in place. In many cases the garden itself could be observed between the pillars of the further gallery, and rich paintings on the wall beyond that.
>
> But how far removed those little palaces of Pompeii were from our notion of well-being is scarcely to be understood by one who has not seen them. It is a question strange in all points of view where the family slept in the houses, nearly all of which had no second story. In the most graceful villas the three to five sleeping chambers round the atrium and four round the peristyle were rather ornamental cupboards than aught else. One did not differ from another, and if these were devoted to the household the slaves, male and female, must have slept on the floor outside. The master, his family and his guest used these small, dark rooms,

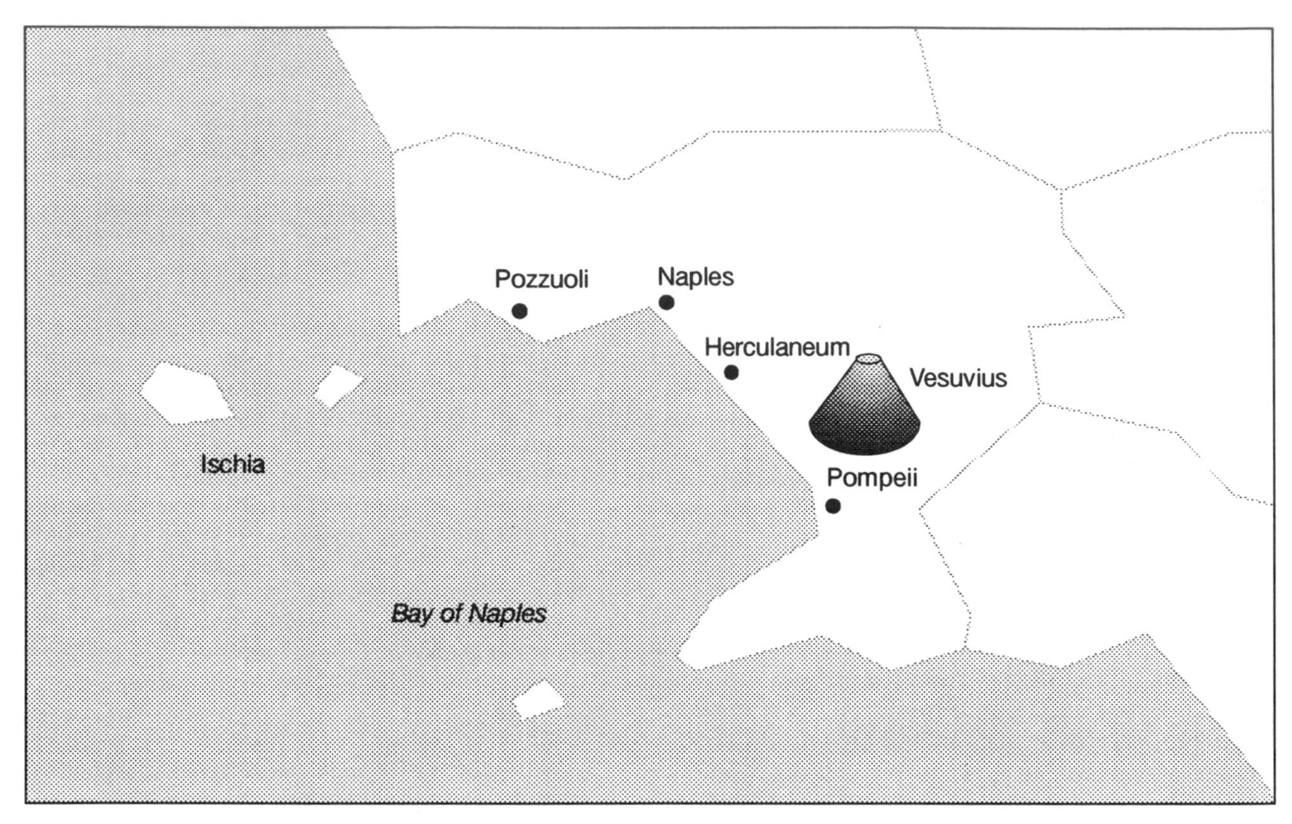

The Bay of Naples and vicinity, including the volcano Vesuvius and the destroyed cities of Pompeii and Herculaneum (© Facts On File, Inc.)

> which were apparently without such common luxuries as we expect in the humblest home. All their furniture could hardly have been more than a bed and a footstool . . . The kitchen of each villa certainly was not furnished with such ingenuity, expense or thought as the stories of Roman gourmandising would have led us to expect. In [one] house . . . the cook seems to have been employed in frying eggs at the moment when increasing danger put him to flight. His range, four partitions of brick, was very small; a knife, a strainer, a pan lay by the fire just as they fell from the slave's hand.

Pompeii was buried by falling TEPHRA, whereas Herculaneum was covered by mudflows.

Popocatépetl, volcano, Mexico The stratovolcano Popocatépetl stands south of Mexico City and has undergone explosive eruptions, most of them small, on more than a dozen occasions since 1512. There was a major eruption in 1720. Popocatépetl was one of the first volcanoes into which European explorers actually made a descent, to collect sulfur for gunpowder.

Port Royal, Jamaica The earthquake that hit Port Royal on June 7, 1692, struck about 11:43 A.M. (In 1959 divers brought up the remains of a watch that apparently stopped at that moment.) The earthquake caused the waterfront—built on sandy soil—to fall into the sea. A tavern, a warehouse and other buildings slid seaward and crumbled into the waters. Approximately half the town was wiped out.

Puerto Rico Puerto Rico experiences numerous earthquakes, but highly destructive ones are rare. However, the Caribbean basin, in which Puerto Rico is located, is an area of high-intensity seismic and volcanic activity.

On October 11, 1918, an earthquake of Richter magnitude 7.5, with Mercalli intensities in the range of VIII–IX, caused property damage estimated at about $4 million and killed more than 100 people. The earthquake was centered in the northwestern Mona Passage and was accompanied by a tsunami that caused widespread destruction and loss of life. The sea withdrew before the tsunami's arrival, then returned and reached heights possibly greater than 19 feet. A church in Aguada was destroyed, and all brick buildings were ruined in Anasco. Structures of wood or reinforced concrete reportedly came through the earthquake with little damage where building materials were sound. At Aguadilla, the earthquake caused serious damage to buildings constructed on alluvium. The collapse of a factory at Mayagüez killed several people. Damage to railways, bridges, pipelines and chimneys was widespread, and two cable links in the Mona Passage were broken. Strong aftershocks followed the October 11 earthquake on October 24 and November 12.

Puget Sound, Washington, United States An inlet of the Pacific Ocean between the Olympic Peninsula to the west and the mainland to the east, Puget Sound is an area of intense seismic activity. (See also CASCADIA SUBDUCTION ZONE; CASCADE MOUNTAINS; WASHINGTON.)

pumice An extremely lightweight, "foamy" volcanic rock, pumice forms when magma bearing a high concentration of dissolved gases solidifies just as the gases are bubbling out of solution. Pumice is cast out in large quantities during the eruptions of some volcanoes and is so lightweight that it may float on water. In eruptions of certain volcanoes on islands or near seacoasts, pumice may accumulate on the water and form great floating "rafts" capable of supporting a man's weight. Such rafts of pumice may impede navigation.

Purace, volcano, Colombia The stratovolcano Purace has undergone explosive eruptions on more than two dozen occasions since 1827. Two of these eruptions caused fatalities.

Puu Oo, volcanic cone, Hawaii, United States This cone, whose name means "hill of the Oo Bird," was produced by an eruption of KILAUEA that began in 1983 and came to be known as the Puu Oo eruption. Puu Oo underwent several episodes of eruption in the 1980s and reached a height of more than 750 feet. Although episodes of eruption became generally shorter, the extrusion rate of lava increased, and lava fountains rose higher. Some lava fountains rose approximately 1,200 feet into the air, approaching the height of the Empire State Building in New York City.

P-waves See SEISMOLOGY.

Ash clouds rise above a pyroclastic flow moving down the side of Mount St. Helens, August 7, 1980 (USGS)

pyroclastic Made up of fragmented pieces of rock, often those produced in an explosive volcanic eruption. An ignimbrite, for example, is composed of pyroclastic material.

pyroclastic flow More commonly known as an ash flow, a pyroclastic flow is made up of clastic material included and carried along in a flow of ash. The pyroclastic flow behaves as a fluid, hence its name. An ignimbrite is an example of a pyroclastic flow. Such a flow may extend outward many miles from the volcano that produced it. A nuée ardente is a fast-moving cloud of gas released in a volcanic eruption and containing pyroclastic material.

Q

quartz One of the most common minerals, quartz is found in many igneous rocks. Quartz also is found in the form of massive crystals.

R

Rabaul, volcanic cluster, Papua New Guinea Located at the end of the Bismarck volcanic arc, Rabaul is thought to lie near the meeting point of three plates of the earth's crust: the Bismarck, Solomon Sea and Pacific crustal plates. Two collapse events are thought to have occurred at Rabaul within the last 3,500 years.

Earthquakes occur often in the vicinity of Rabaul, and the caldera has a history of eruptive activity starting in the 18th century. An eruption was recorded in approximately 1767, probably either at Tavurvur, on the eastern side of the harbor, or at Rabalankaia, about two miles north of Tavurvur. Another eruption in 1791 occurred at Tavurvur. In 1850, an eruption at Sulfur Creek, at the northern end of the harbor, followed a time of powerful earthquakes and extensive uplift. Over the next 28 years, uplift in the vicinity of Vulcan, on the western side of the harbor, raised reefs out of the water. This period of uplift preceded a series of powerful earthquakes that started early in 1878 and were accompanied by tsunamis (on February 4) and dramatic uplift. Along the coast near Tavurvur, the ground reportedly was elevated by about 20 feet in some places. At the same time, subsidence occurred at Davapia Rocks north of Vulcan and put some homes underwater. Steam and pumice rose to the surface of the harbor near Vulcan at a spot that soon would become the site of Vulcan Island. Several hours after this eruption of pumice and steam, Tavurvur underwent a phreato-magmatic eruption. Rabaul remained quiet for the next several years, but unrest resumed in 1910 when a strong earthquake shook the area and was accompanied by a smell of sulfur. In 1916, another powerful earthquake occurred, along with considerable subsidence. Uplift during an earthquake in 1919 again raised some of the land that had subsided in 1916, and in 1919 fumaroles became highly active at Tavurvur and at Matupit Island, about two miles from Tavurvur. After this eruption, a causeway connected Matupit Island with the mainland.

Earthquakes resumed in 1937, including one strong event on May 28 that destroyed at least two buildings and set off landslides and possibly a tsunami as well. Dramatic uplift was observed at this time. Earthquakes continued, especially near Vulcan, on the following day. In between the stronger earthquakes, a steady vibration, possibly volcanic rather than tectonic in origin, was felt. On May 29, an eruption began. Vulcan Island was uplifted about six feet, and cracks appeared on Matupit Island. Just prior to the eruption, rumbling sounds were heard, and gas was seen rising to the surface of the waters. An area at the southern end of Matupit Island was uplifted several feet, exposing a reef. Another reef near Vulcan Island was lifted out of the water, submerged again and reexposed within 10 minutes. Some 500 people were killed by pyroclastic flows from this eruption. Another, less powerful eruption occurred at Tavurvur. This eruption was phreatic and generated destructive rains of mud. Great amounts of carbon dioxide emanated from vents along the eastern side of the caldera and killed animals there for months after the eruption. Earthquakes stopped with the onset of the eruption.

In 1938 emissions of sulfur dioxide and hydrogen chloride increased at Tavurvur and then increased again in 1940. Minor steam explosions occurred at Tavurvur in 1940, and a powerful tectonic earthquake shook Rabaul that same year. Late in 1940, fumaroles at Tavurvur showed a marked increase in temperature, and in January 1941, a major earthquake occurred in the vicinity of the caldera. Fumarole temperatures increased sharply just before the earthquake and continued rising afterward. Meanwhile, hydrogen chloride and sulfur dioxide output also increased. These changes apparently were restricted to Tavurvur, because fumaroles on the opposite side of the harbor at Matupit Harbor were reportedly unaffected. Earthquakes occurred often at Rabaul in the latter half of 1941 and early in 1942 and again in late 1943. Strong tectonic earthquakes occurred east of Rabaul again in 1967, but no changes were observed in local volcanoes. In 1971, earthquake swarms resumed, and the caldera floor started rising. Earthquakes became more frequent (increasing from approximately 2 per day to 10 per day in some periods) between 1971 and 1983. After a strong tectonic earthquake about 120 miles east of Rabaul in 1983, earthquake activity diminished in 1984 and 1985. Emissions of carbon dioxide from a crater on Tavurvur killed birds in 1981, and fresh activity from fumaroles destroyed plants at a site slightly more than one mile northwest of Tavurvur in 1983.

In view of the recent unrest at Rabaul, there is a chance that new eruptions will occur there in the near future. Calderas apparently were formed by eruptions there some 3,500 years and 1,400 years ago, and eruptions of comparable magnitude must be considered possibilities for the years ahead. With such events in mind, scientists and public officials have prepared an emergency plan that would respond in several stages, up to and including evacuation of the local population, to a threatened eruption at Rabaul. (See also PHREATIC ERUPTION; PLATE TECTONICS.)

radioactivity The loss of particles and energy from the nucleus of an atom, resulting in a more stable product.

radionuclides Radioactive atoms that generate heat during their decay. Radionuclides are believed to be an important source of internal heat for the earth. They include isotopes of such elements as uranium and thorium and are commonly found in igneous rocks such as granite and basalt.

Rainier, Mount, volcano, Washington, United States One of the most famous volcanoes on earth, Mount Rainier stands near Seattle and is part of the CASCADE MOUNTAINS, a volcanic range associated with the CASCADIA SUBDUCTION ZONE off the Pacific coast of the northwestern United States. Mount Rainier is a stratovolcano and has been dormant since the mid-1800s, when a minor eruption reportedly took place.

The most recent major eruption of Mount Rainier is thought to have occurred perhaps 2,000 years ago and laid down a thick layer of tephra east of the mountain. This eruption also is believed to have produced a lava cone at the summit of the volcano, as well as large mudflows. One mudflow estimated to have occurred almost 3,000 years ago traveled down the South Puyallup River and Tahoma Creek valleys as a mass of mud hundreds of feet thick; this same flow is thought to have exposed the core of the volcano. In any future eruptions, mudflows may recur on a large scale, through melting of glacial ice on the peak, and may present a threat to nearby communities. The potential for devastating avalanches from Mount Rainier is considerable because internal heat is believed to have turned the mountain's core into soft material that might fail easily during a powerful earthquake (a distinct possibility in the seismically active Pacific Northwest), releasing overlying material to fall downslope in vast amounts. Such an avalanche, in turn, might allow an explosion comparable to the one that destroyed the area around Mount ST. HELENS during the eruptions of 1980. The White River Valley is believed to be especially vulnerable to major avalanches from Mount Rainier, should conditions ever permit them to occur. Rockfalls said to have exceeded 10 million cubic yards occurred in the White River Valley in 1963.

Mount Rainier is noted for its "steam caves," a network of ice tunnels and chambers about 1.5 miles long altogether. The steam caves are produced by heat from the volcano acting on firn, a moderately dense form of ice that results when snow accumulating on the volcano is compressed to a density not quite that of the crystalline ice found in glaciers. Heat from the volcano melts spaces between the ice and the rock of the mountain. Warm air circulating in these spaces enlarges them. The result is a network of passageways within and beneath the ice. The steam caves represent a delicate equilibrium between fumarole activity and the accumulation of snow and ice on the volcano. The caves were discovered in 1870 when climbers took shelter in them, although there is evidence that Native Americans knew of the caves' existence long before European-descended settlers arrived in the Northwest. Most of the tunnel network exists in the East Crater atop Mount Rainier, although the steam caves extend also to the adjacent West Crater. Similar systems of steam caves are found at nearby Mount Baker and also at Wrangell in Alaska.

Ranau, caldera, Sumatra, Indonesia Located on the Semangko Fault Zone, the Ranau caldera is occupied in part by Lake Ranau. The caldera has undergone some unrest, but no major eruptions, during the late 19th and early 20th centuries. On two occasions in 1887 and 1888, large numbers of dead fish were discovered on Lake Ranau and on its beaches. The lake water was discolored and had an odor of sulfur. The deaths of the fish were attributed to volcanic activity in the lake. A powerful earthquake in 1895 reportedly produced SEICHES in Lake Ranau; dead fish were noted again on the surface of the lake in December of 1903. This fish kill was accompanied by discoloration of the lake waters, as in 1887 and 1888, and a smell of sulfur was detected near the shore.

Rasshua, volcano, Kuril Islands, Russia The Rasshua volcano has been active on an intermittent basis since the early 19th century. There appears to have been some unrest at the site in 1810, but details are unavailable. In 1846 a volcano on the island was reported steaming and giving off flame. Increased steaming activity at Rasshua in 1946 preceded a similar increase in activity at nearby Sarichev volcano, which erupted five days later. Fumaroles at Russhua became considerably more active in 1957, and small explosions were reported. Young lava flows extend down the eastern, northwestern and southern flanks of the volcano.

Rayleigh waves These earthquake waves are characterized by retrograde motion (that is, they orbit in an elliptical pattern in the direction opposite to that in which vibrations from the earthquake's EPICENTER are moving) and occur at a right angle to the earth's surface. Rayleigh waves, also known as ground roll, are named after John William Strutt, Lord Rayleigh (1842–1919), British mathematician and physicist.

Raymond Fault, California, United States The Raymond Fault is about five miles long and is located several miles southeast of Pasadena. The fault has exhibited surface rupture within historical times and is suspected of being the source of the Los Angeles earthquake of 1855.

Redoubt, volcano, Alaska, United States The stratovolcano Redoubt, located some 110 miles southwest of Anchorage, has undergone major eruptions several times in the 20th century. Major explosive eruptions also have occurred in this century at nearby Spurr and Augustine volcanoes. Redoubt's eruption of December 14–19, 1989, produced columns of ash that rose to altitudes exceeding 40,000 feet and had serious effects on airline travel in the vicinity of the volcano. The ashfalls from this series of explosive incidents extended over the heavily populated Cook Inlet area and inland to the vicinity of Fairbanks. A period of lava-dome building activity followed these ashfalls and produced more than a dozen such domes over four months. Each of the domes was destroyed by explosive eruptions that produced ash clouds, although the clouds did less to disrupt air travel than the clouds from the early phase of the volcano's eruption. More than 20 significant falls of tephra were observed at Redoubt between December 14, 1989, and the end of April 1990. A fall of tephra on December 15 included coarse-grained deposits rich in pumice, but subsequent tephra falls were generally characterized by ash-sized material.

Recent eruptions of Redoubt have allowed few direct observations because of poor weather conditions and short periods of sunlight in the winter. Studies of these eruptions have yielded considerable information, however, about lightning in the clouds surrounding the volcano. Ordinarily, lightning is rare in the vicinity of Redoubt, but lightning discharges have been observed during most of the eruptions of Redoubt that released ash. A commercial lightning detection system deployed by the Alaska Volcano Observatory detected lightning in 11 of 12 eruptions and located the lightning strikes in nine of those eruptions, following the activation of the system early in 1990. Most of the strikes between clouds and the ground occurred close to the mountain, and the number of cloud-to-ground discharges was found to be related to the amount of ash released. Lightning discharges in the early stages of the eruption had a negative charge, whereas discharges later in the eruption were positive. This pattern indicates that coarse particles of ash, expelled early in the eruption, carry a negative charge, and finer particles expelled in later stages of the eruption carry a positive charge. "Intracloud" strikes of lightning, within the volcanic cloud itself, tend to be more numerous than cloud-to-ground strikes and to occur even at great distances from the volcano. Intracloud discharges during the February 15, 1990, eruption of Redoubt occurred up to about 80 miles from the mountain.

Intracloud lightning may prove useful in developing systems capable of tracking airborne volcanic clouds. Satellites have been used to track ejecta from Redoubt in more than 20 eruptions since late 1989. Environmental observation satellites in geostationary and polar orbits have monitored emissions of ash and volcanic debris. Data provided by the advanced very high resolution radiometer (AVHRR) on two U.S. weather satellites in polar orbit were used to detect a large mudflow along the Drift River near Redoubt as the event occurred during the February 15, 1990, eruption, and to provide warnings to Anchorage. Studies of AVHRR images of eruptions of Redoubt and Augustine indicate that volcanic clouds can be detected under various weather conditions, whether during day or night or over water or land. Multiple infrared channels are useful here, but there is as yet no single algorithm capable of detecting all eruption clouds. (See also AVIATION AND VOLCANOES.)

reef A reef is a barely submerged expanse of rock or coral, often found in shallow water near a continent or island. Coral reefs are often associated with volcanic islands and seamounts, when a submarine volcano's summit rises close enough to the surface for light to penetrate and allow the growth of coral. When coral builds up to the surface from the summit of a seamount and forms a ring of dry land, the resulting formation is called an atoll. Reefs can serve as useful indicators of uplift or subsidence at a volcanic site. On some occasions, reefs may be uplifted, dropped, and then lifted again very rapidly. At Rabaul, for example, a reef on one occasion was seen to appear and disappear, then reemerge from the water, within only about 10 minutes.

Réunion Island See PITON DE LA FOURNAISE.

Reventador, volcano, Ecuador A stratovolcano, Reventador is believed to have been active as early as the mid-1500s but was not explored until the 1930s.

Rhode Island, United States Rhode Island has not been the site of numerous strong earthquakes. It belongs to seismically active New England, however, and has been affected on occasion by earthquakes elsewhere in the region. Two notable earthquakes did occur in the Narragansett Bay area in the 1960s. The earthquake of December 7, 1965, rattled doors and windows and caused widespread alarm in Warwick; this earthquake was felt also in Massachusetts. The earthquake of February 2, 1967, registered 2.4 on the Richter scale of earthquake magnitude and frightened residents of communities including Newport and North Kingstown.

rhyolite A category of volcanic rock with high silica content, rhyolites are commonly tan or light gray, although obsidian, or volcanic glass, a widespread rhyolitic rock, is dark brown or black.

Richter, Charles (1900–1985) An American geologist and physicist, Richter's name is applied to a widely used scale of earthquake magnitude, indicating energy released in an earthquake. The logarithmic Richter scale was first proposed in 1927, and Richter refined it with the help of Beno Gutenberg. Richter magnitude is expressed in Arabic numerals and decimal (for example, 5.5). The Richter scale is different from but complements the Mercalli scale of earthquake intensity. (See also SEISMOLOGY.)

rift A breach or "crack" between two masses of crustal rock that once were joined together.

rift valley A long, linear crack in the earth's crust that occurs where magma rising from below reaches the surface, solidifies and moves away to either side from its zone of formation.

Rincón de la Vieja, volcano, Costa Rica Rincón de la Vieja is the biggest volcano in the northern part of Costa Rica and appears to be located inside nested calderas. The volcano is made up of several volcanic centers arrayed along a ridge. Explosive eruptions were recorded between 1860 and 1863, but details are unavailable. Explosive activity was observed again at the volcano in April of 1922, together with evidence of a very powerful recent eruption. A strong eruption occurred on June 4, 1922, preceded by sulfur dioxide emissions and loud noises from fumaroles. In 1963 the volcano was reported to be steaming and giving off sulfur gases. Vapor emissions from the volcano in 1965 preceded an eruption of ash in 1966. On a single day (February 23, 1968), more than 20 eruptions occurred in half an hour. These eruptions required that villages in the area be evacuated. Steam blasts from the volcano occurred in 1983 that affected locations more than a mile from the crater. Another eruption took place in 1984, and more eruptive activity is thought to have occurred in the autumn of 1985. Minor explosions were recorded in 1986 and 1987.

"Ring of Fire" This is the popular name for a narrow belt of intense earthquake and volcanic activity that follows approximately the borders of the Pacific Ocean basin and includes most of the active volcanoes on earth. The "Ring of Fire" extends from Tierra del Fuego at the southern extremity of South America, northward along the Andes Mountains of South America, through Central America, along the Cascade Mountains in the Pacific Northwest of the United States and south-

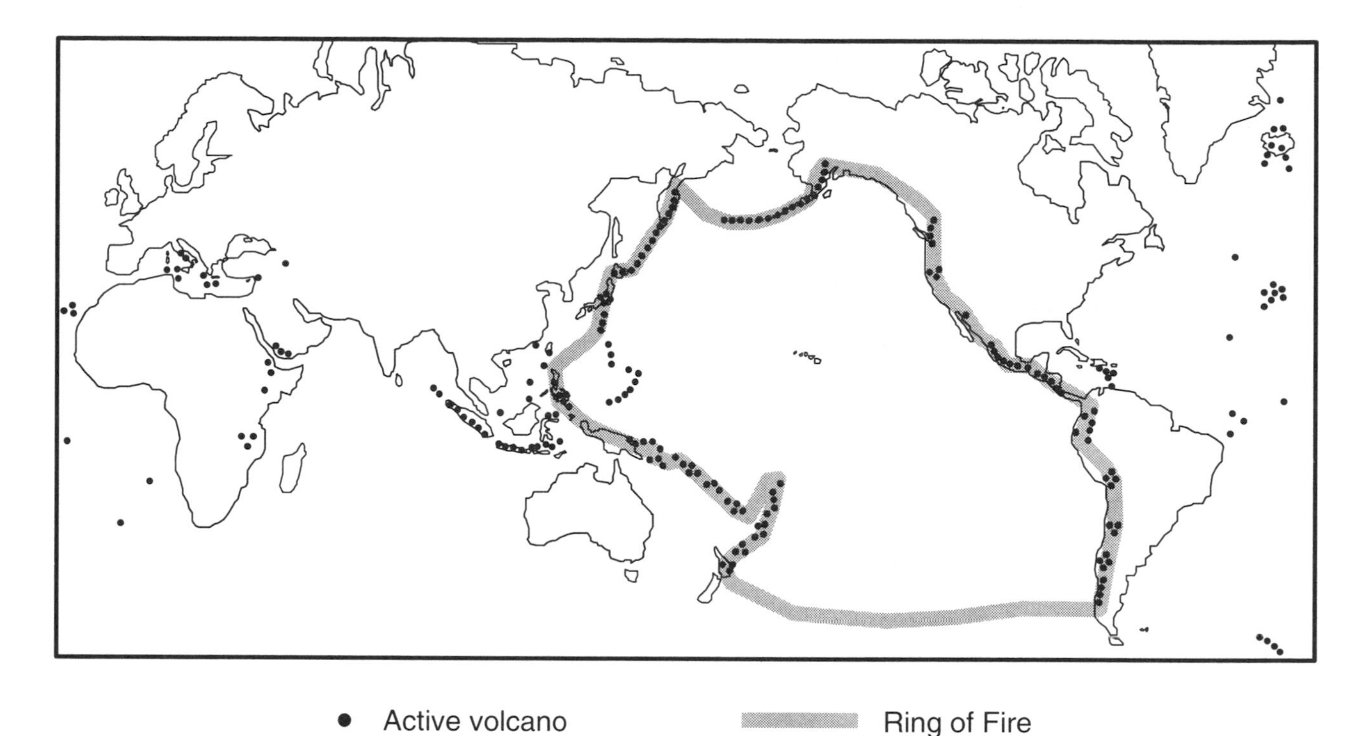

The "Ring of Fire," the popular name of a narrow belt of volcanic and earthquake activity that encircles the Pacific basin (© Facts On File, Inc.)

western Canada, down the ALEUTIAN ISLANDS in Alaska, westward to volcano-filled KAMCHATKA Peninsula and southward to JAPAN and eastern CHINA, the PHILIPPINE ISLANDS, INDONESIA, New Guinea and NEW ZEALAND. Altogether, the "Ring of Fire" measures some 30,000 miles in length and encloses an area of approximately 70 million square miles.

Because the "Ring of Fire" contains some of the largest cities in the world, notably LOS ANGELES and TOKYO, it has been the site of numerous fatalities in earthquakes and volcanic eruptions over the centuries, and the potential for loss of life in future earthquakes and eruptions is great, despite all precautions that may be taken. Adding to the potential for damage from earthquakes and eruptions around the "Ring of Fire" is the danger of tsunamis, which can carry destruction all the way across the Pacific Ocean and even farther. Although a full listing of major earthquakes and volcanic eruptions along the Ring of Fire for the past several centuries is beyond the scope of this article, the following selected listings provide some idea of the destructive potential of seismic and volcanic activity along it:

- *China, 1556.* Although little historical information is available about this earthquake, it is thought to have affected three provinces and killed more than 800,000 people.
- *Japan, 1596.* An earthquake offshore reportedly generated a tsunami, or seismic sea wave, that destroyed the island of Uryu-Jima completely and caused more than 4,000 deaths.
- *Philippine Islands, 1616.* An eruption of the volcano MAYON buried numerous villages near the mountain under ash, although casualty figures are unavailable. A later eruption in 1766 is said to have killed some 2,000 people and produced large flows of hot mud, possibly from lakes that spilled out of the caldera of the volcano. An additional 2,000 people were reported killed in an 1814 eruption of Mayon. The volcano has erupted frequently since then, but casualty figures have been comparatively low.
- *Japan, 1737.* A tsunami said to have been more than 200 feet high struck the northern shores of Japan and also Russia's Kamchatka Peninsula.
- *Chile, 1757.* A major earthquake occurred at Concepción, allegedly accompanied by a tsunami. Some 5,000 people perished, and 10,000 were injured.
- *Mexico, 1759.* The emergence of JORULLO volcano is thought to have killed some 200 people. Jorullo's activity continued for approximately 40 years.
- *Java, Indonesia, 1772.* Powerful earthquakes at Papandayang generated a huge depression,

approximately six miles wide and 15 miles long. An entire town was destroyed, and some 2,000 people were killed.

- *Japan, 1793.* The volcano Unsen exploded, killed some 50,000 people and deposited pumice on the sea in layers thick enough to support a person.
- *Venezuela, 1812.* An earthquake at Caracas killed about 10,000 people there and thousands more in nearby communities.
- *Indonesia, 1815.* The eruption of TAMBORA is estimated to have cast out more than 30 cubic miles of solid material in a single week.
- *Java, Indonesia, 1822.* The volcano GALUNG GUNG erupted, killing some 4,000 people.
- *Chile, 1822.* An earthquake at Valparaiso killed some 10,000 people and uplifted the shoreline several feet, exposing shipwrecks on the ocean floor.
- *Chile, 1835.* One of the most calamitous earthquakes in history destroyed Concepción and Santiago on February 20 and was accompanied by a tsunami that caused widespread destruction along the shore.
- *New Zealand, 1855.* A very strong earthquake in southeastern New Zealand uplifted the Rimutaka mountain range by several feet.
- *Japan, 1857.* An earthquake and subsequent fire destroyed Tokyo and killed an estimated 100,000 people.
- *Ecuador, 1877.* The most violent recorded eruption of the volcano COTOPAXI killed some 1,000 people and destroyed much of the mountain's summit.
- *Indonesia, 1883.* The eruption of the volcanic island KRAKATOA in the Sunda Strait generated a huge tsunami that killed thousands of people along shorelines near the volcano. The tsunami circled the globe several times before diminishing completely. Ash ejected into the upper atmosphere from this eruption made sunsets redder in the following year.
- *Alaska, 1891.* A series of extremely powerful earthquakes near Yakutat Bay raised a nearby mountain range almost 50 feet and affected an area of some 200,000 square miles.
- *Guatemala, 1902.* An earthquake and subsequent fire destroyed Guatemala City and killed more than 12,000 people.
- *Formosa (Taiwan), 1906.* An earthquake reportedly destroyed more than 6,000 buildings and took some 1,300 lives.
- *California, United States, 1906.* The SAN FRANCISCO earthquake of this year is perhaps the most famous seismic event in United States history.
- *Chile, 1906.* An earthquake struck Valparaiso and killed approximately 1,500 people.
- *Java, Indonesia, 1919.* Keluit volcano erupted and generated a vast flow of hot water and mud that is thought to have killed more than 5,000 people.
- *China, 1920.* Kansu Province experienced a powerful earthquake that reportedly killed about 200,000 people.
- *Japan, 1923.* The "great Kanto earthquake" destroyed much of Tokyo, killing approximately 143,000 people and leaving half a million more homeless.
- *Japan, 1927.* An earthquake almost as powerful as the Kanto earthquake struck the Tango Peninsula on Japan's western shore, killing some 3,000 people and destroying approximately 14,000 buildings.
- *Netherlands East Indies (now Indonesia), 1928.* The volcano Rokotinda erupted and killed more than 200 persons with landslides and falling rocks.
- *Java, Indonesia, 1931.* MERAPI volcano's eruption lasted three weeks and took more than 1,000 lives.
- *California, United States, 1933.* The LONG BEACH EARTHQUAKE killed more than 100 people and caused extensive damage.
- *Chile, 1939.* An earthquake destroyed Concepción, killing some 50,000 people and leaving perhaps 750,000 more without shelter.
- *Japan, 1946.* An undersea earthquake generated large tsunamis in the Inland Sea that obliterated some 50 communities along the shore. About 2,000 people are thought to have died in the tsunami, and possibly 500,000 were left homeless after the waves had passed.
- *Ecuador, 1949.* An earthquake in central Ecuador killed some 6,000 people and injured perhaps 20,000 others.
- *Japan, 1952.* The eruption of a submarine volcano at Myozin-Syo in the Bonin Islands destroyed a ship, leaving no survivors, when the vessel passed over a vent during the eruption.
- *Hawaii, United States, 1960.* A tsunami that accompanied an earthquake in Chile caused extensive destruction in Hawaii. Some 60 people died in Hilo City. More than 400 other people were killed when the wave reached Japan and the Philippine Islands.
- *Alaska, United States, 1964.* The GOOD FRIDAY EARTHQUAKE of 1964, with its accompanying

tsunami, killed more than 100 people and carried destruction down the Pacific coast to Crescent City, California.

- *Chile, 1968.* Almost 500 people died in an earthquake that struck central Chile. The earthquake itself was responsible for only a few of these deaths. Most resulted from the failure of dams that released their impounded water and buried communities in mud.
- *California, United States, 1971.* The SAN FERNANDO EARTHQUAKE was small by some standards (fewer than 100 persons killed) but caused heavy damage to property and became one of the most intensively studied earthquakes in history.
- *Mexico, 1985.* The devastating MEXICO CITY EARTHQUAKE demonstrated that city's vulnerability to seismic events. Much of the city is built on unconsolidated sediment, which reacted to passing seismic waves in such a way that numerous buildings collapsed. The earthquake originated not in the Mexico City area but rather along the Pacific coast of Mexico.
- *California, United States, 1989.* The Loma Prieta earthquake that hit SAN FRANCISCO damaged the Bay Bridge between San Francisco and Oakland and killed motorists whose vehicles were crushed under a collapsing segment of highway in the East Bay region.

Rio Grande Rift, New Mexico, United States The Rio Grande Rift borders the Colorado Plateau and runs through the San Luis Valley between the Sangre de Cristo Mountains and the San Juan Mountains. The northern part of the rift runs near Taos and Albuquerque, New Mexico. The lower portion of the rift is characterized by flat territory, such as the Plains of Saint Augustin. Signs of volcanic activity are abundant along the Rio Grande Rift. An example is the Valles Caldera, in the Jemez Mountains near Los Alamos. This is one of the largest CALDERAS known and is believed to have formed through the eruption and eventual collapse of a volcano whose output of tephra filled in large portions of the rift in the vicinity of Los Alamos. Cinder cones in the Valles Caldera area indicate that volcanism was active here only a few thousand years ago. Numerous lava flows are visible along the rift, notably the Malpais lava beds near Carriozo. This lava flow may be seen at Valley of Fires State Park. Molten rock may still underlie the rift; there is evidence of a pool of magma between 10 and 15 miles deep beneath the rift near Socorro. The Rio Grande Rift has been investigated as a possible source of geothermal energy.

Roccamonfina, volcano, Italy The stratovolcano Roccamonfina is located some 30 miles north of Vesuvius on the western shore of the Italian peninsula. The volcano is about 11 miles in diameter and has two CALDERAS. The main caldera is approximately four miles in diameter and occupies the summit of the mountain, while a second, smaller caldera is located on the northern side of the volcano. Roccamonfina has many parasitic cones. The volcano has generally been quiet within historical times but has had one recorded period of unrest, around 270 B.C., when a flame reportedly rose up and burned for several days. What exactly happened in this case is uncertain. It may or may not have been an actual eruption.

Rocky Mountain Trench, Alberta and British Columbia, Canada The Rocky Mountain Trench is a RIFT VALLEY running roughly northwest to southeast, parallel to the Canadian Rockies, through the provinces of Alberta and British Columbia. The rift valley stops in northwestern Montana, just north of the Idaho Batholith.

Rotorua, caldera, North Island, New Zealand The Rotorua caldera is located in the Taupo volcanic zone and is noted for its variable HYDROTHERMAL ACTIVITY. In 1886 there apparently was a failure for several weeks in a hot-water supply to baths at Rotorua, although this change may or may not have been connected with activity at the caldera. In 1932 a strong hydrothermal explosion took place in Lake Rotorua.

Ruapehu, volcano, North Island, New Zealand Adjacent to the volcanoes NGAURUHOE and Tongariro, Ruapehu has a crater lake at its summit whose eruptions cause occasional mudflows. A mudflow in 1953 destroyed a railroad bridge and brought about the deaths of more than 150 people in a train wreck.

rupture Slippage along a FAULT PLANE. Surface rupture does not necessarily coincide with the EPICENTER of an earthquake. (See also SEISMOLOGY.)

S

St. Helens, Mount, volcano, Washington, United States The 1980 eruption of Mount St. Helens caused $1.5 billion or more in total damage, destroyed more than $100 million in crops and demolished some 150 square miles of timber. More than 100 people were reported dead or missing following the eruption. The first indications of an impending eruption that year began on March 20 when a seismograph at the University of Washington in Seattle detected an earthquake approximately 20 miles north of Mount St. Helens. Earthquake activity near the volcano increased over the following several days. On March 27, the volcano emitted a loud explosion along with a plume of ash and vapor that rose to more than 20,000 feet in altitude. This eruption left a crater some 250 feet wide on the summit. Another cloud of vapor and ash two days later reached Bend, Oregon, some 150 miles south of Mount St. Helens.

By April 7, tremors indicated that magma was flowing beneath the volcano. An eruption on April 8 lasted more than five hours. Soon afterward, the north flank of the mountain was noticed bulging outward, in much the same manner as a blister on an overinflated tire. The blister indicated that magma and gas released from it were building up pressure within the peak. Governor Dixy Lee Ray declared a state of emergency on April 9, and the National Guard took steps to prevent onlookers from approaching the volcano. The bulge remained intact for more than another month, despite a violent eruption of May 7. Eventually the bulge grew to be about 2,000 feet long and 500 feet high. The bulge disintegrated at about 8:30 A.M. on May 18, 1980, when an air-blast explosion from Mount St. Helens created a blast wave that moved at an estimated velocity of 200 miles per hour and knocked down trees some 20 miles away. A plume of ash rose from the mountain to an altitude of about 12 miles. The mountain's north flank, which had been destroyed in the explosion, became an avalanche that flowed into the southern fork of the nearby Toutle River and Spirit Lake, displacing water from the lake and sending it into the northern fork of the Toutle River. The resulting flood of debris-laden water carried away virtually everything in its path, including a railroad bridge. Much of the fine sediment from the avalanche and ash cloud made its way into the Columbia River, where it interfered with navigation.

One of the personal stories associated with the eruption concerned one Harry Truman, proprietor of a lodge at Spirit Lake near the volcano. Truman refused to leave his lodge as earthquakes and eruptions became more frequent. He claimed the mountain "didn't dare" harm him. Truman and his lodge were buried under volcanic debris during the eruption of May 18.

The 1980 eruptions of Mount St. Helens made the mountain one of the most intensively studied volcanoes. Mount St. Helens has added to the knowledge of many phenomena of volcanology, notably air-blast explosions.

St. Lawrence Valley, Canada and United States The St. Lawrence river valley, which forms the boundary between a portion of New York and eastern Canada, has been the location of numerous earthquakes, some of them severe, over the past several centuries. It would be impractical in this space to provide a complete listing of earthquake activity in the St. Lawrence Valley since settlement of the area by Europeans began, but descriptions of several notable earthquakes may convey a sense of the seismic potential of this region.

An earthquake on June 11, 1638, for example, is believed to have occurred in the St. Lawrence Valley. This very strong earthquake affected Trois-Rivières (Three Rivers), Quebec and was felt so strongly in New England that people in Massachu-

Satellite photography captured images of the eruption of Mount St. Helens on May 18, 1980 (USGS)

setts had trouble standing upright and portions of chimneys were shaken down in the vicinity of Lynn, Plymouth and Salem, Massachusetts. Connecticut and Narragansett, Rhode Island were also affected by the earthquake, as were ships off the coast. Lesser shocks occurred for some three weeks after this earthquake.

Another powerful earthquake on September 16, 1732, caused substantial damage in Montreal and reportedly killed seven people there. Houses in New Hampshire were damaged, and the earthquake was perceived in Boston and as far away as Annapolis, Maryland. On October 20, 1870, a powerful earthquake was felt strongly at Montreal and Quebec and was reported over a vast area of the eastern United States as well in such widely separated states as Iowa, Michigan and Virginia. (There is some question, however, whether certain reports, from Virginia and Iowa, are accurate.) The shock was strong enough in Maine to break windows and was felt along the northeastern coast of the United States from Portland, Maine to New York. The vicinity of Lake George in eastern New York state was also affected. The earthquake lasted one minute and was accompanied by a rumbling sound at Albany, New York. Clocks reportedly stopped in Cleveland, Ohio.

The 20th century has also seen some notable earthquakes in the St. Lawrence Valley. An earthquake on February 28, 1925, was centered near Laurentides Provincial Park, between La Malbaie (Murray Bay) and Chicoutimi. The exact epicenter was difficult to determine because instrumentation in the region at the time was inadequate for that task. Although this was a powerful earthquake, measuring about VIII on the Mercalli scale in the vicinity of its epicenter, and was perceived over a great area of eastern Canada and the northeastern United States (it was felt as far away as Boston, Massachusetts and New York City), it appears to have caused little significant damage, and that was confined to a narrow zone on either side of the St. Lawrence River. More than a hundred aftershocks were experienced at Chicoutimi in the week following the earthquake. For some months, shocks continued, including strong shocks on April 10 at La

Devastated area near Mount St. Helens (*Earthquake Information Bulletin*/USGS)

Malbaie and April 25 at Chicoutimi. (There also appears to have been a foreshock on September 30, 1924, felt in Maine and Vermont and estimated at magnitude 6.1 on the Richter scale.) Reportedly, no one was killed directly by the earthquake, although several deaths from shock were attributed to its effects. Among the intriguing effects of this earthquake was the rotation of monuments. Statues and furniture at La Malbaie rotated clockwise during the earthquake.

This earthquake distributed its energy in ground motion of moderate intensity over a very wide area, with considerable variations in intensity at separate locations equally distant from the epicenter in different directions. The earthquake had interesting effects on buildings in Quebec. In some cases one structure was damaged while another standing nearby was relatively untouched. For example, sheds and elevators for handling grain underwent significant damage at one location, but a nearby hotel built on rock rode out the earthquake so easily that there was hardly any sign an earthquake was occurring. Near the epicenter, there was apparently little damage to buildings because there were few structures of any great size except for churches. A church at Saint-Urbain underwent heavy damage, and the bells were shaken out of place and the organ pipes were bent at another

Timber blown down by eruption of Mount St. Helens, 1980 (Earthquake Information Bulletin/USGS)

church at Rivière-Quelle. As a rule, buildings situated atop rock came through the earthquake well. Vertical ground motion near the epicenter must have been considerable; at Rivière-Quelle, a 200-pound rock was reportedly thrown off its foundation, and a chimney was destroyed by vertical thrust from below. Cracks formed along the southern shore in a grid pattern. One such crack near Rivière-Quelle was several inches wide and deeper than two feet. In some areas, frozen earth near the surface appears to have moved atop the unfrozen portion and cracked, producing fissures through which wet sand and water made their way to the surface.

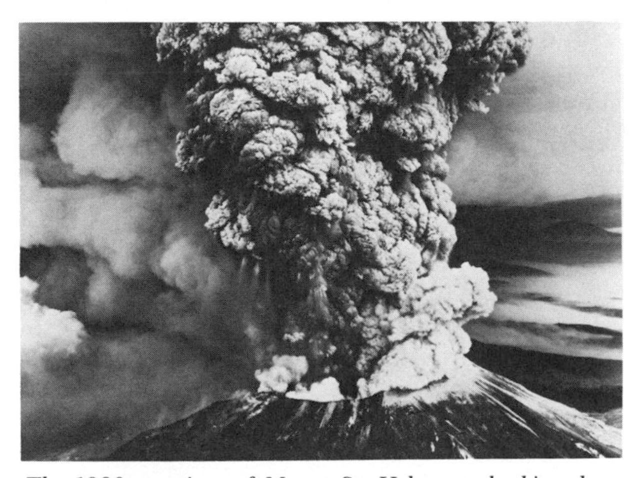

The 1980 eruptions of Mount St. Helens made this volcano in Washington one of the most intensively studied in the world (USGS)

The Grand Banks earthquake of November 18, 1929, estimated at magnitude 7.2 on the Richter scale, was felt all through New England and through parts of Canada south of the St. Lawrence River and the Strait of Belle Isle. The earthquake and subsequent undersea landslide broke submarine communications cables that traversed the area of the epicenter. A tsunami, or seismic sea wave, associated with the earthquake was responsible for loss of life and property damage at Placentia Bay, Newfoundland. In Maine, objects were knocked from shelves, and clocks stopped. Ships at sea also felt the shock strongly.

St. Vincent See SOUFRIÈRE.

Sakura-zima, volcano, Japan Located on Kagoshima Bay, Sakura-zima erupted with great violence in 1476, 1779 and 1914. Thousands of smaller eruptions have been recorded since the early eighth century A.D. An account of the 1779 event says that strong earthquakes of varying duration preceded the eruption, which began when the volcano's summit exploded and expelled a great white cloud. Hot stones fell from the volcano and reportedly set houses on fire. Some of the stones were so large that they were compared to birds in flight even when viewed from a distance of several miles. Darkness covered the area around the volcano, accompanied by thunder and lightning. Pumice reportedly covered the waters to such a depth that one could walk on it, and two men allegedly used this route to escape from the volcano. Estimates placed the number of dead at approximately 150. One group of refugees hid in a cave and perished there when an avalanche buried the entrance. Snow-white tephra accumulated up to five feet deep in places. Following the eruption, authorities ordered everyone, regardless of wealth or social station, to work on cleanup.

The 1914 eruption was preceded by earthquake activity and several violent eruptions from Kirishima, a volcano about 30 miles away. Settlements around Sakura-zima were evacuated, and not one life was reported lost as a direct consequence of the eruption. Thirty-five people were reported killed and more than 100 injured, but as a result of earthquakes, not the eruption. The 1914 eruption started on the morning of January 12. A large cloud burst from a fissure on the volcano's western flank shortly after 10 A.M. and rose about five miles. A powerful earthquake in the early evening generated a tsunami that damaged the waterfront area. A spectacular display of gas-rich lava emerged from the volcano. Lava flows from this eruption covered some 10 square miles and joined the volcanic island to the mainland, thus converting the island into a peninsula.

Sakura-zima has been the subject of intensive research into predicting volcanic eruptions. Japanese researchers report that explosions at the crater on the summit can be predicted using data on tilt and strain at the summit and on earthquake swarms occurring at shallow depths below the crater.

The dark line indicates the path of the San Andreas Fault along the San Francisco Peninsula in California (*Earthquake Information Bulletin*/USGS)

San Andreas Fault, California, United States Perhaps the most famous fault line in the world, the San Andreas Fault runs along the California coast for about 500 miles and marks the approximate boundary between the North American crustal plate and the Pacific plate. The San Andreas Fault is not a single fault but rather a pair of faults joined end-to-end. The northern half of the San Andreas Fault, which runs past San Francisco, is the less active of the two. The southern portion of the fault, which transverses the Los Angeles area, generates more earthquakes than the northern portion does, especially where a northward-moving piece of crustal rock collides with the Sierra Nevada range. This meeting of crustal block and mountain range has caused numerous small fragments of crust to chip off from the northward-moving block and make their way around the western side of the Sierra Nevada before continuing on their path northward. Los Angeles is built on one such fragment that has not quite completed its turn, so to speak, around the Sierra Nevada. This situation has given the LOS ANGELES area a complex geology. The San Andreas Fault in southern California has numerous tributaries, including the Garlock Fault, Elsinore Fault, White Wolf Fault and San Jacinto Fault. The NEWPORT-INGLEWOOD FAULT ZONE (actually a string of several shorter faults) also runs through greater Los Angeles. The 1857 FORT TEJON EARTHQUAKE occurred where the Garlock Fault meets the San Andreas Fault.

Despite its fearsome reputation, the San Andreas Fault is difficult to see in many locations because development has erased its traces on the surface. Near San Francisco, for example, the fault is clearly visible only in a few places, such as Lake San Andreas, which runs parallel to the fault just south of the city, and Mussel Rock, where the fault runs out to sea several miles away.

The San Andreas Fault was named not for a saint of the Roman Catholic Church but rather in honor of Andrew Lawson, a geologist whose study of the 1906 San Francisco earthquakes is considered one of the classics in the literature of geology. (See also PLATE TECTONICS.)

Earthquake-generated fault scarps at Point Reyes, California along the San Andreas Fault, 1906 (*Earthquake Information Bulletin*/USGS)

San Fernando earthquake, California, United States The earthquake that struck California's

A view of destruction at Olive View Hospital following the 1971 San Fernando earthquake (*Earthquake Information Bulletin*/USGS)

San Fernando Valley near LOS ANGELES on February 9, 1971, is one of the most famous and most intensively studied earthquakes in history. The earthquake registered about 6.9 on the Richter scale, killed 64 people and injured about 2,400 others. Most of the fatalities (47) occurred at the Veterans Hospital by Pacoima Canyon, much of which collapsed completely. At another hospital, Olive View, only three people were killed, one by falling debris and two others by a power failure that cut off their life-sustaining equipment. The Olive View facility was damaged so badly, however, that it was almost a total loss. The extent of damage was remarkable because the building was said to be earthquake-resistant. Because the earthquake occurred at 6:01 A.M., injuries and fatalities were few compared to what might have happened if the quake had struck several hours later when more people would be expected at the hospital.

The earthquake came within a narrow margin of destroying the Lower San Fernando Dam and releasing a torrent of water on some 80,000 residents of the valley downstream. Part of the dam collapsed into the reservoir and reduced the freeboard, or distance between the reservoir level and the top of the dam, to only a couple of feet. Had the earthquake lasted a few seconds longer, the dam might have failed. A curious effect of the earthquake was seen near the intersection of Rajah and Wallaby Streets in Sylmar. Here the ground motion was so intense that it plowed grass under, so to speak, and brought up to the surface fresh mineral soil from below. After the earthquake had done its work, only 10% of the surface soil at this location was made up of grassy soil clods. The other 90% consisted of fresh mineral soil churned up from underground. Surface ruptures occurred in a zone roughly 10 miles long, between San Fernando and Big Tujunga Canyon. Displacement involved slip of more than six feet in some places along the San Fernando Fault Zone.

San Francisco, California, United States Widely known as "Earthquake City," San Francisco sits beside the San Andreas Fault and close to other active faults, including the HAYWARD FAULT just

Tracks buckled by 1906 San Francisco earthquake (G. K. Gilbert, USGS)

across the bay. Numerous earthquakes have shaken San Francisco since its founding. Unreinforced adobe houses reportedly suffered heavy damage from a series of earthquakes that hit the city in June and July of 1808. Mission Santa Clara and Mission San Jose were damaged by a strong earthquake in 1822, and two powerful earthquakes struck San Francisco in 1836 and 1838. A big quake in 1851 damaged some buildings severely, and another earthquake the following year is said to have cracked the ground so severely that the waters of Lake Merced drained into the sea. An 1856 earthquake was felt as far away as Stockton and disturbed the waters of the bay. In 1864 an earthquake broke store windows in San Francisco, and another in 1865 destroyed numerous buildings and created a fissure two blocks long in the earth along Howard Street. An earthquake in 1868, centered near San Leandro on the east shore of the bay, damaged every building in adjacent Hayward and was felt more than 150 miles away. Mark Twain experienced this earthquake while walking down a street in San Francisco. He described the event as "terrific" and noted an interesting detail: Many people suffered motion sickness from the earthquake, and some remained incapacitated for hours and even days later.

The destruction of San Francisco in 1906 by earthquake and subsequent fire is one of the most widely known catastrophes in history, although earthquakes elsewhere have caused greater loss of life. Fire consumed almost 500 blocks (or about 2,800 acres) of the city. Roughly 30,000 buildings were destroyed, about 3,000 of those by fire. The surface rupture from this earthquake extended almost 300 miles. Thirty schools and 80 churches and convents were listed as destroyed, and approximately a quarter million men, women and children were left homeless. How many died in the earthquake and subsequent fire is still uncertain. Initial reports put the number of dead at 315, plus 352 missing and unaccounted for, but the total of casualties has risen to more than 2,000 as information continues to accumulate about victims of the earthquake. The earthquake, estimated at 8.2 on the Richter scale, struck San Francisco at 5:13 A.M. on April 17, 1906. The first tremor lasted some 40 seconds and was followed by another shock lasting a minute and a half. Falling chimneys damaged or destroyed numerous wood-frame homes. As many as 80 guests were reported killed at one particular hotel, though not all of them by the earthquake itself; many of them are said to have drowned when a ruptured water main flooded the first floor. Reports indicate that a cooking fire in the area south of Market Street burned out of control, and the fire spread quickly through the city, which was built without close attention to fire safety and therefore was soon ablaze. The army tried to slow or halt the fire's spread by using dynamite to blow up buildings and create firebreaks, but the powder they used may itself have ignited fires and contributed to the conflagration. The army also used artillery to generate firebreaks by shelling houses to the ground, but this measure too proved ineffective. San Francisco was rebuilt on much the same layout as before, but this time planners and builders incorporated into the city an extensive system of reservoirs, secondary water mains and cisterns to provide an adequate water supply for fighting future fires.

Almost equal in fame to the 1906 catastrophe was the less powerful but nonetheless highly destructive earthquake that struck the San Francisco Bay area on October 17, 1989. This earthquake was the largest in northern California since the San Francisco earthquake 83 years earlier. In terms of damage, the earthquake is thought to have been the single most expensive natural disaster in U.S. history up to that time. Estimates of damage are

San Francisco had to be rebuilt after the calamitous earthquake and fire of 1906 (*Earthquake Information Bulletin*/USGS)

as high as $10 billion. Also known as the Loma Prieta earthquake after a mountain near its epicenter, the 1989 disturbance was centered just south of San Jose, in the vicinity of Gilroy, Watsonville and Santa Cruz, and was caused by movement along the SAN ANDREAS FAULT at a depth of more than 11 miles, as the rock along the western side of the fault moved northward several feet. The earthquake occurred at 5:04 P.M., measured 7.1 on the Richter Scale, lasted some 15 seconds and is thought to have been the most powerful earthquake in California since 1952.

Although media coverage concentrated on destruction in San Francisco, where structures built on unconsolidated material suffered serious damage, destruction from the Loma Prieta earthquake was greatest within a radius of about 20 miles from the epicenter. Buildings collapsed in Hollister, Gilroy and Santa Cruz, and major damage was reported in Watsonville as well. A portion of Highway 101 collapsed over a slough near Watsonville. The earthquake reportedly shifted hundreds of frame homes in Santa Cruz County from their foundations and blocked highway access to the community of Santa Cruz, which was cut off by rockfalls and damaged bridges. Collapsing walls of aged buildings allegedly killed three people in Santa Cruz. Damage was light at the University of California at Santa Cruz but relatively heavy at nearby Stanford University, where destruction was estimated at $165 million. The ground cracked in many locations near the epicenter in the Santa Cruz Mountains, but no signs of surface rupture along the fault itself were found, although one estimate put the length of the underground rupture from this earthquake at about 30 miles. One crack near Highway 17 measured more than 600 yards long and several feet wide. Sixty-seven people were reported killed by the Loma Prieta earthquake.

Loss of human life was concentrated in western Oakland, on the east side of the bay, where a portion of Interstate 880, some 1.5 miles long and called the Cypress Viaduct, collapsed and crushed dozens of vehicles and their occupants as the upper level of the two-level highway fell on the lower level. Survivors on the lower level were few; although one man was found alive 89 hours after the collapse, he died soon afterward. Another collapse occurred on the Bay Bridge between San Francisco and Oakland, where a 50-foot-long segment of the bridge's upper deck fell onto the lower. Heavy damage was also reported to the Embarcadero Freeway in downtown San Francisco, and damage including landslides and harm to bridges forced the closing of Highways 1 and 101 south of San Francisco.

The earthquake interrupted the third game of the World Series at Candlestick Park. Fires from a ruptured gas main consumed portions of the Marina district, but the fires were brought under control within hours. The Marina district also exhibited widespread damage from liquefaction because the district was built on fill laid down early in the century. Other areas of heavy damage included portions of the Mission district and the area south of Market Street. Significant damage also occurred in the Mission, Haight, Sunset and Tenderloin neighborhoods.

Sangay, volcano, Ecuador Sangay has erupted frequently, though on a small scale, from the early 18th century to the present.

San Jacinto Fault Zone, California, United States The San Jacinto Fault Zone includes a number of faults in the greater LOS ANGELES area and is responsible for most of the strong earthquakes there. The fault zone extends for some 200 miles along the California coast and includes the San Jacinto, Glen Helen, Lytle Creek, Claremont, Casa Loma, Hot Springs and Clark faults. The potential for powerful and highly destructive earthquakes appears to be greater, however, along the nearby SAN ANDREAS FAULT because the fault segments in the San Jacinto Fault Zone are comparatively short (about 50 miles long or less) and separated from one another.

Santa Maria, volcano, Guatemala Santa Maria erupted explosively in 1902. A LAVA DOME called Santiaguito formed in 1922 in the CRATER left by the earlier explosion. The volcano has remained active.

Santorini, caldera, Greece Located on the island of THERA (now known as Thira), Santorini was the site of the catastrophic explosion that destroyed the Minoan civilization and much of the island around 1470 B.C. and is thought to have given rise to the myth of Atlantis. Over some 2,000 years following that eruption, a dome complex emerged in the center of the underwater caldera and became the Kameni islands, Micra Kameni and Palaea Kameni. Palaea Kameni may have appeared in 197 B.C. and was enlarged in A.D. 46 and 726, and the dome of Micra Kameni appeared between 1570 and 1573. Although there were reports of an eruption in 1457, it appears that a cliff on Palaea Kameni merely collapsed.

Seismic activity in 1649 required residents of Santorini to evacuate, and very powerful earthquakes occurred in 1650, followed several days later by an eruption that formed a cone. The cone was eroded by wave action and became Colombo Banks. Apparently subsidence followed this eruption because the sea reportedly moved inland in at least two locations on Thera, the main island. In May of 1707, earthquakes preceded the emergence of a shoal some 20 feet high, called Nea Kameni, off the northeast shore of Micra Kameni. One account says that mariners, curious to see what was happening, went to examine the shoal and came back to report that a new shoal was rising from the sea floor. The following day, more investigators set out for the shoal and managed to land on it, even though the shoal was growing perceptibly under them. The explorers brought back souvenirs, including some large oysters that were found clinging to a rock and were said to have an excellent flavor. In July of 1708, the water grew hotter and more turbulent at the site over the next several weeks, and the smell of sulfur intensified. During an explosive event that month, Nea Kameni continued rising, while Thera and Micra Kameni reportedly subsided about six feet. The new island rose to a height of about 30 feet and a diameter of approximately three miles by November of 1708 and attained a height of almost 250 feet by 1711. Micra Kameni subsided during this period of Nea Kameni's growth.

In the early 19th century, copper-bottomed ships took advantage of the submarine volcanism at Nea Kameni to clean their hulls. The ships simply anchored in the bay for a while and let the sulfur-laden waters clean and polish the metal sheathing on their hulls. Late in January of 1866, fissures appeared between Nea Kameni and Micra Kameni. Earthquakes occurred, the water grew hotter, and gas rose through the bay water south of Nea Kameni. The southern side of Nea Kameni subsided dramatically. An eruption began around the first of February and a cone perhaps 15 feet high reportedly formed on the sea floor. Earthquakes continued all through the eruption. The lava dome rose, while other nearby areas subsided; at one place along the shore of Nea Kameni, a pier sank almost six feet between May 1866 and March 1867. Several new domes had grown on the southern side of Nea Kameni by 1870.

The temperature rose in the waters around Nea Kameni in 1925, and local earthquakes were reported. At one point, the heat and turbulence were so great that navigation ceased in the area. An explosive eruption in August 1925 led to the formation of lava domes and lava flows. Seismic and eruptive activity continued in the late 1920s. An eruption occurred in 1939, and a dome formed in 1950.

satellite observations Earth satellites have proven to be useful tools for monitoring volcanic eruptions. Satellites cover large areas of the globe in a single field of view and can provide multispectral images that can reveal images invisible in a single band of the spectrum. A famous example of satellite monitoring of a volcanic eruption occurred in 1980 when the U.S. Geostationary Operational Environmental Satellite (GOES) photographed the ongoing eruption of Mount St. Helens. In the United States, satellite data have also been used to monitor eruptions of Augustine and Redoubt volcanoes in Alaska. Weather satellites have monitored ash clouds from other eruptions, including those of Galung Gung in 1982, Mayon in 1984 and Kelut in 1990. There are limits to what satellite imagery can reveal; it can be difficult for satellites to distinguish between the cloud from a volcanic eruption and ordinary clouds in the surrounding atmosphere. Satellites can, however, help monitor ash clouds from major eruptions that pose hazards to aircraft. Australia's Bureau of Meteorology uses images from the Japanese Geostationary Meteorological Satellite to help in providing warnings to aviators about clouds of volcanic ash over Australia and in areas to the north, where volcanic eruptions are commonplace. The Australians base their warnings also on reports from aircraft and from information supplied by the country in which the eruption occurs.

scoria A lightweight, basaltic volcanic rock similar to PUMICE, scoria may accumulate around a volcano in quantities great enough to form a cone several hundred feet high.

seamounts Seamounts are volcanoes rising from the ocean floor. Some seamounts reach above the ocean's surface and form islands, of which the Hawaiian Islands and Aleutian Islands are good examples, or atolls, where the summits are covered with coral reefs. If the seamount does not quite reach the surface and has a flattened crown indicative of erosion by wave action in the past, the mountain is called a guyot. Seamounts may occur in chains or clusters. Seamounts spaced very near to one another are known as aseismic ridges.

sediment See SEDIMENTARY ROCK; SEDIMENTATION.

sedimentary rock Rock that was deposited in the form of sediment, such as sand, silt or clay, and then transformed into rock. The material making up sedimentary rock was originally part of another rock and was deposited by wind, water or glaciers. Sedimentary rocks vary greatly in composition and texture. Examples of sedimentary rock are sandstone, shale and limestone.

sedimentation The process by which clastic material is laid down before forming sedimentary rock. Sedimentation may involve deposition by air and/or water and generally produces horizontal layers of sediment, although angled beds of sediment may occur in some situations.

Segara Anak, caldera, Lombok, Indonesia Located on the island of Lombok, this caldera includes Gunung Rinjani, a stratovolcano on its eastern rim, and Gunung Baru, a cone within the caldera. A strong earthquake in 1884 preceded an eruption of Gunung Rinjani, which is said to have expelled smoke and fire for several days. Another earthquake in 1898 was suspected of having originated at Gunung Rinjani. Two strong earthquakes occurred under Gunung Rinjani in November of 1903, and the volcano was reported smoking in November of 1905. Several weeks after an earthquake early in 1966, a climber reported seeing a flow of lava from Gunung Baru and detecting a strong smell of sulfur.

seiche A periodic oscillation in an enclosed, or nearly enclosed, body of water. Earthquakes may contribute to producing seiches.

seismic gap This expression refers to a recently "quiet" portion of an active earthquake-generating fault. This section forms a gap in a map of recent earthquake epicenters along the fault and may be a likely spot for strong earthquakes in the future because the forces that generate earthquakes may be building up along the gap and "waiting" to be released.

seismograph A device for detecting and recording the vibrations of earthquakes. See SEISMOLOGY.

Seismometers and seismographs at National Earthquake Information Service, Golden, Colorado (Earthquake Information Bulletin/USGS)

seismology The scientific study of earthquakes. Earthquakes generate several kinds of waves. These are sometimes classified in two categories: surface waves and body waves. Surface waves, or S-waves, are confined to the surface of the earth and show "sinusoidal," or S-shaped, patterns of motion. These sometimes are also known as transverse waves because they exhibit transverse motion, at a right angle to the direction of propagation. Two kinds of surface waves, named Love waves and Rayleigh waves after their discoverers, have periods of perhaps half a minute. Love waves exhibit horizontal shear, meaning a side-to-side shaking motion. Rayleigh waves involve an elliptical motion like that seen in ocean waves. Body waves are also called P-waves, the *P* standing for "pressure." These waves travel deep into the earth's mantle and core and move by compression and subsequent rarefaction of the rock media through which they move, in much the same manner as sound waves traveling through the air. P-waves have much shorter periods than S-waves and move more rapidly—more than 8 miles per second in some cases, as compared to perhaps 2 to 2.5 miles per second for S-waves.

Factors Affecting Destruction from Earthquakes The extent of destruction from an earthquake depends on several factors:

- *Magnitude.* Earthquake magnitude refers to how much energy the earthquake released. Magnitude is not always proportionate to damage; under some conditions, such as high population density in the affected area, a relatively minor earthquake may do far more damage than a more powerful quake in a sparsely populated locale.

  Two scales are used to measure earthquake magnitude: the Richter scale and the Mercalli scale. The Richter scale, named after Dr. Charles Richter of the California Institute of Technology, is a logarithmic scale (expressed in powers of 10) that reflects the intensity of ground movement. Several steps are involved in establishing an earthquake's magnitude on the Richter scale. First, the origin of the quake must be determined. This is done by noting the time a wave passed two widely separated locations and inserting the data into a set of equations to determine how far each station was from the origin of the quake. The next step is to estimate the intensity of ground motion at the earthquake's origin by using a correction factor that introduces some error into the calculation of magnitude. Measuring magnitude on the Richter scale does not yield precise results, but within certain limitations, it does provide a useful estimate of how strong an earthquake was. The Richter scale starts at zero and is open-ended. An earthquake of magnitude 1 on the Richter scale probably would go unnoticed by anyone but an observer with a seismograph. Magnitude 4 earthquakes are mild but still perceptible, and earthquakes of magnitude 8 and above are devastating if they strike densely populated areas. The San Francisco earthquake of 1906 is estimated to have been about 8.2 on the Richter scale, and two other earthquakes of the 20th century—one beneath the Pacific Ocean near Japan in 1933 and the other off the coast of Ecuador in 1933—are thought to have ranked around 8.9. In theory, earthquakes of Richter magnitude 11 or 12 are possible, but extraordinary circumstances would be required to create such a disturbance.

  The second scale for measuring earthquake magnitude is the Mercalli scale. It uses Roman numerals to avoid confusion with the Richter scale and is based on easily observable effects, such as swaying of buildings. Here is a condensed version of the Mercalli scale:

Mercalli Intensity	Observed Effects
I	Not felt at all
II	Felt by only a few individuals
III	Felt indoors by many persons, but not necessarily identified as an earthquake
IV	Felt both indoors and outdoors and comparable to vibrations produced by a passing train or heavy truck
V	Strong enough to awaken a sleeping person; small objects fall off shelves, and beverages may splash out of cups or glasses
VI	Perceptible to everyone; pictures fall off walls, and plaster may fall from ceilings.
VIII	Difficult to stand upright; ornamental masonry falls from buildings; waves may occur in ponds and swimming pools.
VIII	Mass panic may occur; chimneys and smokestacks may fall; unsecured frame houses slide off their foundations
IX	Heavy damage to masonry structures and underground pipes; large cracks open in ground; panic is general
X	Numerous buildings collapse; water splashes over riverbanks
XI–XII	Virtually complete destruction

Seismic moment is another measure of the size of an earthquake. Seismic moment is the product of the average amount of slip, the area of the rupture and the shear modulus, or strength, of the rocks affected. An expression of magnitude related to seismic moment is moment magnitude, which is approximately the same as Richter magnitude in range 3–7.

- *Duration.* Some earthquakes last only a few seconds, but others have been known to last for a minute or two. As a rule, damage increases with duration.
- *Local geology.* The character of underlying material does much to determine the extent of damage from an earthquake. The safest material on which to build, generally speaking, is solid rock, which transmits vibrations from an earthquake but maintains its structural integrity. More dangerous is unconsolidated sediment, which may exhibit liquefaction and extensive consequent damage where groundwater rises close to the surface. Fractured rock, as found in many parts of the western United States, may absorb energy from earthquake waves and thus reduce the geographical extent of damage from a major earthquake. By contrast, in areas where geological formations stretch unbroken for hundreds of miles, as in portions of the eastern United States, vibrations may travel great distances with much of their initial destructive potential intact. The 1906 San Francisco earthquake, for example, was barely felt at all outside California, but the New Madrid, Missouri earthquakes approximately a century earlier were perceived over much of the United States east of the Mississippi River. The irony here is that the same processes that predispose a given part of the earth's crust to major earthquakes may also confine damage from earthquakes through extensive fracturing, whereas comparatively earthquake-free regions may harbor much greater potential for destruction because of the relatively undisturbed, contiguous rock that underlies them.
- *Time of day.* This is an important factor in determining the number of casualties from an earthquake. Deaths and injuries from collapsing buildings in rural areas, for example, are likely to be more numerous if an earthquake occurs at night or in the early morning when the population is indoors sleeping or having breakfast. In cities, on the other hand, deaths and injuries may be more likely during the working day, when large numbers of pedestrians are exposed to harm from falling walls and ornamental masonry.
- *Architecture.* Wood-frame buildings have a reputation for standing up well to earthquakes because wood can bend and sway under earthquake vibrations, thus improving a building's chance of survival. Buildings of brick and stone construction, by contrast, are more likely

to crumble and collapse because they lack wood's resiliency, although careful design and construction may increase the survivability of a nonwooden structure. Possibly the worst building material in regard to earthquake resistance is adobe, which was responsible for numerous fatalities during earthquakes in the early years of Euro-American settlement of the western United States. (For an illustration of the dangers of adobe construction, see FORT TEJON EARTHQUAKE.)

Predicting Earthquakes Earthquake prediction has been a goal of scientists for centuries, and efforts in this direction have often been colorful, even when unsuccessful. Charles Mackay, in his 1841 book *Extraordinary Popular Delusions and the Madness of Crowds,* tells how one man named Bell apparently used a primitive mathematical model to predict the date of an earthquake in England during the 18th century.

> In the year 1761, the citizens of London were alarmed by two shocks of an earthquake, and the prophecy of a third which [supposedly] was to destroy them altogether. The first shock was felt on the eighth of February and threw down several chimneys in the neighborhood of Limehouse and Poplar; the second happened on the eighth of March and was chiefly felt in the north of London . . . It soon became the subject of general remark, that there was an interval of exactly a month between the shocks; and a . . . fellow named Bell, a soldier in the Life Guards, was so impressed with the idea that there would be a third in another month that he lost his senses altogether and ran about the streets predicting the destruction of London on the fifth of April [that is, four weeks after the previous earthquake].

Mackay tells how Bell's prediction affected the public:

> There were not wanting thousands who confidently believed the prediction and took measures to transport themselves and their families from the scene of the impending calamity. As the awful day approached, the excitement became intense, and great numbers of credulous people [migrated] to all the villages within 20 miles, awaiting the doom of London. Islington, Highgate, Hampstead, Harrow, and Blackheath were crowded with panic-stricken fugitives who paid exorbitant prices for accommodation to the housekeepers of these secure retreats. Such as could not afford to pay for lodgings at any of these places remained in London until two or three days before the time [of the prediction], then encamped in the surrounding fields, awaiting the tremendous shock which was to lay their high city all level with the dust . . .
>
> [The] fear became contagious, and hundreds who had laughed at the prediction a week before packed up their goods when they saw others doing so and hastened away. The [Thames] river was thought to be a place of great security, and all the merchant vessels in the port were filled with people who passed the night between the fourth and fifth on board, expecting every instant to see St. Paul's totter and the towers of Westminster Abbey rock . . . and fall amid a cloud of dust.

Even when the predicted earthquake failed to occur on April 5, the public's fears took days to subside completely.

> The greater part of the fugitives returned on the following day . . . but many judged it more prudent to allow a week to elapse before they trusted their dear limbs in London . . . [Bell] lost all credit in a short time and was looked upon by even the most credulous as a mere madman. He tried some other prophecies, but no one was deceived by them; and . . . a few months afterwards, he was confined in a lunatic asylum.

Such "earthquake prophets" have remained active from Bell's day to ours, but only in recent years has knowledge of earthquakes and their origins advanced to the point where accurate prediction of earthquakes, or some quakes at least, has become a possibility.

Determining the potential for future rupture along a fault can be difficult because faults differ in their behaviors. One fault may show no activity for many centuries, then suddenly move 30 feet or more, whereas another fault may exhibit virtually continuous, gradual movement and many small earthquakes. This variability in behavior means that geologists cannot make accurate predictions of a fault's future activity merely by establishing that the fault is active. Two selected faults, though both active, may move at greatly different intervals and exhibit equally great differences in rates of slip. Also, the rate of slip along an individual fault may change. The slip rate along California's SAN JACINTO FAULT ZONE, for example, is believed to have fluctuated by more than 1,200%.

Another source of uncertainty in determining slip rate and history of activity is difficulty in estimating the ages of offset deposits or other features along a fault; in the Los Angeles area, for example, such estimates are unreliable in many or most locations where faults are active. Yet another problem in determining the true slip rate for a fault is the partial nature of information on components of slip. Data may be restricted to either the vertical or horizontal component. Such limited data may yield an unreliable estimate of slip along a fault, although the horizontal or vertical component alone may prove useful if the ratio of vertical to horizontal motion along a fault is known already. (In situations involving a strike-slip fault, where motion along a fault is primarily horizontal, the horizontal component alone may provide a reasonably accurate measure of the slip rate. This is the case along the SAN ANDREAS FAULT in California. Vertical component data reportedly have provided an approximation of true slip rates along the reverse-slip faults in California's TRANSVERSE RANGES.) The best information that can be supplied, in many cases, is an average figure for slip rate over a long period of time. As a rule, however, the higher the average slip rate, the more active the fault and the more closely it bears watching as a potential source of future earthquakes. Only in a few areas—namely, boundaries between major plates of the earth's crust, as along the San Andreas Fault—do faults show very high average slip rates, perhaps half an inch or more per year. Active faults in other areas generally show less slip.

The future behavior of a fault may be inferred, to some extent at least, from evidence including the rate of slip, the size of earthquakes and intervals between them and the amount of slip in each incidence of movement. No one set of criteria exists, however, for determining how active a fault may be in the future.

Complicating such analysis further is the fact that some active faults do not extend to the surface of the earth and, on the surface, may display only faint evidence of recurring seismic activity. A highly damaging earthquake that struck Coalinga, California in 1983 involved such a "hidden" fault. Moreover, simply establishing that a fault is active does not allow geologists to make accurate predictions of its future activity. Ongoing measurements of seismic activity are useful tools for estimating the likelihood of earthquakes along a given fault in the future. Seismic data may be misleading, however, because ongoing earthquake activity, or the absence of it, along a fault does not necessarily reflect the potential of that fault for generating destructive earthquakes. An active fault may have very little potential to cause destructive earthquakes because it releases its energy through CREEP, without actually generating earthquakes. Other active faults may be too shallow to reach rocks capable of holding great amounts of elastic strain and thus are unlikely to produce strong earthquakes, even though movement along such faults may rupture the surface of the ground. Also, data on seismic activity along a particular fault may span too short a time to provide a useful means of predicting the behavior of earthquake-generating faults over a long period. Even when historical records are more extensive, as in parts of Asia, great uncertainties remain in reconstructing the seismic history of a given locality or region.

Despite these limitations, certain methods exist for estimating the size and frequency of possible future earthquakes along a fault. These methods involve, but are not limited to, analyzing the earthquake history of a region to find the biggest seismic event linked to a given fault and comparing a given fault's history of quake activity with that of other faults similar in structure and tectonic characteristics to it. Another approach is to use empirically-established relations between earthquake magnitude and length of faults. It has been known for decades that the greater the dimensions of the fault surface involved in an earthquake, the greater magnitude the earthquake will have. One widely used method for estimating the most powerful earthquake that is likely to occur on a given fault rests on the assumption that half the total length of the fault may rupture in a particular earthquake. This approach is not fully reliable, however, because experience has shown that a major earthquake may involve rupture of anywhere from only a small percentage to almost the whole extent of a fault surface that existed prior to the earthquake. Another drawback to this method is difficulty in measuring the length of a given fault accurately. Much of a particular fault may lie hidden under sediment and water. In greater Los Angeles, the 1857 Fort Tejon Earthquake is thought to represent about the most powerful earthquake that might affect that area, although an earthquake of still greater magnitude is possible in theory. It has been suggested that the whole San Andreas Fault might rupture in southern California, producing a single gigantic earthquake; but there are questions about whether or not the geology of the region would allow such a single catastrophic event.

Semeru, volcano, Java, Indonesia One of the world's most active volcanoes, Semeru has produced lava flows and nuées ardentes. Semeru is a stratovolcano.

Semisopochnoi, volcanic island, Alaska, United States Semisopochnoi Island is located in the Aleutian Island arc and consists largely of a single volcano, Pochnoi. The central portion collapsed and formed a caldera following the eruption

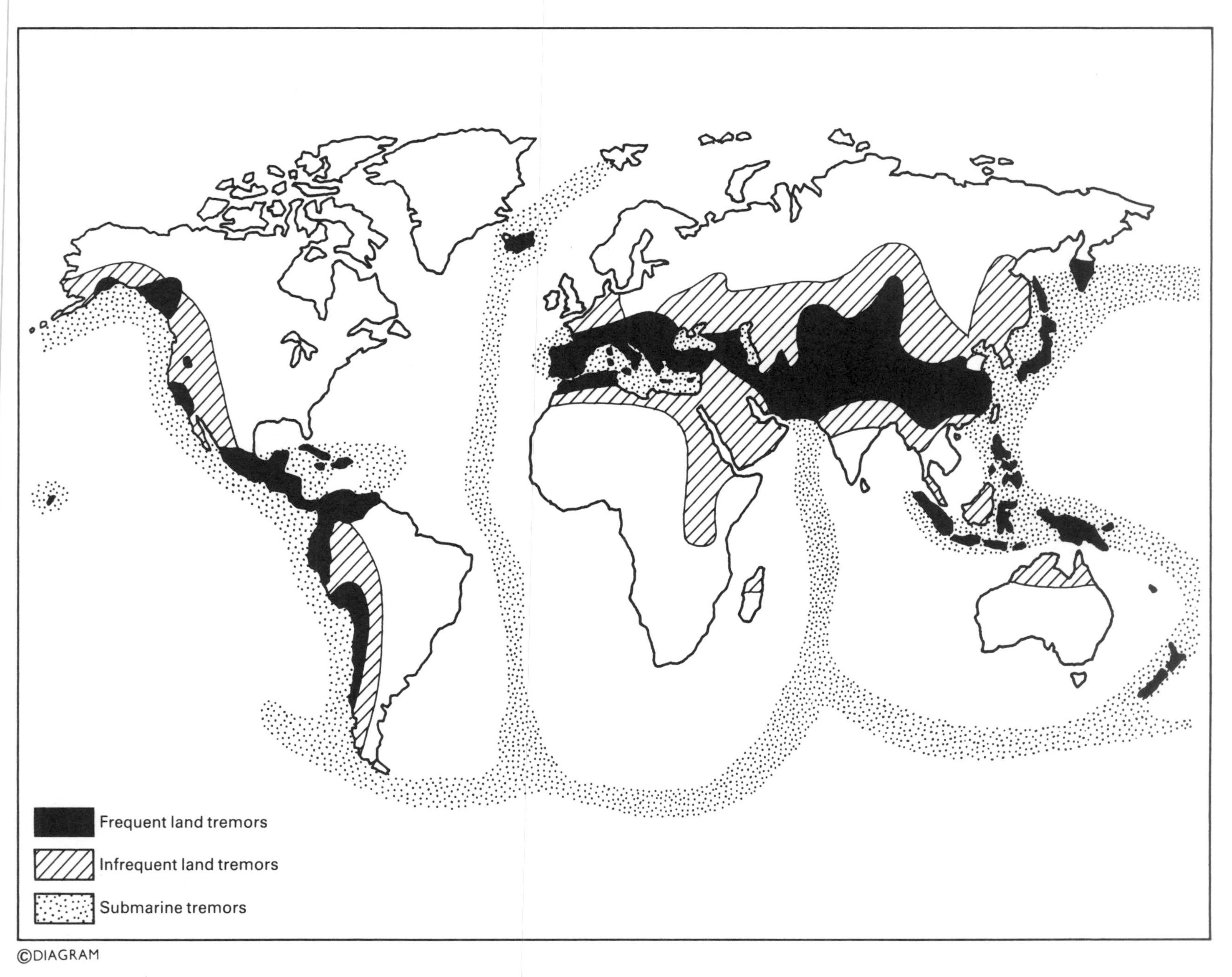

A map of the world showing the frequency and distribution of earthquakes (© Diagram)

of a great volume of pumice, although the date of the eruption and collapse is uncertain. After the caldera formed, several new cones formed. Accounts of eruptive activity from the 18th and early 19th centuries are spotty, but there is reason to think the volcano was active in or around 1772, 1790, 1792 and 1830. Some kind of eruption appears to have occurred in 1873, but details are unavailable. A plume appears to have arisen from the vicinity of the island in 1987. Several days after this event, a pilot reported seeing snow darkened by ash.

Sete Cidades, caldera, Azores The Sete Cidades CALDERA on San Miguel Island has undergone eruptions on several occasions within historical times. An eruption occurred around the year 1444 from a vent on the southwestern floor of the caldera. In 1638 very strong earthquakes in May and early June were followed in July by an underwater eruption from a vent to the west of the caldera. A submarine eruption in 1682 followed strong earthquake activity, and an earthquake swarm in 1713 may have accompanied an underwater eruption west of Sete Cidades. Strong earthquakes in 1810 and early 1811, including one that caused widespread destruction on June 24, 1810, were followed in 1811 by a submarine eruption several hundred yards southeast of the site of the underwater eruption of 1638. This 1811 eruption produced an islet about a mile and a half off the western shore of the island. The eruption subsided in early February but resumed in June, when powerful earthquakes shook the island and a strong odor of sulfur was detected. An underwater eruption began on June 14, 1811, near the site of the eruption in February. Further submarine eruptions were recorded in 1861 and 1880.

shatter cones These are conical structures associated with IMPACT STRUCTURES such as the craters blasted out by the fall of giant meteorites. Shatter cones are thought to be generated by the passage of powerful shock waves through rock as a result of these impacts and are used to identify structures created by those impacts.

shear Shear is a measure of the "shaking power" of an earthquake. It is defined as motion at right angles to the direction in which an S-wave, or surface wave, from an earthquake is advancing. Shear is responsible for much of the destruction that accompanies major earthquakes. Shear waves are also known as transverse waves. (See also SEISMOLOGY.)

shield volcano A dome-shaped volcano with a gradual slope, formed by eruptions of basaltic lava. OLYMPUS MONS on Mars and the HAWAIIAN ISLANDS are examples of shield volcanoes.

Shikotsu, caldera, Japan The Shikotsu caldera appears to have been the site of a great flow of ash and fall of tephra in prehistoric times. Lake Shikotsu occupies much of the interior of the caldera. The volcano Tarumai is located on the caldera's southeastern edge. In historical times, the caldera has been active on numerous occasions. Major eruptions occurred in 1667 and 1739 at Tarumai and left a caldera approximately one mile wide on the volcano's summit. A vigorous eruption of Tarumai in 1874 preceded by about three weeks a powerful tectonic earthquake along the northwestern coast of Hokkaido, the northernmost major island in the Japanese archipelago. Small emissions of ash occurred in 1894. Tarumai underwent a strong eruption in 1909, approximately a year after a major tectonic earthquake off the northern coast of Hokkaido. This 1909 eruption began in late April with loud detonations, shaking of the ground and a tremendous column of dark smoke. Eruptive activity occurred again in May, apparently with great emissions of smoke and vigorous seismic activity. Minor eruptions occurred every few months from 1917 to 1923, and eruptive activity continued on an intermittent basis through 1936. A powerful explosion occurred in 1920. Another eruption, on December 1, 1933, preceded a powerful earthquake on December 5 and a renewal of volcanic activity on December 11. Explosive eruptions, and possible PHREATIC ERUPTIONS as well, occurred between 1944 and 1955. Earthquake swarms started increasing in 1967, and minor swarms of earthquakes were reported again in 1975, 1978, 1979 and 1981. Eruptions resumed in 1978, and small eruptions continued through 1981. A minor eruption in February of 1981 marked the end of earthquake swarm activity that had been occurring since 1967, although some earthquakes took place after this eruption. Inflation appears to have occurred at the caldera during the 1980s.

Ship Rock, New Mexico, United States One of America's most spectacular and famous landforms, Ship Rock is a VOLCANIC NECK with DIKES radiating from it.

Shishaldin, volcano, Alaska, United States A stratovolcano in the eastern Aleutian Islands, Shi-

Mauna Loa, Hawaii. Note the gently sloping profile of a shield volcano (Hawaiian Volcano Observatory/USGS)

shaldin has erupted intermittently since the late 18th century.

Sierra Negra, volcano, Galapagos Islands, Ecuador A large shield volcano, Sierra Negra occupies Isabela (Albemarle) Island and has a shallow caldera on its summit. An explosive eruption was reported in 1813, and other eruptive activity of unknown character was recorded in 1817. A lava flow occurred during an eruption in 1844, and an explosive eruption was reported in 1860. Some kind of unrest appears to have occurred at Sierra Negra in 1911 or 1912, but details are unavailable. An explosive eruption and associated flow or flows of lava were reported in 1948 and again in 1953–54. Unrest of unknown character was reported in 1957. Explosive eruptions with lava flows occurred in 1963 and again in 1979–80. An eruption in 1979 injected large amounts of sulfur dioxide, detectable on a global scale, into the atmosphere.

Sierra Nevada, California, United States Approximately 350 miles long and 60 miles wide, the Sierra Nevada range runs along the western edge of North America's Great Basin and is bordered on the north by the CASCADE MOUNTAINS. The Sierra Nevada comprises a huge block of granitic rock with faults on the eastern side and tilted westward. In some places, magma has made its way to the surface and formed volcanoes atop the granite block. The Sierra Nevada range is noted for seismic activity, especially along its southern border, where a northward-moving block of crustal rock encounters the deep roots of the mountain range and, being unable to overcome this obstacle, generates numerous earthquakes in southern California as the block of crust grinds against, and sends smaller pieces of crust moving around the southwestern edge of, the Sierra Nevada. The TRANSVERSE RANGES have arisen along the zone of collision. Veins of gold were deposited in the rocks along the Sierra Nevada and became known as the mother lode. Fragments of gold eroded from the mother lode were laid down with sediments in placer deposits that, when discovered during the early settlement of California by Americans of European descent, led to the gold rush of the mid-19th century. Glaciation has done much to form the topography of the Sierra Nevada. The Sierra Nevada range exhibits a curious inverted topography in places, where lava flows that once filled canyons now form ridges because erosion has worn away the granitic rocks on either side of the lava flows, leaving the former valley (topped with lava) in separate streams. The basins parallel to the Sierra Nevada on the western side of the range contain tremendously deep deposits of sediment eroded from the mountains. These deposits are five miles deep in places. They are thought to have formed as the basins sank relative to the mountains, giving sediment an opportunity to accumulate to great depths.

silica Silicon dioxide, or SiO_2. Silica is an important component of many minerals. Quartz crystals are familiar examples of silica.

silicates Silicate minerals incorporate the silicon tetroxide molecule, which consists of a silicon atom and four oxygen atoms arranged in a tetrahedral structure. Other atoms such as potassium, iron and magnesium may occupy spaces within tetrahedra. Silicate tetrahedra may be arranged in various patterns. In minerals where tetrahedra occur in isolated form, tetrahedra are packed together closely; garnet and olivine are examples of minerals with this structure. An example of a mineral having rings of six tetrahedra is beryl, commonly found in the form of six-sided crystals. Tetrahedra arranged in single chains (pyroxenes, for example) and double chains (amphiboles) are moderately hard and tend to separate in a regular fashion, parallel to the chains. In sheet form, tetrahedra are linked together to form a planar structure; mica is a familiar example of a mineral with this structure. In frameworks, where tetrahedra join to form a three-dimensional lattice comparable to the framework of a building, one finds feldspars, the most common silicate minerals, and quartz. (See also CRYSTALS.)

silicic Rich in silica.

sill Similar to a DIKE, but running in a horizontal or nearly horizontal plane, a sill is a tabular body of intrusive igneous rock.

Skaptar Jökull, volcano, Iceland The eruption of Skaptar Jökull in 1783 was one of the most powerful eruptions in history. The eruption started on June 11 and followed a long series of earthquakes, which became very violent just before the eruption. Lava issued from the mountain in numerous locations (although volcanic gases had been vented starting several days earlier), and the separate streams of lava united to form a stream of molten rock that blocked the gorge of a nearby river, which was some 200 feet wide and 400 to 600 feet deep. The lava not only filled this gorge but also overflowed it on either side. The lava also filled the bed of a lake along the river. Continuing on its way, the lava flow encountered an area of ancient volcanic rocks honeycombed with caverns. The lava flowed into these caverns and sent fragments of old lava flying some 150 feet into the air.

On June 18, another flow of lava advanced rapidly over the hardened surface of the first flow. On August 3, more lava flowed from the mountain. The fresh lava moved southeastward and spread out over the plain. Eruptions of lava continued intermittently for two years and, by one estimate, amounted to a mass as great as that of Mont Blanc. Individual currents averaged some 100 feet deep, but in some places depths of 600 feet were attained. Vapors were still rising from these lavas as late as 1794. Partly because of melting snow and ice and partly because of blocked rivers, great floods of water occurred and caused extensive damage. Lava overwhelmed 20 villages. Ashes covered the entire island and the waters offshore. Winds carried ashes from the eruption over the European continent on several occasions, creating effects similar to those following the eruption of KRAKATOA. Of the 50,000 people inhabiting Iceland at this time, more than 9,000 are thought to have been killed, along with more than 11,000 head of cattle, some 28,000 horses and more than 190,000 sheep. This loss of life resulted from streams of lava, from vapors, from floods of water, from destruction of plants by ashfalls, and from a shortage of fish, which disappeared from the coastal waters during the eruption. Previously, fish had supplied much of the food of the people.

In 1784, Benjamin Franklin noted that the year 1783 had been unusually cool and that sunlight reaching the ground seemed diffused. He observed that when the rays of the sun "were collected in the focus of a burning glass, they would scarcely kindle brown paper." He also reported that a strange "dry fog" hung over the land in the summer of 1783 and that this fog appeared to cut off some incoming sunlight. Franklin suggested that the "vast quantity of smoke" from Skaptar Jökull's eruption in 1783 had created the cloud, blocked part of the incoming solar radiation and thus given the northern latitudes a colder than usual year. The eruption of the volcano ASAMA-YAMA in Japan that same year also may have contributed to the effects observed by Franklin.

Snake River Plain, Idaho, United States The Snake River Plain in southern Idaho is part of the Columbia-Snake River Plateau, an expanse of lava flows overlying great rhyolitic ash deposits extending through much of the states of Idaho, Washington and Oregon, with portions in Utah, Nevada, California and Wyoming. The Snake River Plain stands about 3,000 feet high at the west and rises to about one mile in elevation in the east. The plain exhibits many volcanic cones, particularly at CRATERS OF THE MOON MONUMENT, about 50 miles north of American Falls. Two impressive volcanic cones, Menan Buttes, may be seen near Idaho Falls. The plain is not noted for frequent or strong earthquakes but borders on areas of considerable seismic activity. Earthquakes around the edges of the Snake River Plain indicate the plain is descending with respect to the areas around it. The rate of descent is believed to be a few inches per century on the average.

One scenario for the formation of the Snake River Plain says that the plain and the Columbia Plateau both had their origins in a giant meteorite impact that produced first the flood basalts of the Columbia Plateau, then—as the North American crustal plate moved westward across the point of impact—the volcanic structures of the Snake River Plain. The rhyolitic ash deposits underlying the basalts of the Snake River Plain are believed to have been formed when granitic continental crust moving over the "hot spot" melted and generated vapor-rich magma that emerged in explosive eruptions, creating a series of calderas that became a series of basins stretching through the plain. The overlying basalts presumably did not form during these explosive eruptions but instead were deposited later from fissures. The postulated "HOT SPOT" created at the impact site now lies beneath YELLOWSTONE NATIONAL PARK. It has been suggested that this "hot spot" is also responsible for the rifting that gave rise to the northern BASIN AND RANGE PROVINCE of the western United States. Dramatic evidence of rifting activity may be seen at Great Rift National Landmark near American Falls. (See also COLUMBIA PLATEAU; PLATE TECTONICS.)

Socorro, caldera, New Mexico, United States The Socorro caldera is associated with the RIO GRANDE RIFT and has been the site of considerable unrest, although no actual eruptions, since the mid-1800s. Earthquake swarms were reported at the caldera in 1849–50, 1904 and 1906–07 and from the early 1960s on. Earthquakes at Socorro occur within several miles of the surface. Uplift also has been noticed at Socorro. The uplift and seismic activity are thought to be linked to a large body of magma that is located at a depth of perhaps 10 miles and causing uplift in an oscillating or episodic pattern.

solfatara A volcanic fissure or vent that emits only vapor, notably sulfurous gases.

Soufrière 1. A volcano on the island of St. Vincent in the Caribbean Sea, south of Jamaica. The volcano's name means "sulfur" or "brimstone." The volcano erupted in 1718 and again in 1812, but its most famous eruption took place in 1902. Earthquakes began to shake the island in April 1902, and in May the peak began to emit puffs of steam. The crater lake started bubbling on May 5, and large amounts of steam rose from it. On the afternoon of May 7, the volcano released a nuée ardente, or "fiery cloud," a mixture of ash and superheated gases. The cloud was heavier than air and spilled downslope, killing some 1,350 people on the island. Some escapes from death were remarkable: About a hundred people, for example, are said to have avoided destruction by hiding in a rum cellar.

The eruption laid waste about a third of the island and removed more than 700 feet from the summit of the volcano. The earlier eruption in 1718 is said to have released lava, and the violent eruption in 1812 reportedly lasted three days and caused extensive loss of life. A new crater about half a mile wide and 500 feet deep formed in the 1812 eruption, northeast of the original crater (which was about the same size). Barbados, some 95 miles distant, received an ashfall of several inches.

Before the 1812 eruption, a conical hill stood in the middle of the crater, beside which were two lakes, one sulfurous and the other taste-free. Vegetation made the site beautiful, and white smoke, with an occasional touch of light blue flame, emanated from fissures on the cone. The eruption of April 27, 1812, began with a strong earthquake accompanied by what one historian describes as a "tremulous" noise, as well as by a column of dense black smoke. The first signs of the eruption were almost comical. A boy herding cattle on the mountainside noticed a stone fall near him, then another. He supposed that some other boys were tossing rocks at him from the cliff above, and he began throwing stones upward in reply. He soon saw that an eruption was starting, however, and he ran for his life. The eruption continued for three days and nights. Then, on the 30th, lava flowed from the crater's rim downward to the sea before the eruption stopped. This eruption appeared to be the culmination of almost two years of earthquake activity, especially in the vicinity of CARACAS, Venezula. On March 26, 1812, a great earthquake struck Caracas as large numbers of residents were gathered in churches. The earthquake killed some 10,000 people in the collapse of churches and homes and was felt up to 180 miles away. Just over one month later, the 1812 eruption of Soufrière occurred. On April 30, just as the eruption was ending, a loud underground noise resembling cannon fire was heard at Caracas and other locations, although no shock was reported. The noise was said to be as loud along the shore as it was hundreds of miles inland, and at Caracas preparations were made to defend the city against what sounded like invading troops with heavy guns.

In the 1902 eruption, Soufrière erupted simultaneously with Mount PELÉE. Soufrière's eruption occurred at the north end of the island of St. Vincent and reportedly wiped out most of the native Carib population of the island. In April, unrest was noted at Soufrière, and on May 5, the crater lake started bubbling and releasing great clouds of steam. Earthquakes followed, some of them very strong. At noon on Wednesday, May 7, 1902, Soufrière suddenly sent six separate streams of lava down its sides. On the night of May 7, Soufrière's eruption was visible from St. Lucia. The eruption lasted all night, and the day and night afterward. On Thursday morning, a tremendous black column rose to an estimated height of eight miles. Ashes and rock fell from the column for miles around. On Thursday night. a steamship on the way to Kingstown encountered a floating mass of ashes and was beset for three hours by a cloud of sulfurous gas. When the ship reached Kingstown at dawn, the streets were covered to a depth of two inches with ash and rock. Lava flowed down the side of Soufrière.

The eruption abated slightly on Friday. Showers of rocks ceased falling, although lava continued to flow from the volcano. Corpses of humans and livestock added to the pollution generated by the eruption. Hundreds of bodies remained unburied. In one ravine, 87 bodies were found together. Nearby lay the carcasses of hundreds of cattle. The bodies were destroyed by quicklime, and the cattle were burned.

Soufrière **2.** A volcano on Gaudeloupe Island in the Caribbean, with a history of explosive eruptions dating back to the 15th century. A minor eruption in 1976 led to the evacuation of thousands of people.

South Carolina, United States South Carolina's reputation as a high-intensity earthquake area is due largely to the great CHARLESTON earthquake of 1886. This is not, however, the only strong earthquake in South Carolina's history. A powerful earthquake accompanied by a loud roar occurred in Pickens County on October 20, 1924, for example. Coastal areas of South Carolina are highly susceptible to liquefaction and resultant damage in the event of future strong earthquakes.

Spatter cone in yard of home at Kilauea, Hawaii, March 14, 1955 (G. A. Macdonald, USGS)

South Dakota, United States South Dakota does not have an outstanding history of strong earthquakes, but some such events have occurred since the state was settled by Americans of European descent. The James River Valley earthquake of June 2, 1911, for example, was felt in Iowa and Nebraska as well as in South Dakota.

spatter cones Volcanic structures resembling giant anthills, standing several feet high and formed by eruptions of highly fluid lava.

stock A large, approximately circular igneous INTRUSION extending horizontally.

stratovolcano A steep-sided VOLCANIC CONE formed by eruptions of LAVA and PYROCLASTIC material. (See also VOLCANO.)

Stromboli, volcano, Tyrrhenian Sea, near Sicily Known as the "Lighthouse of the Mediterranean," the volcanic island of Stromboli rises slightly more than 3,000 feet above sea level and has erupted on numerous occasions over many centuries. Stromboli's eruptions have varied in output and intensity from mild activity of fumaroles to violent outbursts such as those of 1907, in which explosions shattered windows in villages on the island and incandescent material was cast out of the crater, and 1930, when two strong explosions on September 11 followed an earthquake that produced a minor tsunami. Four people were killed in an avalanche during the 1930 eruption.

Strombolian eruption See ERUPTION.

subduction zone An area where a plate of crustal rock is being overridden by another plate and is moving downward into the asthenosphere, where the descending plate melts and undergoes chemical fractionation, releasing its lighter chemical fractions. Those fractions rise back toward the surface and provide magma for volcanoes. Subduction zones are characterized by deep ocean trenches with curved or "arcuate" shapes as well as by volcanic activity and earthquakes.

submarine eruptions Eruptions that occur beneath the sea may be modest in scale, but in some cases they are much more powerful and have provided some of the most spectacular events in the history of volcanology. The eruption of SURTSEY near Iceland and MYOZIN-SYO near Japan are recent examples of such eruptions. Iceland also has a vivid history of submarine eruptions, located as it is on a MID-OCEAN RIDGE, along which volcanic activity is abundant. The name *Reykjanes* means "smoky cape," a reference to submarine eruptions near that Icelandic cape. Several islets were formed here during an eruption in 1240; most of them vanished later, but others reappeared to take their places. In 1783, about a month before the volcano SKAPTAR JÖKULL erupted, a volcanic island appeared, but it too was soon reduced to a mere reef by wave action. Yet another submarine eruption in 1830 produced an island. (This eruption was notable for apparently destroying the skerries known as the Geirfuglaska and with them the great auks that had bred on them up to that time.) Another island emerged from the sea approximately ten miles off Reykjanes in July of 1884.

The ALEUTIAN ISLANDS of Alaska are monuments to submarine volcanism, having arisen from the ocean floor in eruptions that continue into the present. These eruptions have been observed frequently since Americans of European descent began to explore and settle Alaska. In 1856, for example, one Captain Newell of the whaling bark *Alice Fraser,* together with men from several other vessels, watched a spectacular submarine eruption in the Aleutians. No island appeared during this eruption, but huge jets of water rose up from the sea, and the waters were strongly agitated. Volcanic ash and stones soon emanated from the eruption site, and pumice was strewn over the sea for great distances. Loud noises accompanied the eruption, as did shocks like those of earthquakes.

A submarine eruption like this one near Japan can be dangerous to ships (USGS)

An eruption that produces large amounts of PYROCLASTIC materials may build up an island in time, but as described above, wave action quickly reduces such an island to sea level unless lava flows occur to give the new island greater resistance to erosion. When magma is relatively gas-poor and flows freely, as in the case of the great shield volcanoes of the HAWAIIAN ISLANDS, lava flows occurring over many years can build up a huge island. Volcanoes that exhibit more explosive eruptions may also emit enough lava to produce large islands, but the explosive character of their eruptions may destroy those islands soon after they are formed; the island may simply blow itself to pieces, as in the famous case of KRAKATOA.

An undersea volcano that produces a mountain whose top has not yet reached the surface is called a seamount. A seamount with a flattened top, indicative of wave action and subsequent submergence, is known as a guyot.

Submarine eruptions occur frequently, though not always with spectacular results, along midocean ridges, where magma rises from the earth's interior to the surface. This process creates new crustal rock, which moves outward on either side, away from the ridge, in somewhat the same manner as goods on a conveyer belt. Not all undersea volcanism, however, occurs along such ridges. "Hot spots" within the boundaries of crustal plates may produce whole chains of undersea volcanoes far from plate edges. The Hawaiian Islands are familiar examples of volcanic islands that appear to have arisen from submarine eruptions at a "hot spot" within a crustal plate.

Sukaria, caldera, Flores Island, Indonesia The boundaries of the Sukaria caldera are uncertain, but it may be as much as 11 miles in diameter. Likewise, the geologic history of the caldera is largely unknown. Several lakes are located at the caldera. The color of one or more of these lakes changed in 1937–38 because of sulfur and sulfuric acid in the water. A fountain of water some 300 feet high arose from one of the lakes in 1968. On April 27, 1986, one crater lake showed an increase in gas bubbling, and an earthquake occurred the following day. Evidently some phreatic explosive activity occurred at the caldera in the 1860s, but details are unavailable.

sulfur A mineral commonly associated with volcanic eruptions, sulfur in its pure, native form is a yellow mineral that may be found in crystals or in irregular masses. Sulfur may join with various metals to produce sulfides, some of which are widely mined as ores. Galena (lead sulfide) and pyrite (iron sulfide) are familiar examples of such ores.

Sunda, caldera, Java, Indonesia The Sunda caldera on the island of Java may be only about 2,000 years old, according to one estimate. The diameter of the crater is not known with certainty but is thought to be between 3.5 and 5 miles. The volcano Tangkubanparahu occupies part of the caldera. Although a stratovolcano, it is basaltic and has a structure like that of a shield volcano. There are several large craters—Kawah Baru, Kawah Ratu, Kawah Upas—and a number of lesser craters on the summit of the volcano.

Within historical times, eruptions have tended to be minor and phreatic. Carbon dioxide and hydrogen sulfide gases may gather in craters in concentrations high enough to endanger humans and other animals. An apparent eruption was reported in 1829 after several days of "violent" noise. An eruption in 1846 was stronger than most at this volcano and is thought to have involved a PYROCLASTIC surge that destroyed a whole forest and broke off trees several feet above the ground. There is also a report of a flow of ash and mud from this eruption that moved eastward from the volcano and destroyed trees and shrubbery. In 1896 a renewal of activity at Kawah Upas involved minor explosions and the formation of a new crater, Kawah Baru. The volcano apparently emitted rumbling noises on several occasions in 1908, but no sign of activity was observed in the craters. Underground noises and an ash column from Kawah Ratu were reported in 1910, and a strong earthquake occurred on April 22, followed by a less powerful earthquake on April 30. Ash eruptions were observed in May.

In 1913 solfataras became more active; subterranean rumbling was reported, along with vigorous steaming from vents involved in the 1896 and 1910 eruptions. In 1926 PHREATIC ERUPTIONS produced a new crater some 150 feet wide, Kawah Ecoma, inside Kawah Ratu. Solfataric activity intensified in 1967, and a SOLFATARA cast out mud in May of that year. Small eruptions of ash also occurred in 1967, and certain solfataras showed a sharp increase in temperature in September of the same year. Phreatic eruptions in 1969 deposited a small layer of ash on the volcano. In 1971 the volcano emitted a small quantity of mud, and thermal activity intensified.

Seismic activity started increasing in June of 1983 and showed a dramatic increase later that year, from several earthquakes per day in June to more than 1,000 per day on some dates in September. Most of these earthquakes occurred near the surface and in the vicinity of the Baru fumarole field, although only a very slight increase in fumarole temperatures was recorded. Temperatures at fumaroles at Kawah Baru increased sharply in later 1985. Slight earthquake activity was reported in 1986. A magma reservoir approximately one mile under Kawah Upas is thought to have increased by more than 70 million cubic feet of magma in the 1980s, and measurements taken at Tangkubanparahu indicate that some inflation occurred at the volcano in the early 1980s.

Suoh depression, Sumatra, Indonesia The Suoh depression is located along the Semangko Fault Zone on the island of Sumatra. There is disagreement about the possible origin of the Suoh depression. It has been argued that the depression is purely tectonic in origin and, alternatively, that it formed by a combination of tectonic and volcanic activity. Many hot springs occur within the depression. A strong earthquake in 1933 was followed a few hours later by increased steaming in the Suoh depression. Two weeks later, a large PHREATIC ERUPTION took place in the depression, forming two explosion craters and making a noise that was heard approximately 400 miles away. A PYROCLASTIC surge may have been associated with this event, but this is not certain. Another strong earthquake occurred in 1985, but there was no sign of increased volcanic activity.

Surtsey, volcanic island, near Iceland The volcano Surtsey emerged from the northern Atlantic Ocean south of Iceland in 1963. The crew of a fishing boat noticed a cloud rising from the sea on November 14 and recognized the cloud as coming from a volcanic eruption. Apparently the eruption had been under way beneath the ocean's surface for some time, unnoticed except for a few small indications such as minor earthquakes and venting of hydrogen sulfide gas, which produced a characteristic odor of rotten eggs, perceptible on the Icelandic shore nearby. Within 10 days of its discovery, the eruption had produced an island about half a mile long, almost that wide and about 300 feet high at the rim of the principal crater. Made up of fragments of volcanic rock, the island grew in size despite the erosive effects of waves and measured almost a mile wide at maximum by early February. Steam explosions sent volcanic BOMBS flying out from the volcano. As steam explosions diminished, lava outflows from the volcano, after cooling and hardening, made the island more secure against erosion. Surtsey's eruption continued for more than three years before finally subsiding in 1967. Volcanoes in shallow water, of which Surtsey is an example, tend to erupt explosively because of lava's contact with relatively cold seawater. The eruption becomes less explosive when the summit rises above sea level and direct contact between sea and magma is reduced.

S-waves Sinusoidal (that is, S-shaped) waves that travel outward across the earth's surface from the epicenter of an earthquake. See SEISMOLOGY.

T

Taal, volcano, Luzon, Philippines The volcano Taal, on an island inside Lake Bombon, a CALDERA lake, has a history of eruptions dating from the late 16th century. Records of eruptions in the 16th and 17th centuries indicate that destruction was widespread. An eruption in 1707 is said to have brought about a spectacular display of lightning but caused no damage to nearby communities. In another eruption in 1719, eruptive activity is said to have progressed down the side of the volcano and into the lake; on this occasion tsunamis appear to have occurred and fish reportedly were killed by heat. Evidently no one was killed in several subsequent eruptions in the early 18th century. Even a vigorous eruption in 1754 that expelled great amounts of material and crushed a church under the weight of the ashfall is said to have taken only a dozen lives, although it seems likely that other deaths from the eruption went unreported. Eruptions continued at intervals through the 19th century, but again, no great loss of life appears to have occurred. The most destructive eruption of Taal occurred in 1911, when activity thought to have centered around Green Lake, one of two crater lakes, resulted in an outburst of mud that killed most inhabitants of the island. The waters of Green Lake were said to be highly acidic, and the acid content of the resulting mud is believed to have been responsible for many of the deaths that occurred. At other locations, asphyxiating gases appear to have been responsible for numerous deaths. More than 1,300 people were reported killed in this eruption. Another, similar (though comparatively minor) eruption in 1965 killed approximately 500.

table mountains These curious tabular formations in Iceland represent volcanoes that erupted under cover of glacial ice, forming mountains topped with lava flows that exhibit gentle sloping. Table mountains exhibit some of the same features as submarine eruptions, such as PILLOW LAVA formed when molten rock encounters cold water or ice. Herdubreid in Iceland is a good example of a table mountain.

Tambora, volcano, near Java, Indonesia The eruption in 1815 of the 13,000-foot volcano Tambora (also known as Tomboro) near Java was remarkable in three respects. It emitted an unusually large volume of ash and other solid material, the eruption was recorded in detail by European observers in the vicinity, and Tambora's ash cloud was invoked later to account for a history-making change in global climate. Tambora was believed to be extinct during the early years of European settlement, but the volcano began emitting small showers of ash in 1814. A strong earthquake on the night of April 5, 1815, was followed by a series of explosions and an eruption of ash and smoke that darkened the sky for some 300 miles' distance.

Sir Stamford Raffles, founder of the British Colony of Singapore, served as military governor of Java when Tambora erupted on this occasion and reported that on Java the sky was darkened at noonday with ash clouds. An ashfall covered the land, while explosions like the sound of artillery or faraway thunder could be heard from the eruption. The sound of some explosions was heard in Sumatra, more than 900 miles away. Tambora's eruption was most violent on April 11–12 but did not end until July, Raffles wrote. He reported that glowing lava appeared to cover the volcano, and stones the size of a person's head fell in the vicinity of Tambora. Twelve thousand natives perished in this eruption, Raffles wrote.

Considerable subsidence was reported; for example, the site of the village of Tomboro was said to be covered by 18 feet of water following the eruption.

This eruption produced a deep caldera at the summit of the volcano. The collapse of this caldera may have been at least partly responsible for earthquakes that were felt up to approximately 300 miles away. Between 1847 and 1913, a small cone and lava flow developed in the caldera, possibly in coincidence with a powerful earthquake, centered near the volcano, that occurred on January 13, 1909.

Tambora after the eruption of 1815 stood almost a mile shorter than before. Much of the estimated 36 cubic miles of solid material cast out from the volcano remained airborne in the form of very fine ash that formed a cloud in the upper atmosphere. This high-altitude cloud intercepted incoming sunlight. The resulting drop in insolation, or solar radiation reaching the earth's surface, was implicated in a dramatic change in climate and weather patterns in the northern hemisphere during the following year. The year 1816 is known as the "Year Without a Summer" because there was no warm season over much of the northern hemisphere that year. Although the winter of 1815–16 apparently was not unusually bitter, the cold season lasted well into the summer of 1816, with subfreezing temperatures and several inches of snowfall recorded in New England in June. Nonetheless, New England's harvest appears to have been adequate for that summer, partly because much of the harvest consisted of crops that can endure, and even thrive in, cool and moist weather.

Other parts of the northern hemisphere, however, experienced such poor harvests that starvation became a major cause of death among poor families in Canada, and in some corners of Europe, the human population was reduced to eating rats. The economic effects of poor harvests were devastating. Grain prices reportedly rose fourfold in Switzerland; and when hungry Europeans and Canadians turned to the relatively well-off United States for help, the foreigners bought up so much American grain that the price of everything connected with grain rose sharply and increased inflation in the United States. The "year without a summer" was accompanied by political turmoil in France, where an incipient famine, coming just after the devastation of the Napoleonic wars, strained the social fabric to the point of rupture. Many farmers were afraid to take their produce to market for fear of being robbed and murdered by famished mobs along the way. Farmers who dared take their crops to market required government troops in some instances to protect them from half-starved hordes who fought to reach the food.

Tambora's eruption has not been identified conclusively as the cause of the "year without a summer" because the unusual cold spell of 1816 is believed to be within the range of normal fluctuations in weather and climate, meaning that a summerless year such as that one may happen even in the absence of a major volcanic eruption the year before. This is because volcanoes are not the only known influence on weather and climate. Many other factors have been identified or at least implicated in weather and climatic change, from sunspots to irregularities in the earth's motion. Yet a U.S. Weather Bureau historical study of temperature as a function of solar radiation reportedly indicated that lesser eruptions than that of Tambora in 1815 have preceded significant drops in insolation and surface temperature. For example, total measured heat received from sunlight fell to about 88% of the normal figure, a reduction of 12% from that value and more than 16% from the previous year, immediately after the eruption of Krakatoa in 1883. Insolation diminished 4% following the eruption of Alaska's Bogoslov volcano and several other volcanoes in 1889. Insolation fell about 13%, from 101% to 88% of normal, after the eruptions of Mount Pelée and Soufrière in 1902. A similar drop in incoming solar radiation, from 101% of normal to 84%, was recorded after the Mount Katmai eruption in Alaska in 1912. There is strong reason to suspect, then, that the eruption of Tambora in 1815 was a factor in bringing about a memorable year in both world climate and human history, long after the eruption itself was over.

Tanaga, volcano, Aleutian Islands, Alaska, United States Tanaga volcano is located at the northern tip of Tanaga Island in a structure that has been interpreted as a caldera. Activity was reported on the island in 1763–70, possibly in 1791 and again in 1829 and 1914. In the 1914 activity, a lava flow reportedly formed.

Tango Peninsula, Japan The Tango Peninsula borders Wakasa Bay on the western coast of Japan and was the site of a 1927 earthquake comparable in magnitude to Tokyo's KANTO EARTHQUAKE of 1923.

Tangshan, China The Tangshan earthquake of July 28, 1976, is perhaps the most destructive earthquake in history, whether destruction is measured in terms of lives lost or property damaged. Estimates put fatalities from the earthquake from about 650,000 to 750,000, with some 800,000 people injured. The city of Tangshan, approximately 85 miles southeast of Beijing, was simply demolished. The physical damage at Tangshan has been compared to the destruction of Hiroshima. Two major earthquakes struck Tangshan on this occasion. The first, reported as magnitude 8.2 on the Richter scale, took place at 3:45 A.M. and lasted some two minutes. An aftershock of magnitude 7.9 occurred several hours later. Each shock was roughly equivalent in magnitude to the earthquake that destroyed San Francisco in 1906. Reports of the disaster told of widespread subsidence caused by the collapse of mine tunnels underground. The subsidence is said to have done particularly great damage to railway facilities. The Tangshan disaster occurred so close to Beijing that officials in the capital ordered the public to move outdoors in case an earthquake struck the city. Some six million people are believed to have slept outdoors in temporary shelters for more than two weeks. One intriguing aspect of this earthquake was the reported occurrence of EARTHQUAKE LIGHT. The red and white light is said to have been seen up to 200 miles from the epicenter of the quake and, as viewed from Beijing, allegedly lit up the sky as brightly as daylight in the direction of Tangshan.

Tao-Rusyr, caldera, Kuril Islands, Russia The Tao-Rusyr CALDERA is located on Onekotan Island in the Kuril chain and has an intracaldera cone, Krenitzyn Peak. Surrounding the cone is a crater lake that does not freeze in winter. In 1846 and 1879, solfataras were observed in the caldera, although it is not known whether this activity was merely ordinary or remarkable. An eruption occurred in 1952 at Krenitzyn Peak following, and possibly the result of, a powerful tectonic earthquake that took place several days earlier. Three days before the volcano erupted, there were reportedly disturbances in the magnetic field in the vicinity of the caldera. A magnetic compass is said to have behaved in a strange fashion, decelerating slowly and unevenly in one direction.

Tarawera, volcano, North Island, New Zealand An eruption of Tarawera along a fissure in 1886 reportedly killed more than 150 people. An eruption on the night of June 9 of that year destroyed two native communities near the volcano. Early in the morning of the following day, strong earthquakes preceded a series of explosions from Wahanga, one of the peaks of Tarawera, that generated a huge cloud of ash and vapor. Apparently incandescent material then burst from the crater, creating an appearance of flame leaping from the volcano's throat. In quick succession, other volcanoes nearby erupted, until a chain of mountains some nine miles long was in eruption. Blasts of steam are also said to have occurred from Lake Rotomahana, which the eruptions caused to merge with adjacent Lake Rotomakiriri. These eruptions destroyed the beautiful lakeside geyserite formations known as Pink Terrace and White Terrace. A great fissure called Tarawera Chasm, more than a mile long and about a thousand feet deep, opened during Tarawera's 1886 eruption. The violent eruptions of June 10 lasted only about two hours, although lesser activity was observed for days afterward. Vigorous geyser activity has been recorded along the fissure involved in that eruption.

Tarumai, volcano, Japan The stratovolcano Tarumai has undergone explosive eruptions on more than 30 occasions since the mid-17th century. An especially large eruption in 1909 released great quantities of ash and BOMBS.

Taupo, caldera, North Island, New Zealand The Taupo caldera is located in the Taupo Volcanic Zone near Wairakei and is partly occupied by Lake Taupo. The boundaries of the caldera are poorly defined, but it is thought to be about 24 miles wide. Roughly 2,000 years ago, during a series of violent eruptions that occurred here, a set of fissures opened, then filled with IGNIMBRITES. Slightly more than a mile outside the caldera is Tauhara volcano, a complex of domes of undetermined age. Tauhara is also the site of a geothermal field. In 1895 a powerful earthquake occurred that caused landslides and fissures in the vicinity of Wairakei. Earthquake activity was frequent in 1922 and involved displacement of approximately 10 feet in places along fault lines. Subsidence of more than six feet occurred along the northern shores of Lake Taupo. Several years later, further subsidence was reported, this time more than 10 feet. Fissures developed along the Kaiapo Fault, along the northeastern rim of the caldera at the edge of the Wairakei field, and numerous fountains of water were reported seen along the fault.

A moderately strong earthquake occurred immediately west of Lake Taupo in 1953 and was followed by aftershocks. Earthquakes occurred again to the west and southwest of Lake Taupo in 1956–57. Late in 1964, a swarm of more than 1,000 earthquakes greater than magnitude 2.5 started under Lake Taupo and continued through January 1965. This swarm is thought to have accompanied several inches of uplift along the eastern shore of the lake. In 1974 and 1981, hydrothermal explosions took place in the Tauhara geothermal field. An earthquake occurred several days before the 1981 explosion, but there is no proof that the earthquake was the cause of events leading to the explosion.

Very slight subsidence was noticed on the north shore of Lake Taupo in the early 1980s, and in 1983 earthquake swarms occurred to the north of the lake, followed by several inches of uplift along the northern shore. Earthquake activity intensified in mid-1983, followed by fast subsidence northwest of the Kaiapo Fault and uplift southeast of the fault. Slight subsidence preceded an earthquake swarm at the southern side of the lake in 1984, but an equivalent amount of rebound occurred over the following months. The northeast shore was uplifted slightly in the latter part of 1984, then subsided after a series of minor earthquakes in 1985.

In 1986, it appeared that an eruption might be imminent, because pumice was seen in the lake and gas bubbles were reported rising to the surface. Fears of eruption, however, proved to be groundless. The pumice was thought to have fallen into a stream that carried the material to the lake, and the gas bubbles evidently were no different from others that occur often in the lake. It is hard to predict eruptive activity at the Taupo caldera because of difficulty in separating magmatic from tectonic unrest.

Taupo Fault Zone, North Island, New Zealand The Taupo Fault Zone on North Island has been the site of intensive volcanic activity both before and after European settlement of the islands in the late 18th century. Great quantities of volcanic material emerged from fissures in the Taupo Fault Zone in prehistoric times and were deposited as IGNIMBRITES. Similar conditions gave rise to the VALLEY OF TEN THOUSAND SMOKES in Alaska, although the New Zealand deposits are far more voluminous than those in Alaska—some 50,000 square miles in area and thousands of feet deep (estimated at almost 200 cubic miles total) in the Taupo Fault Zone, compared to only about 50 square miles and several hundred feet deep (some seven cubic miles total) at the Valley of Ten Thousand Smokes. Eventually, eruptive activity was confined to a few locations but became more explosive. The Taupo Fault Zone was the site of the famous eruption of TARAWERA in 1886.

tectonics The study of the large-scale features of the earth and how they form. (See also PLATE TECTONICS.)

Tengger, caldera, Java, Indonesia The Tengger caldera played an important part in debate during the early 20th century about how calderas originate. It was determined that the caldera formed after an eruption of vast amounts of tuff and formed during two separate episodes of collapse. The first episode formed the Ngadisari depression immediately to the northeast of Tengger, and the second episode of collapse created the Tengger caldera itself. Tengger is occupied partly by a lake, and a small cone called Bromo is located in the center of the caldera. Bromo undergoes frequent eruptions of the Strombolian type. The historical record of eruptions dates back to the early 19th century. After Bromo erupted in 1835, a small lake formed in the crater. The lake grew in size following another eruption in 1838, then began changing color from blue to green between 1838 and 1841. Sometime in the early 1840s, the floor of the crater appears to have bulged upward and given off gases. Several weeks later, the swelling subsided, and the crater floor collapsed. This swelling and subsequent collapse evidently were confined to the crater at Bromo.

In March of 1858, Bromo became active again. This eruption was preceded by loud underground noises. Such noises also occurred before another eruption in 1858; before this eruption, earthquakes were felt up to a distance of about a mile from the volcano. Bromo was blamed for a very strong earthquake that shook the island of Java in May of 1865. Although the volcano was active for part of that year, there is no proof that the eruption and the earthquake were related, but some connection between the seismic and volcanic activity is possible. In 1888 Bromo emitted steam and sulfur gas. Several days before an eruption in 1980, fumaroles became more active.

Tennessee, United States Tennessee has an extensive history of strong seismic activity, partly because of its proximity to the NEW MADRID FAULT ZONE, which was responsible for a series of extremely powerful earthquakes during the winter of 1811–12. Another severe earthquake occurred in western Tennessee on January 4, 1843, and affected a wide area including portions of Kentucky, Missouri, Alabama, Indiana, Iowa, South Carolina and Georgia. Western Tennessee could undergo severe damage in the event of any future earthquakes comparable in intensity to the 1811–12 events because the area is settled much more heavily now than it was at the time of the great earthquakes of the early 19th century.

tephra Geologists define *tephra* as all solid particles emitted by a volcano during an eruption. Tephra particles range in size from extremely fine ash with approximately the consistency of confectioner's sugar to large pieces several feet in diameter. Tephra forms when magma containing dissolved gases nears the surface; the gases bubble out of solution, forming cavities in the magma as it solidifies. The resulting, fragmented solid material is cast out of the volcano's throat and, under the right conditions, may travel for thousands of miles before settling to the surface. Almost all volcanoes release tephra when erupting, although some eruptions produce far more tephra than others. Tephra comes in various forms, including bombs, masses of fluid magma that solidify in flight, and lapilli, gravel-like bits of lava. Tephra may mix with extremely hot gases and flow downslope from the volcano as a nuée ardente (fiery cloud), a phenomenon that can destroy a landscape as effectively as a nuclear explosion. A nuée ardente caused much of the destruction at Mount Pelée in 1902. A high-altitude cloud of fine tephra from KRAKATOA appears to have altered global climate by intercepting incoming sunlight and lowering surface temperatures. A similar cloud from the 1815 eruption of TAMBORA has been implicated in the dramatic drop in temperatures the following year, the "YEAR WITHOUT A SUMMER."

Tephra's constituents are known collectively as PYROCLASTICS. Pyroclastic deposits may occur either as pyroclastic flows (that is, as nuées ardentes) or as airfall deposits, which accumulate as tephra settles out of the atmosphere. Each kind of pyroclastic deposit shows a characteristic pattern when exposed vertically. In a pyroclastic flow, particles of many different sizes, large and small, are mixed together with little or no evidence of sorting or layering. Airfall deposits, by contrast, exhibit distinct layering and sorting into fine and coarse layers. Mudflow deposits consist of gravel of various sizes deposited amid fine silt. These deposits occur when large quantities of water flow down a volcano's flanks, picking up tephra along the way and finally depositing the mudflows in nearby stream valleys. Mudflows may occur when an eruption melts ice and snow on a volcano's summit or casts water out of a lake in the crater. Heavy rains may also generate mudflows. A mudflow may overwhelm an entire community without any advance warning.

terrane A piece of the earth's crust that differs greatly in history and composition from adjacent areas. Terranes are commonplace in the western United States because of the collision of several large tectonic plates and numerous smaller plates there.

Terror, volcano, Antarctica See EREBUS, MOUNT.

Texas, United States Earthquake activity in Texas has provided interesting discoveries for geologists on occasion. For example, the July 30, 1925 earthquake in the state's panhandle region was a strong earthquake, felt as far away as Roswell, New Mexico, 225 miles away, and Leavenworth, Kansas, 400 miles distant. The area was not thought to be given to earthquakes (the only previous recorded earthquake in the Texas panhandle occurred in 1917), and for a while, oil drilling was suspected as a reason for the earthquake. The earthquake affected such a wide area, however, that drilling was eliminated as a possible cause. Oil-well borings revealed that the earthquakes occurred in the vicinity of a buried mountain with slopes of 1,500 to 2,000 feet over a distance of just a few miles. Although the area looked quiet on the surface, it appeared that the area near the buried mountain once was highly active seismically, and earthquakes were still occurring from time to time. Another earthquake, at Mount Livermore in western Texas on August 16, 1931, caused heavy damage at Valentine and would have been extremely destructive if the event had occurred in a heavily settled area.

Thera, volcano, Mediterranean basin, circa 1470 B.C. The island of Thera (also known as Santorini or currently Thira), between Greece and Turkey, is believed to have been almost totally destroyed in a volcanic explosion that wiped out the Minoan civilization. Thera became active after long dormancy around 1500 B.C. and appears to have erupted at intervals over some 30 years before a final, explosive eruption destroyed much of the island. Archeological evidence indicates that powerful tsunamis generated by the explosion caused widespread destruction on shores in the eastern Mediterranean. Apparently Minoan civilization never recovered from the natural disasters associated with this eruption of Thera. The destruction of Thera is believed to have given rise to the legend of Atlantis, the ancient civilization supposedly destroyed in a single day when the island of Atlantis sank in a great natural cataclysm. Initially, the legend placed the demolished island accurately in the vicinity of Crete, but the Greeks later imagined that Atlantis had been located farther west, in the great ocean beyond what we know as the Straits of Gibraltar. The Atlantic Ocean was named for the fictional island. Thera's eruption is sometimes compared to that of KRAKATOA. As in the case of Krakatoa, the explosion of the volcanic island left behind a caldera that filled with water. A group of small volcanic islands arose from the caldera and erupted intermittently from 1938 to 1941. (See also SANTORINI.)

thermal gradient The rate at which the earth's temperature increases with depth. Although an average figure of 30°C per kilometer is sometimes given for the thermal gradient, the rate of increase actually varies from one level of the earth's internal structure to another. The increase is believed to be rapid within the crust, where high concentrations of radionuclides in granitic rock generate large amounts of heat by their decay. The thermal gradient is thought to diminish as the crust gives way to the outer layer of the mantle. In this layer, some 60 to 100 kilometers (about 36 to 60 miles) beneath the surface, temperatures are estimated at approximately 800° to 1200°C. (See also EARTH, INTERNAL STRUCTURE OF.)

Thira See THERA.

Three Sisters, Oregon, United States A cluster of volcanic peaks in western Oregon, the Three Sisters are part of the Cascade Mountains and are located near Mount Hood and Crater Lake.

tidal wave See TSUNAMI.

Tien-Chi, caldera, China Located on the border between China and Korea, the Tien-Chi caldera has a spectacular crater lake some three miles wide and more than 1,000 feet deep. The caldera is thought to have formed less than 1,000 years ago, during or after an eruption that deposited a huge volume of pumice over an area some 24 miles wide around the caldera. Eruptions were reported at the caldera in 1597 and 1702, but these reports have not been confirmed. Tien-Chi is located near the volcano P'aektu-san, which is either a stratovolcano or a dome surmounting a vast lava shield.

Toba, caldera, Sumatra, Indonesia The Toba caldera is a volcano-tectonic depression. Although it has been argued that the depression is tectonic in origin, the vast quantities of tuff expelled from eruptions here indicate to many geologists that volcanic activity was involved in the caldera's origin. The Toba caldera marks the point where a chain of andesitic volcanoes undergoes a change in relation to Sumatra's Semangko Fault Zone. To the south of the Toba caldera, active volcanoes are located very close to the fault zone, but north of Toba the volcanoes are situated more to the east of the fault zone and are distributed over a broader area. Toba is thought to occupy a boundary, running north to south and characterized by high seismicity, between two subducted plates of the earth's crust. Volcanism at Toba may be the result of increased production of magma along this boundary zone.

The caldera has not experienced any eruptions within historical times, but earthquake activity is frequent in the vicinity of Toba. Powerful earthquakes have occurred at Toba on several occasions since the late 19th century. A very strong earthquake on May 17, 1892, centered about 75 miles south of Toba, demolished numerous buildings and accompanied several feet of displacement along the Semangko Fault. A slump took place around the year 1914 at the upper end of the Asahan Valley on the southeastern side of the caldera and made Lake Toba rise several feet, although no particular importance is attached to this event in terms of earthquake or volcanic activity. An earthquake occurred in November of 1920 and was followed by earthquake swarms in 1922. Damage occurred along the southern shore of Lake Toba, where subsidence took place at several locations. Fumaroles to the south and west of Lake Toba became more active after powerful earthquakes. A moderately strong earthquake in 1987, centered on the southern shore of the lake near the community of Muara, caused significant damage.

Tofua, volcanic island, Tonga Tofua Island is the site of Tofua caldera and has a history of eruptions and other unrest dating back to the late 18th century. One eruption in 1958–59 was severe enough to require the evacuation of the island's population for several months.

Tokai earthquake See KANTO EARTHQUAKE.

Tokyo, Japan The capital of Japan is known for its history of frequent and highly destructive earthquakes. The 1923 KANTO EARTHQUAKE and subsequent fire destroyed much of Tokyo and nearby Yokohama. The earthquake was estimated at a magnitude of up to 8.3 on the Richter scale. A similar earthquake today would have a much more devastating effect than the 1923 earthquake, for various reasons. The city is larger than in

1923, and therefore opportunities for damage are greater. Also, Tokyo now has numerous hazardous materials stored in it in large quantities, and those materials might pose serious dangers in the event of a major earthquake. The economic effects of a major earthquake in Tokyo would reach to numerous nations other than Japan, notably the United States, and might cause serious hardship on an international scale because of Tokyo's importance as a financial center. Tokyo is located near the volcano Mount Fuji, which has become a symbol of Japan. (See also ECONOMIC EFFECTS; FUJI, MOUNT.)

Tolbachik, volcano, Kamchatka, Russia Although the stratovolcano Tolbachick has erupted often since the mid-18th century, it is most famous for a major eruption in 1975 that was predicted so accurately that television crews were able to film its onset. Earthquake activity just before the eruption made the prediction possible.

Toledo, caldera, New Mexico, United States The Toledo caldera is thought to have produced an estimated 50 cubic miles of tephra in an eruption some 1.3 million years ago. The Toledo caldera lies buried under a heavy ashfall from the nearby VALLES CALDERA.

Tomboro See TAMBORA.

Tondano, caldera, Celebes, Indonesia The Tondano caldera occupies the northern portion of Celebes and has a record of activity within historical times dating back to the early 19th century. Eruptions have been explosive, with some PHREATIC ERUPTIONS. Several andesitic stratovolcanoes have formed along the rim of the caldera. A new crater called Tompaluan appeared in 1829 and started smoking in 1893. A fissure some 300 feet long that emitted smoke is thought to have formed in the caldera around 1898. Earthquakes in 1901 were ascribed to increased activity at the Soputan volcano on the southern rim of the caldera; two mud pots formed at the foot of the volcano, and there was a report of ash ejection and steaming from Soputan at this time. About a month later, a new solfatara had appeared near the warm spring at Rumerega and was casting up boiling mud to a height of more than 700 feet, but the crater of Soputan reportedly showed no increase in activity.

In 1906, the new crater Aeseput formed at Soputan. That same year, earthquake activity increased, accompanied by fumes from the volcano and a glow at night. Soputan erupted on June 17, 1906, the same day an earthquake shook Tomohon, some 18 miles north-northeast of the volcano. Soputan continued erupting at intervals through 1913. A new crater called Aeseput-weru formed at Soputan in 1915. A powerful tectonic earthquake occurred in 1932 but apparently did not affect the volcanoes. In 1958 and 1959, minor eruptions of ash occurred, and another series of minor eruptions took place from 1961 to 1962. Small explosions were reported between 1963 and 1968, and lava flows and PYROCLASTIC FLOWS occurred at Soputan between 1966 and 1968. Small explosions occurred again in 1969–71. Gray gas, possibly containing ash, emerged from Soputan in 1970, and fumarolic activity increased through a lava flow that had been laid down in 1966.

There was an increase in seismic activity at the caldera in 1976, and the temperature of the crater lake increased sharply in early 1978. A strong explosive eruption occurred in August 1982 without any premonitory signs, but earthquakes did precede another eruption in November of that year. In April of 1984, minor eruptions of gas through the lake killed plants as far as two miles downwind, possibly from the action of acid rain or acid gases. An eruption in May of 1984 showed no premonitory activity, but earthquake activity increased before an eruption in August of 1984. Tectonic earthquake activity appears to have increased dramatically at the caldera in early 1985. In May of 1985, the steam plume that normally rose from the volcano increased in height, and by May 19 the plume, now containing ash as well as steam, rose to an altitude of about three miles. In March of 1986, the Tompulan crater sent up puffs of steam to an altitude of more than 1,000 feet, and phreatic explosions produced lahars. In the second half of 1986, dozens of explosions occurred daily. Explosions continued, though less frequently, through May of 1987. Between April and May of 1987, light-colored fumes emerged from the Magawu crater, where the crater lake rose in temperature and increased in volume.

Tongariro, volcano, North Island, New Zealand Tongariro is adjacent to Ngauruhoe and Ruapeho volcanoes and has a summit crater so large that it might be considered a small caldera.

Torfajökull, volcano, Iceland Located in southeastern Iceland, the Torfajökull volcano is associated with a caldera and is in an area of frequent earthquake activity. The volcano erupted around the year A.D. 900 and emitted both lava and tephra. Another eruption occurred around 1480, again involving both tephra and lava. Both these eruptions occurred during episodes of rifting on the Veidivotn fissure swarm, which runs from northeast to southwest through the middle of Iceland. Tephra and lava in these eruptions were rhyolitic. Earthquakes have occurred at the caldera in recent years, sometimes at a rate of several dozen per day, but there have been no further eruptions since the 15th century.

Towada, caldera, Japan The Towada CALDERA is located in the northern portion of the island of Honshu. The caldera contains a lake and is thought to have formed from the collapse of several small stratovolcanoes. Towada has been quiet for approximately the last 1,000 years, but an eruption in approximately A.D. 915 is thought to have deposited airfall tephra and PYROCLASTIC FLOWS over the surrounding area.

Toya, caldera, Japan Located on the island of Hokkaido, the Toya CALDERA lies immediately north of Usu volcano and contains a lake, Lake Toya, in the middle of which a group of lava domes has formed an island. The historical record of activity at the caldera dates back to the 17th century. Eruptions were recorded in 1663, 1769, 1822 and 1853. An eruption in 1910 was preceded by several days of earthquake activity, which was accompanied by the appearance of cracks and faults aligned mostly east to west along the northwestern foot of Usu volcano. Small craters formed at the north foot of Usu, and numerous small mud cones formed along the seashore nearby. The explosive phase of the eruption was largely finished by early August.

In May 1932, an area of subsidence some 200 feet wide was noticed off the north shore of Lake Toya; although this area subsided by more than 30 feet in places, no eruption followed, although lake temperature rose considerably. Earthquakes started late in December of 1943. An explosive eruption began in June of 1944 and lasted through late October. Earthquake activity again showed a sharp increase in August of 1977, followed by an explosive eruption of pumice. Magmatic explosions continued through August 14 on an intermittent basis. PHREATIC ERUPTIONS occurred on occasion in 1977 and 1978. Activity at Usu volcano has produced new hot springs, which in turn have generated opportunities for tourism, because hot springs are popular attractions in Japan.

transform fault A fracture in the crust characterized by lateral movement. Transform faults are found along MID-OCEAN RIDGES. (See also PLATE TECTONICS.)

Transverse Ranges, California, United States These mountains run along an east-west line through southern California and include the San Bernardino, San Gabriel, Santa Susana, Santa Monica and Santa Ynez mountains. The Channel Islands off Santa Barbara are also seen as an extension of the Transverse Ranges. The Transverse Ranges appear to mark the zone where the Los Angeles basin is being driven under the crust to its northwest. Compression is evident in the Transverse Ranges and is thought to originate in this case from the movement of the Pacific plate to the northwest. Uplift along the Transverse Ranges is thought to have been responsible for the highly destructive SAN FERNANDO EARTHQUAKE of 1971. Spectacular displays of exposed sedimentary rock may be seen along U.S. 101 in the Santa Ynez Mountains near Santa Barbara and Buellton. (See also PLATE TECTONICS.)

Trident, Mount, Alaska, United States Mount Trident is a volcano located in the Aleutian Range near Mount KATMAI. Movement of magma from under Katmai toward Trident is believed to have played a part in the collapse of Katmai during its 1912 eruption.

Tristan de Cunha, volcanic island group, South Atlantic Ocean The islands of Tristan de Cunha, part of a MID-OCEAN RIDGE, made news worldwide in the early 1960s when earthquakes and an eruption of lava required the evacuation of the local population. The residents of Tristan de Cunha remained in exile for about a year and a half, then chose to return to their island home.

tsunami A tsunami is a wave generated in a body of water by a physical disturbance. The wave may be produced by an earthquake or volcanic eruption or by several other processes. Tsunamis are known, erroneously, as "tidal waves," although tides are not a factor in their creation. The classic description of a tsunami goes roughly as follows. First the waters withdraw from the shore, leaving the seabed exposed. Then the waters return as a breaker that may reach heights of 40 feet. The wave carries away virtually everything in its path and may travel a mile or more inland before returning to the sea.

Actual tsunamis may not follow that exact pattern. Although some tsunamis are preceded by a withdrawal of water from the shore, other tsunamis may rise up from the sea with little or no warning. Nor does the wave always take the form of a curling breaker. A tsunami may manifest itself instead as a sudden upsurge of the sea. A train of

Trees mark the trimline of a giant wave that swept this inlet in the Lituya district of Alaska in 1936 (D. J. Miller, USGS)

tsunamis characteristically has such a great wavelength (the distance from crest to crest) that the waves may be imperceptible as they pass under ships in midocean. As the waves approach land and "touch bottom" in shallow coastal waters, however, the tsunamis rise and take on a familiar wavelike form. In open oceans, the waves may travel at several hundred miles per hour, but near shore their velocity drops to only perhaps 50 or 60 miles per hour so that the waves crowd together and may be spaced only a few thousand feet apart. This compressed spacing means that several towering waves may hit the shore in quick succession, with devastating effect. The exact characteristics of a tsunami depend on many factors, including the geometry of the seabed and the shoreline. A narrow inlet may "squeeze" a tsunami into a relatively small area and thus multiply its destructive power manifold. Tsunamis are associated closely with Pacific Ocean shores because the rim of the Pacific coincides roughly with the so-called "Ring of Fire," a belt of intense earthquake and volcanic activity that exists where the Pacific crustal plate encounters adjacent plates. Along the "Ring of Fire," earthquakes and volcanic eruptions have generated numerous destructive tsunamis.

The National Oceanic and Atmospheric Administration (NOAA) Coast and Geodetic Survey's Pacific Tsunami Warning System was instituted following the tsunami that devastated Hilo, Hawaii in 1946. At the system's headquarters in Honolulu, Hawaii, seismic data and information on tides are monitored to provide early warning of tsunamis and potentially tsunamigenic events. If and when evidence indicates a tsunami poses a danger to shores around the Pacific Ocean, a warning goes out from Honolulu. The Coast and Geodetic Survey also maintains the Regional Tsunami Warning System in Alaska. The survey's seismological observatory at Palmer, Alaska maintains a watch on seismic data and issues a tsunami warning if a sufficiently powerful earthquake occurs in the Aleutian Islands or along the southeastern shores of Alaska.

tuff A rock made of PYROCLASTIC fragments fused together by the heat of an eruption.

turbidity current Loosely defined, a turbidity current is a submarine landslide. When some disturbance, such as an earthquake, sets a turbidity current in motion, the current flows downslope as

Tuff beds at Red Hill on Oahu, Hawaii. Note lava projectile embedded in tuff at right (H. T. Stearns, USGS)

a dense suspension of mud, silt and sand (though the composition of the sediment in it varies with location). The deposit of sediment laid down by a turbidity current is called a turbidite. As a rule, turbidity currents have little effect on human activities, but in at least one celebrated case they knocked out a deep-sea cable system. An earthquake beneath the Grand Banks off Newfoundland in 1929 set off a massive turbidity current that snapped cables on the seabed and interrupted telegraph communications. The maximum velocity of the current was estimated, by timing the breaks in various cables, at about 40 miles per hour. At the epicenter, the earthquake caused a mass of sediment some 60 miles long and 1,200 feet thick to slide away. Altogether, the earthquake displaced perhaps 20 cubic miles of sediment spread over an area of roughly 36,000 square miles.

U

United States The United States of America has a long record of earthquake and volcanic activity. Seismic activity is most pronounced on the Pacific coast of the United States, notably in CALIFORNIA, where ongoing movements among plates of the earth's crust have resulted in numerous strong earthquakes within the 20th century alone, most notably the SAN FRANCISCO earthquake of 1906, which destroyed much of the city through ground motion and a subsequent fire. Southern California, including the LOS ANGELES area, is extremely susceptible to earthquakes and is considered perhaps the most likely candidate for the next major earthquake (comparable in destructive power to the 1906 San Francisco earthquake). In the 20th century alone, the Los Angeles area has undergone several strong and destructive earthquakes, such as the Long Beach earthquake of 1933 and the SAN FERNANDO EARTHQUAKE of 1971. The SAN ANDREAS FAULT, which runs several hundred miles along the California coast, has been responsible for some of the most powerful earthquakes to strike California in the last two centuries, but many other active faults in the state have generated, and remain capable of generating, strong earthquakes. (An example is the HAYWARD FAULT underlying the densely populated East Bay area of San Francisco.)

Other areas of strong earthquake activity in the United States include New England, where the area around Boston, Massachusetts has been struck by major earthquakes on several occasions; upstate New York, which includes part of the earthquake-prone ST. LAWRENCE VALLEY; South Carolina, where one of the most powerful U.S. earthquakes of the 19th century hit CHARLESTON; the Mississippi Valley, site of the New Madrid, Missouri earthquakes of 1811–12, thought to be the most powerful earthquakes in the history of the nation; ALASKA (site of the GOOD FRIDAY EARTHQUAKE), also subject to frequent volcanism; the HAWAIIAN ISLANDS, where earthquakes have accompanied the volcanic activity that created the islands; and OREGON and WASHINGTON, where strong earthquakes have been rare within historical times but have the potential to occur and cause great destruction nonetheless. Very few portions of the United States are completely free from earthquakes, although some states, such as Florida, are generally quiet from a seismic standpoint, compared to such active areas as California and Alaska.

Although California is the most seismically active state of the United States, the potential for highly destructive earthquakes actually may be greater in other portions of the nation, where major earthquakes are few and preparations for them are lacking. The danger from major earthquakes is especially great in portions of the eastern United States where active faults may underlie thick layers of sediment so that the faults give no sign of their activity until a powerful earthquake occurs. Adding to this threat is the fact that many large urban areas in the eastern United States are built on sediments in which ground water rises close to the surface. Under these conditions, liquefaction may cause soil to behave as a liquid during the passage of earthquake waves and cause massive destruction of buildings atop the liquefied soil. Liquefaction could cause tremendous property damage and loss of life in future earthquakes in such places as Boston, where much of the city is built on geologically unstable landfill, and cities in the Mississippi Valley around Memphis, Tennessee and St. Louis, Missouri, which are built on the floodplain of the Mississippi River and are situated near the NEW MADRID FAULT ZONE, thought to be responsible for the tremendous earthquakes of 1811–12. The New Madrid Fault Zone is believed to have its origins in a great RIFT VALLEY that developed in the middle of what is now the United States and was buried under thick layers of sediment that today fill the Mississippi Valley.

From an economic standpoint, earthquakes are a subject of some concern to the United States. It may be only a matter of time before one or more major U.S. cities will be destroyed, either in part or in whole, by an earthquake. Property damage alone in such an event would reach tremendous values, and the economic impact of a major earthquake would be amplified by America's growing dependence on services and products furnished by such cities, from financial transfers to electronic components. Total damage in the hundreds of billions of dollars, or even into the trillions, is conceivable as an outcome of a "superquake" in the densely settled areas of the United States, particularly the California coast.

In modern times, the danger from volcanism is largely confined to the Pacific Northwest. Here an encounter between the North American crustal plate and the JUAN DE FUCA CRUSTAL PLATE beneath the Pacific Ocean has generated an active volcanic mountain range, the Cascade Mountains, reaching from northern California through Oregon and Washington state into the Canadian province of British Columbia. Among the famous and spectacular volcanoes of the Cascades are Mount BAKER; Mount HOOD; Mount RAINIER; and Mount ST. HELENS, which underwent an explosive and highly destructive eruption in 1980. In recent geologic time, the eruption and subsequent collapse of MAZAMA, a huge volcano in what is now Oregon, generated a caldera that filled with water and became CRATER LAKE, a popular tourist attraction. Volcanic activity has continued at Crater Lake since the destruction of Mazama and has created a small new volcano, Wizard Island, within the lake. The Cascade volcanoes are associated with an offshore trench belonging to the CASCADIA SUBDUCTION ZONE, where the Juan de Fuca plate is believed to have descended into the earth's mantle, melting along the way and producing magma that rises back to the surface through eruptions of the volcanoes. In California, Mount Shasta and LASSEN PEAK represent very recent volcanism on the geologic time scale (Lassen Peak erupted early in the 20th century), and renewed volcanic activity is a possibility for the near future in the vicinity of MONO LAKE, in eastern central California near the Nevada border.

Although modern volcanism is largely restricted to the far western United States, evidence of ancient volcanic activity is abundant in many parts of the nation. In New Hampshire, for example, ancient volcanism produced peculiar circular ridges visible today in the WHITE MOUNTAINS. Volcanic activity has produced spectacular calderas and other formations at Long Valley, California, site of one of the most powerful eruptions in the history of North America; YELLOWSTONE NATIONAL PARK in Wyoming, where HYDROTHERMAL ACTIVITY continues to this day, producing (among other phenomena) the famous geyser "Old Faithful"; and the VALLES CALDERA in New Mexico. In the Northwest, vast flows of lava cover the COLUMBIA PLATEAU and SNAKE RIVER PLAIN. These flows may have resulted, in part at least, from the impact of a giant meteorite in prehistoric times. Volcanism in Alaska is frequent, widespread and spectacular. Alaskan eruptions tend to be highly explosive, in contrast to the relatively tranquil eruptions of Hawaii's volcanoes. In this century, the eruption of Mount KATMAI in Alaska resulted in the collapse of the volcano's peak, the formation of a huge caldera and the generation of a plain of fumaroles, the VALLEY OF TEN THOUSAND SMOKES.

Volcanism and its associated phenomena have had considerable economic and historical importance for the United States. The deposits of gold that set off the gold rush of the mid-19th century, thus accelerating the settlement of the western United States, are thought to have their origins in chemical processes associated with hydrothermal activity. Also tied in one way or another to volcanism in the western United States are numerous deposits of rich metal ores that have yielded great quantities of silver, tin, copper and other economically important metals. In some cases, the mines dug to exploit these minerals have encompassed a cubic mile of rock with their tunnels.

The western coast of the United States is vulnerable to tsunamis, or seismic sea waves, generated by earthquakes along U.S. shores or elsewhere around the Pacific Ocean basin. Perhaps the most famous tsunami in U.S. history accompanied the Alaskan Good Friday Earthquake of 1964. This tsunami devastated shorelines in southern Alaska, wiped out much of the state's commercial fishing fleet and retained enough power to cause widespread damage at Crescent City, California. Numerous other (though less destructive) tsunamis have occurred along the California coast, many of them apparently in connection with earthquakes there. (See also IMPACT STRUCTURES; PLATE TECTONICS.)

Utah, United States Although Utah is not commonly associated with strong earthquakes, as nearby California is, Utah is, in fact, extremely susceptible to strong earthquakes and to damage from them. Earthquakes are common in and near the Wasatch Mountains, and the potential earthquake hazard there puts much of the state's population at risk. A U.S. Geological Survey study of possible dam failure revealed that a powerful earthquake in Utah might kill many people by drowning if dams failed. Dam failure, the study suggested, might kill about 10 times more people than the earthquake itself.

A notable pair of earthquakes in Utah history occurred at Elsinore on September 29 and October 1, 1921. The earthquakes affected a wide area and followed several weeks of preliminary activity. A schoolhouse in Elsinore was damaged severely, and many people could have been killed if the damage had occurred during school hours. Another powerful earthquake took place on March 12, 1934, near Kosmo at the northern end of the Great Salt Lake. This earthquake occurred in a lightly settled area but could have caused extensive destruction in more heavily settled territory. Large quantities of water emanated from fissures, and in some areas, the ground surface was altered extensively. Another strong shock followed several hours after this one.

V

Valles, caldera, New Mexico, United States A huge basin in the Jemez Mountains in northern New Mexico near Los Alamos, the Valles caldera is one of the world's largest calderas and has an area of some 180 square miles. The Valles caldera is inactive, although a body of partly molten rock several in diameter is thought to lie beneath the caldera. Experiments in tapping geothermal energy have been conducted around the edges of the caldera.

Valley of Ten Thousand Smokes, Alaska, United States This area of steaming ground owes its existence to the simultaneous eruptions of Mount KATMAI and Novarupta in Alaska in 1912. Siphoning magma away from Katmai, Novarupta emitted a gigantic ash flow more than 40 square miles in area and hundreds of feet thick. Vapors rising to the surface from still-hot tephra deep below generated a plain of numerous fumaroles. A National Geographic Society expedition in 1916 discovered the fumarole-filled valley, and President Woodrow Wilson soon afterward proclaimed the Valley of Ten Thousand Smokes to be a national monument. Members of one expedition to the valley tried cooking food over one fumarole and discovered that the escaping gas (mostly water vapor, with traces of hydrogen sulfide, hydrochloric acid and other compounds) would blast a frying pan out of one's hands unless one held it tightly.

Satellite view of Valles caldera, about 1978 (*Earthquake Information Bulletin*/USGS)

Venezuela Located in northern South America, Venezuela has a history of highly destructive earthquakes. An earthquake on March 16, 1812, for example, destroyed Caracas and is believed to have killed some 20,000 people. Another earthquake at Cua on May 14, 1878, wrecked the city, caused some 600 deaths and was felt in Caracas.

Veniaminof, caldera, Alaska, United States Located on the Alaska Peninsula, Veniaminof is thought to have formed from the summit of a large stratovolcano between 3,500 and 4,000 years ago. A noisy, explosive eruption occurred in 1938, and ash fell as far away as Mount Katmai. Within historical times, the greatest eruption of Veniaminof occurred in 1892. Skies over southwestern Alaska turned dark during this eruption. Veniaminof is of particular interest to scientists because it is located next to the Shumagin gap, a seismic gap that is under study as a possible site of strong earthquakes in the future. It has been suggested that Veniaminof was active for several years before and after an earthquake in 1938 but comparatively quiet for several decades before and after that span of years. This suggestion helped focus attention on Veniaminof's behavior when eruptive activity resumed there in 1983. There is reason to question whether activity at Veniaminof is a reliable indicator of seismic activity at the Shumagin gap, however, because two eruptions occurred during the "quiet" period, in 1944 and 1956. During the eruption that began in 1983, Veniaminof produced explosions, fountains and flows of lava and the growth of a cone in the caldera. Veniaminof emitted a dark ash cloud during an eruption in 1984, and a small amount of ash on another occasion in 1987.

View of Valley of Ten Thousand Smokes, Alaska (W. R. Smith, USGS)

Vent An opening through which volcanic material escapes to the earth's surface.

Venus The second planet from the sun is known to have volcanoes whose activity appears to have reworked the planet's surface extensively. Images returned in 1991 from the U.S. radar-mapping Magellan probe, in orbit around Venus, revealed evidence of fresh lava flows from the volcano Maat Mons. The summit of Maat Mons was much darker—that is, less radar-reflective—than the summits of other peaks on Venus. This relative darkness gave rise to speculation that Maat Mons might be crowned by recent lava flows. Scientists observing the radar images from Venus believed that the extreme heat and acidic atmosphere on Venus, over long periods, may have produced a layer of highly radar-reflective iron sulfide on mountains other than Maat Mons. Had volcanic activity occurred recently at Maat Mons, this sulfide coating would not yet have had time to form, and the volcano's summit would appear dark in radar images. At this writing, astronomers await additional images of Maat Mons to see if any changes appear that would indicate current volcanic activity. Maat Mons is about five miles high, or approximately the same height as Mauna Kea in Hawaii, the tallest volcano on Earth. Formations resembling LAVA DOMES have also been observed on the surface of Venus.

Vermont, United States Located in New England and close to the ST. LAWRENCE VALLEY, the state of Vermont experiences numerous minor earthquakes and an occasional strong one. Burlington experienced an earthquake on January 29, 1952, that affected an area of approximately 50 square miles and caused slight damage. Cracks in the ground some two miles long were reported in the northern portion of the city. The earthquake of April 23, 1957, at St. Johnsbury, estimated at Mercalli intensity V, caused much fright within an area about 30 miles in diameter. On April 10, 1962, another earthquake estimated at Mercalli intensity V occurred in Vermont and affected much of the rest of New England as well as portions of New York; this earthquake caused considerable damage at the statehouse, as well as minor damage in Barre. Another moderate earthquake occurred in Vermont in the autumn of 1983.

vesicles Cavities within volcanic rocks, formed by the release of dissolved gases within the molten rock immediately before it solidifies. The size of vesicles may vary from one kind and sample of volcanic rock to another. In pumice, a lightweight volcanic rock, vesicles are so numerous and tiny that they form, in effect, a "rock foam" that has had numerous applications over the centuries, from abrasives to building material. The large numbers of vesicles make pumice so lightweight that it can float on water. Vesicles are less numerous but larger in scoria, another volcanic rock. Calculations of the volume of rock released in certain eruptions must allow for vesicles, which can expand the volume of volcanic rock considerably. Vesicles are abundant in pumice and other, similar rocks produced where gas-rich magma rises to the surface from below. This situation occurs in many parts of the world where chains of volcanoes border oceans where subduction zones carry great quantities of sediments from the seabed downward toward the asthenosphere. There, miles beneath the surface of the earth, water in the sediments is thought to be dissolved in the resulting magma and carried back to the surface in the form of dissolved gas. This combination of processes appears to be active in the volcanoes of the Cascade Mountains in the Pacific Northwest of the United States.

Vestmann Islands, volcanic islands, Iceland A short chain of volcanic islands off the southern coast of Iceland, the Vestmann Islands include the sites of two famous eruptions of recent years, those of SURTSEY in 1963 and HEIMAEY in 1973.

Vesuvius, volcano, Italy The only active volcano on Europe's mainland, Vesuvius stands some 4,000 feet high and is located near Naples. The most famous eruption of Vesuvius, in A.D. 79, destroyed the nearby cities of POMPEII AND HERCULANEUM and preserved them under ash and mud until excavation began some 17 centuries later. Perhaps 16,000 people are believed to have died in the two cities. The A.D. 79 eruption of Vesuvius altered the appearance of the volcano considerably. The eruption destroyed much of the cone, except for a semicircular remnant, now know as Mount Somma, on the northeastern side of the mountain. The cone now known as Mount Vesuvius arose from the ruins of this older cone.

Records of activity at Vesuvius are unreliable for centuries after this eruption, and it seems likely that only major eruptions are mentioned. An eruption appears to have occurred around 203 and cast out ash and rocks. Another notable eruption occurred in 472, when ash emissions from the volcano are said to have fallen over much of Europe and the eruption reportedly was detected at Constantinople. Vesuvius erupted again in 512, 685 and 993. An eruption in 1036 was notable for its lava flows, thought to have been the first to occur at Vesuvius within historical times. Eruptions are also recorded in 1049 and 1139. An ash eruption appears to have taken place in 1500.

Mount Vesuvius in eruption, 1944 (USGS)

An eruption in 1631 signified a change in the behavior of Vesuvius. Previously, it had been active only on an intermittent basis, with long periods of repose between eruptions. Now the mountain entered a time of virtually steady activity. The eruption of 1631 commenced in December with vigorous explosions and a cloud that darkened the adjacent land. Ash and cinders fell, and communities immediately around the volcano were evacuated. On December 17, fissures appeared on the southwest flank of the volcano and discharged large quantities of lava. The lava advanced down the side of the volcano rapidly and flowed through several communities, including La Scala, Portici, Pugliano, San Giorgio a Cremano and Torre del Greco. The lava flow divided into many tongues before entering the sea, about two hours after emerging from the fissures. That evening, a mud rain began to fall on Naples, and large mudflows rolled through villages on the sides of the volcano. Lava flowed again from Vesuvius, this time from a fissure on the south flank. This eruption continued intermittently for several weeks before finally subsiding in January of 1632 and caused widespread devastation. Lava and mudflows destroyed more than a dozen towns. The eruption is thought to have killed some 4,000 people. Many deaths are said to have resulted from poor judgment on the part of the governor of Torre del Greco, who waited until approaching lava reached the walls of the community before giving orders to evacuate. Before the residents could flee, lava breached the walls and flowed through the streets. The mountain lost more than 500 feet of its summit, and the crater after the eruption was reportedly more than twice as wide as before.

The great eruption of 1767 was recorded in detail. It began in October with powerful explosions that lasted for two days and were followed by the opening of fissures on the northwest and southern sides of the volcano. Lava flowed from these fissures toward nearby towns. An observer on the side of the volcano, within sight of one of the fissures when the lava emerged, described a fountain of lava that shot up high into the air and was accompanied by ashes and dark smoke. The people of Naples were alarmed and flocked to churches to pray for deliverance from the eruption. One account says that a mob set fire to the gates of the cardinal archbishop's residence after he refused to produce relics of Saint Januarius. (Near the end of the eruption, the cardinal gave in to public pressure and brought out the saint's head.) Three years later, Vesuvius resumed erupting and continued to do so, on and off, for several years.

In May of 1779, another violent eruption occurred. A fissure opened on the northern side of the volcano and emitted lava for weeks afterward. On August 8, strong explosions began, and a column of fire reportedly rose from Vesuvius to an altitude of approximately two miles. This eruption devastated the community of Ottaiano, which lay in the path of the advancing fiery column and was showered with glowing scoria. Stones as heavy as a hundred pounds reportedly fell on Ottaiano, and the population took shelter as best it could, in cellars and in archways, for about 25 minutes while the cloud was passing. The tempest of rock, ash and sulfur stopped abruptly, and the townspeople fled.

After this eruption, Vesuvius remained quiet for about five years. Then eruptive activity began again, culminating in the great eruption of 1793. A fissure between Torre del Greco and Resina opened and emitted lava from several locations. These individual flows joined and became a single flow of lava that overwhelmed Torre del Greco. The advancing front of lava is said to have been more than 1,000 feet wide and extended more than 300 feet into the sea before it stopped. This eruption was remarkable for a report of boiling water about 100 yards offshore; one observer wrote that boats could not remain in the vicinity of this disturbance because the heat melted the pitch in their hulls.

Strong eruptions occurred again in 1822, 1838 and 1850. One of the volcano's most powerful eruptions within historical times took place in 1872. The volcano cracked open from summit to base on its north flank. Lava flowed from this great fissure and rolled over the towns of San Sebastiano and Massa, while ash from the crater fell across the surrounding land. This same year marked the start of an eruptive cycle that lasted until the early 20th century. Lava flowed from the crater in 1881, but the lava was so viscous that it tended to build up and form a dome-like formation rather than flow rapidly downhill. A lava dome formed from 1891 to 1894. Another lava formation, the Colle Umberto, arose near the Vesuvian Observatory in the late 1890s. In the late 1800s, a cone was forming inside the crater, and by 1906 the cone had filled the crater entirely. That year, a fissure appeared on the southern flank of the volcano. Lava flowed from the fissure, and great quantities of solid material were thrown up from the crater, falling back onto the volcano. Thick eruption clouds laced with lightning emerged from the volcano. Ottaiano again was devastated; the weight of the ashfall there crushed many buildings. More than 100 people were reported killed in San Giuseppe when the weight of ash caused the roof of a church to collapse on them. Toward the end of this eruption, a powerful outburst of gas blew away the top portion of the cone and reduced its height by more than 300 feet.

Except for minor eruptions of ash, Vesuvius remained quiet after this eruption until 1913, when eruptive activity began to rebuild the cone. The next powerful eruption occurred in 1944. Lava started flowing out of the crater on March 18, 1944. The lava flow proceeded in the direction of Massa and San Sebastiano and overwhelmed the towns. Huge amounts of scoria and ash were cast out from the crater. Three days later, the volcano began sending up tremendous fountains of lava. An outburst of gas similar to that of the 1906 eruption took place in 1944, again blasting away the upper part of the cone. The 1944 eruption caused extensive damage to an American air base near Naples and interfered with Allied operations in the region.

The character of Vesuvius's eruption has changed since the destruction of Pompeii. The 1036 eruption released lava as well as pyroclastic material, and with the 1631 eruption, a new pattern of activity was observed in which the volcano emitted basaltic lava in cyclic eruptions. Several hypotheses have been put forward to explain the change in Vesuvius's activity. One hypothesis is that a fresh supply of basaltic magma became available underneath the mountain. Another hypothesis is that the eruption of A.D. 79 used up magma rich in dissolved gases that caused explosive eruptions so that basaltic magma could flow in to replace it. A third hypothesis is that contact with sedimentary rock near the surface changed the chemistry of the magma. It has also been suggested that some combination of these factors was responsible for the change in Vesuvius's eruptions.

Vesuvius is located near the PHLEGRAEAN FIELDS, an area noted for strong volcanic activity and for dramatic uplift and subsidence along the shore.

Villarrica, volcano, Chile The stratovolcano Villarrica is located in central Chile and has erupted explosively on some 20 occasions since the mid-16th century. Several lava flows have also been observed. Mudflows associated with an eruption in 1971 reportedly killed more than a dozen people. Villarrica is located where a fault zone running north-northeast to south-southwest intersects with a chain of volcanoes oriented toward the northwest. The beautiful symmetry of the volcano is marred slightly by the rim of a caldera at slightly more than 7,000 feet. Eruptions of Villarrica commonly exhibit a variety of spectacular effects, including underground noises, loud explosive sounds, lofty ash columns, lava flows and fountains of glowing debris. The eruptive history of Villarrica extends back to 1558, when an eruption destroyed the community of Villarrica. The town was destroyed again by another eruption in 1575 that killed more than 300 people. An eruption in 1640 released a flow of lava that extended to Lake Villarrica and the Rio Tolten valley. Thereafter, unrest was reported at intervals of every several years through 1898. Loud noises from underground were noted in 1906, but there appears to have been only a slight emission of lava. An earthquake was reported in 1907. The volcano erupted in spectacular fashion in 1908, releasing an outburst of fire that set off an avalanche of snow, ice and rock from a glacier on the eastern side of the mountain. A small eruption occurred in 1908, and eruptions were recorded in 1909, 1913, 1915–18 and 1919.

A powerful earthquake on December 9, 1920, damaged a house about four miles from the crater and displaced furniture; three strong shocks were followed by numerous minor earthquakes that occurred all night at intervals of several minutes, with a stronger shock every 30 minutes or so. The earthquakes reportedly demolished several structures, including houses and stables. A violent, explosive eruption began on the night of December 10 with a fountain of fiery material. It lasted for a day and a half. In general, this eruption was characterized by explosions, at least one fountain of lava and minor emissions of tephra.

Eruptions in late 1948 and early 1949 caused considerable loss of life. The explosions of October 18, 1948, and January 1 and 31, 1949, produced great avalanches (possibly PYROCLASTIC FLOWS) that are thought to have killed 60 people. Lava flows in January of 1949 reportedly burned large areas of forest around the volcano. Eruptive activity between 1959 and 1961 was relatively minor; small explosions and emissions of tephra were reported. Destructive activity resumed in 1963 when avalanches cut several bridges in the vicinity of Lake Calafquen. In March of 1964, mudflows rolled over part of a village and killed 22 people there. Mudflows and emissions of tephra occurred during eruptions that began in October of 1971 and continued through December, killing 17 people and releasing lava that flowed some 10 miles from a vent. Activity at fumaroles intensified in September of 1979, and minor eruptions of tephra occurred in 1980. A red glow was observed at the summit in 1983, and a small emission of tephra was reported, along with minor explosions.

A short eruption of ash, together with an increase in earthquake activity, occurred in August of 1984. On October 30, minor explosions and emission of lava began, and strong earthquake activity was recorded near the surface on November 4–5. Although instruments recorded no more earthquakes, a strong subterranean rumbling and a tremor were reported. Explosions and emissions of lava occurred between November 10 and 13, and mild unrest continued for several days afterward. A prominent bulge was seen on the southwestern flank of the volcano on November 18, followed by a glow and emissions of ash through April 18. An increase in earthquake activity was reported again between June and November 1985.

Virginia, United States The state of Virginia is located in a region of moderate seismic activity in the southeastern United States but has experienced strong earthquakes on occasion since colonial times. Virtually no portion of the state, from the coastal plain in the east to the mountains in the west, is completely free of seismic activity. For example, central Virginia experienced an earthquake on August 27, 1833, that was felt from Norfolk in the eastern part of the state to Lexington in the west, and from Baltimore, Maryland south to Raleigh, North Carolina. The shock was perceptible in Washington, D.C. and rattled windows energetically in Lynchburg. Public fright at Fredericksburg was considerable, and the shock at Charlottesville is described as "severe."

Like the other states along the Atlantic seaboard of the United States, Virginia has a broad coastal plain of unconsolidated sediment in the east. This area could undergo liquefaction, with a resultant potential for serious damage to property, in any future major earthquake, especially in areas where groundwater rises close to the surface.

vog Short for "volcanic smog," this expression refers to a troublesome atmospheric phenomenon—an acidic haze—sometimes observed during eruptions at Kilauea in Hawaii. Vog is thought to form partly from acid droplets released when lava reaches the sea and generates steam.

volcanic cone A conical formation resulting from expulsion of lava and/or clastic material from a vent. Wizard Island in Crater Lake is a familiar example of a volcanic cone. (See also VOLCANO.)

volcanic dome A mound of thick, slowly flowing lava that has formed over the vent of a volcano. A volcanic dome has a characteristic mounded shape and has a surface covered with large chunks of igneous rock produced by the cooling of the lava. A whole series of domes may form, be destroyed and then form again during explosive eruptive activity. The formation of a dome indicates that gas-rich lava has ceased (at least temporarily) to flow, and more viscous lava with lower dissolved-gas content has replaced it.

volcanic neck A tower-like landform consisting of the igneous core of an extinct volcano whose outer layers have eroded away. Devil's Tower, Wyoming is a familiar example of a volcanic neck.

volcanic tremors See HARMONIC TREMORS.

volcanism Any and all volcanic activity. Volcanism may be spectacular, involving great fountains of molten rock spouting high into the air, or tremendous explosions caused by the buildup of gases within a volcano. Alternatively, volcanic activity may be relatively quiet, releasing only small amounts of lava, or even restricted to minor emissions of vapor. Volcanism at a given site may be almost continuous over a period of centuries, or it may be comparatively infrequent, with eruptions separated by intervals of centuries. Some volcanic events are highly destructive of human life and property; other incidents of volcanism may be dramatic but do little harm. Volcanism may take place in a wide variety of environments, from dry land to the bottom of the ocean. Eruptions have even occurred under glacial ice. Depending on the circumstances of an eruption, products of volcanism may include fine ash, blobs of molten rock solidified in mid-air (called bombs) and floods of lava that can cover thousands of square miles. Not always an isolated phenomenon, volcanism may occur in combination with earthquakes, some of them capable of causing serious damage.

Devil's Tower, Wyoming, a famous example of a volcanic landform (N. H. Darton, USGS)

Volcanism also is thought to have a significant influence on global climate, because fine ash released into the upper atmosphere during a major eruption can intercept enough incoming sunlight to reduce the warmth the earth receives from the sun and thus reduce surface temperatures. In some cases, volcanic eruptions have been correlated with short-term changes in climate that resulted in major hardship for humans over large portions of the earth.

Seismic sea waves, or tsunamis, are also associated with certain eruptions along the seacoasts of the world. A famous case in point is the 1883 eruption of the volcanic island Krakatoa between Java and Sumatra, when the explosion of the volcano generated a tsunami that circled the world several times before finally dissipating and was responsible for thousands of deaths along shorelines near the eruption. Mudflows may cause extensive destruction in the vicinity of a volcano when eruptions melt ice and snow on the summit and send large quantities of mud and rock flowing down the flanks of the mountain. The flows may travel miles from their point of origin and bury the underlying land to a depth of many feet. Other phenomena capable of causing great destruction are the nuée ardente, a fiery mixture of gas and glowing rock capable of incinerating a landscape almost as effectively as a nuclear explosion, and emissions of toxic gases, which may kill by asphyxia anyone caught in them.

The economic effects of volcanism are considerable. Negative effects include the destruction of homes, industrial facilities, communications equipment, vehicles, airports, highways and other property, as well as damage to crops and to health. Pollution of water supplies is also a possibility. The economic impact of even a reasonably small eruption may run into the billions of dollars if the event occurs in or near a densely settled or intensively cultivated area. Cleanup alone may be so expensive that the most practical policy is to abandon the affected property, as the United States military did when a 1991 eruption of Mount Pinatubo in the Philippine Islands buried a nearby air base under ash and made the facility unusable. Although economic damage from many eruptions is largely confined to the area immediately around the volcano, under certain circumstances an eruption may cause a change in the global environment that spreads economic hardship over a wide area. An eruption of the Indonesian volcano Tambora, for example, is thought to have been at least partly responsible for a global cooling that had disastrous effects on agriculture in the northern hemisphere in the early 19th century.

Less directly, but no less significantly, ancient volcanism has influenced economics—and thus the rise and fall of nations and whole empires—by concentrating rich deposits of precious minerals such as gold and silver at certain locations. Volcanism and associated phenomena along the Andes Mountains of South America gave those mountains such a heavy concentration of gold-bearing rocks, for example, that the wealth of the Inca Empire became an attraction for invaders from Spain, who conquered the Incas and plundered the mineral wealth of their empire. In similar fashion, volcanic formations in South Africa have yielded large quantities of diamonds, with consequent effects on the South African economy. Not all rich mineral deposits associated with volcanic activity are so exotic; copper and other minerals have played an equally important, if not greater, role in the wealth of the countries in which they have been mined. Volcanism may create new land that may be settled profitably, and the tephra (ash and other ejecta) from an eruption may itself have considerable economic value, in applications ranging from abrasives to building material. A minor but colorful economic benefit of volcanism in some places is tourism. Certain volcanoes in the United States benefit especially from tourism, notably Oregon's Crater Lake, the spectacular water-filled CALDERA left by an ancient eruption; Mount St. Helens, whose explosive eruption in 1980 made the then little-known volcano famous throughout the world; and the volcanoes of Hawaii, whose relatively safe but dramatic eruptions draw many visitors.

Although volcanism has been studied intensively for thousands of years, the science of volcanology operates under certain limitations. The phenomena under study, as a rule, are difficult to study under highly controlled conditions in a laboratory and can pose significant risk to scientists trying to make observations on the spot. Except in a few locations such as Hawaii, eruptions are only intermittent and cannot be predicted far in advance. Many eruptions occur in isolated areas that are difficult to reach with the personnel and equipment required for on-site studies. Despite these inconveniences, however, the science of volcanology has advanced rapidly in the 20th century, thanks to a variety of new sensing techniques and computer analysis, which together have made possible the collection and processing of great quantities of data on eruptions. The 1980 eruption of Mount St. Helens alone expanded the scientific knowledge of eruptions tremendously by giving volcanologists an active volcano to study within easy reach of numerous well-equipped laboratories and other research facilities. (See also "YEAR WITHOUT A SUMMER.")

volcano A volcano is defined, in a broad sense, as an opening in the earth's crust through which MAGMA escapes to the surface as LAVA. More specifically, the word *volcano* refers to mountains produced by volcanic activity, known as volcanism or vulcanism. (The words volcano, VOLCANISM and vulcanism all are derived from the Latin *Volcanus*, or Vulcan, the god of fire in Roman mythology.) In a volcanic eruption, magma and/or hot gases escape from an underground reservoir of magma to the surface through a relatively narrow vent, or conduit. Eruptions differ greatly in character from one volcano to another and sometimes within the history of the same volcano. Some eruptions are extremely violent and involve great outbursts of ash, gas and lava. These eruptions produce cinder cones and composite cones.

A cinder cone is made up of fragments of rock ejected from the vent. Parícutin in Mexico is an example of a cinder cone. These fragments tend to be low-density rock formed when dissolved gases in the magma bubble out of solution as the rock solidifies, producing what might be called "rock foam," full of small cavities that give the rock a "frothy" texture. Rock fragments between about half an inch and two inches in diameter are called cinders and constitute most of the cinder cone. (Cinders are distinct from ash, finer material that winds may carry for great distances away from the volcano.) Mixed in with the cinders, in some cases, are volcanic bombs formed as large masses of lava are ejected intact. Cinder cones may grow rapidly, rising hundreds of feet in the first few days or weeks of their existence, but seldom reach heights of more than 1,000 feet. Cinder cones are steep-sided; the slope of a newly formed cinder cone is approximately 28°. The CRATER tends to be large and to have a rim higher on one side than the other because of prevailing winds that carry the volcano's output in a given direction. Cinder cones may occur virtually anywhere the appropriate kind of magma rich in dissolved gases can reach the surface. Clusters of cinder cones are commonplace. Eruptions of cinder cones may include lava flows.

A composite volcano is more complex than a cinder cone. Composite volcanoes are made up of layers of cinder and ash alternating with lava. Because of these alternating strata, composite volcanoes are known as stratovolcanoes. A stratovolcano has steeply sloping sides, as cinder cones have, but has greater structural strength due to the rigid lava layers inside it. Stratovolcanoes may reach thousands of feet in height. Examples of stratovolcanoes are Vesuvius in Italy, Mount Fuji in Japan and Mount St. Helens in the United States. Most stratovolcanoes are concentrated in two parts of the world: the "Ring of Fire," a belt of intense volcanic and earthquake activity encircling the Pacific Ocean basin, and the Mediterranean Sea. Eruptions of stratovolcanoes involve release of hot gases, ash, cinders, bombs and lava. Eruptions may occur with such violence that they demolish part of the mountain, as happened in the 1980 eruption of Mount St. Helens. An even more violent eruption, that of the volcanic island Krakatoa in 1883, destroyed most of the island and generated a tsunami, or seismic sea wave, that inundated shorelines in the vicinity and killed thousands of people. Eruptions of stratovolcanoes sometimes produce calderas, when the entire central portion of the volcano is blasted away or, alternatively, collapses as magma within it is drained away.

Young stratovolcanoes are characterized by their conical shape and symmetry. Older stratovolcanoes are less symmetrical because water and ice have eroded their flanks. Erosion produces a characteristic radial drainage pattern down the sides of the volcanoes. Another, separate drainage pattern occurs in some cases within the crater, as water flows downward from the rim to the floor of the crater. As erosion continues, the mountain is worn down gradually until only the volcanic neck, the solidified mass of magma where the vent once extended upward to the summit, remains. Ship Rock in New Mexico is a famous example of a volcanic neck, from which DIKES radiate. Because of their great resistance to erosion, lava flows from stratovolcanoes may endure as mesas rising above the surrounding terrain.

When eruptions are comparatively mild and involve only small, nonexplosive eruptions of lava, the product is a shield volcano, a gently sloping volcano (perhaps 5°) with a wide, flat top. The Hawaiian Islands are shield volcanoes, as is Olympus Mons on Mars. Shield volcanoes build up slowly through repeated outpourings of lava. Their lava has less dissolved gas in it than the lava of stratovolcanoes and does not form great quantities of ash and cinders. Lava from shield volcanoes may solidify as volcanic glass or basalt. Shield volcanoes are also characterized by their large central depressions, which may be miles in diameter. On the floor of these central depressions, one may find fire fountains of lava. (See also PUMICE; SCORIA.)

volcano, active See ACTIVE VOLCANO.

vulcanism See VOLCANISM.

Vulcano, volcano, Italy The volcano that gave its name to volcanoes as a phenomenon, Vulcano is a stratovolcano located west of Italy in the Lipari Islands. The eruptive history of Vulcano is difficult to reconstruct for some periods because mythology

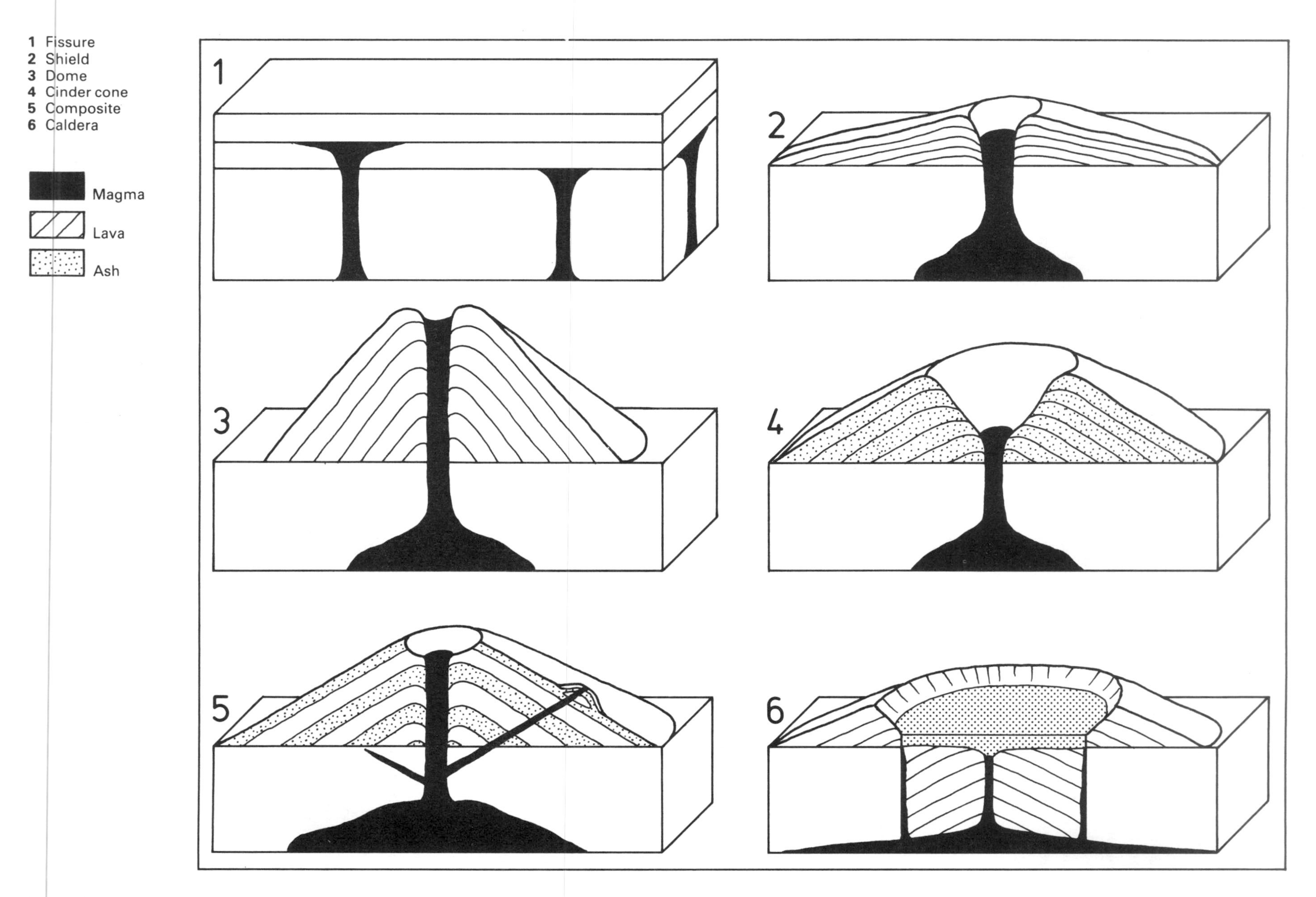

Types of volcanoes (© Diagram)

has influenced accounts. It appears safe to say, however, that Vulcano has been active on a frequent basis for thousands of years. The Greek authors Aristotle and Thucydides both mention eruptions of Vulcano in the fourth and fifth centuries B.C. respectively. Aristotle apparently wrote about an eruption of Vulcano that was occurring at the time of writing. He reports that ash from this eruption reached the mainland of Italy. A small island called Vulcanello (now connected to the northern shore of the island of Vulcano) is thought to have emerged in the second century B.C., although there is some question about this estimated date. An eruption in 126 B.C. is believed to have enlarged Vulcanello greatly. Vulcano is mentioned in the biography of Willebald, the Anglo-Saxon bishop of Eichstatt. Willebald lived in the eighth century A.D. and was reportedly one of the most widely traveled men of his day. On a journey through the Mediterranean, Willebald is said to have stopped at Vulcano and desired to look into the crater during an eruption, but it proved impossible to climb the flanks of the volcano, covered as they were with scoria and ash. So Willebald and his companions had to stand at a distance and watch the flames of the eruption, which also involved a huge column of smoke and loud roaring sounds.

Recorded observations of eruptions at Vulcano are few in the centuries between the start of the Christian era and the early 1400s. This lack of information on Vulcano may be due in part to the sinister reputation that the Lipari Islands once had as a haven for pirates. An eruption in 1444 was documented by a Sicilian who reportedly described underwater eruptions and said that great rocks were cast out of the mountain. Later reports indicated that Vulcano and Vulcanello were separated by water after this eruption and many new rocks had appeared in the waters around the island. A visitor to Vulcano in 1727 mentioned that the noise from a virtually nonstop eruption of the volcano was so tremendous that it prevented him from sleeping at night. At this time, the volcano reportedly had two craters. If that was indeed the case, then some subsequent event apparently destroyed those craters and replaced them with a single crater because a visitor in the late 18th century mentioned only one crater existing in the volcano.

Vulcano is said to have erupted on a nearly continuous basis during the 1730s, but a period of relative quiet appears to have followed because observers of the volcano around 1768 to 1770 reported no dramatic activity. Eruptions occurred in 1771 and 1775. The 1775 eruption is said to have generated a flow of obsidian. A powerful eruption in 1786 lasted more than two weeks and cast out vast amounts of solid material. Underground noises from this eruption are said to have been heard throughout the surrounding islands. The island was evacuated as a result of this eruption and remained uninhabited for decades afterward. Volcanic activity resumed at Vulcano in 1873 when an eruption began that lasted more than a year and disrupted the mining of sulfur from the crater. Another eruption commenced in 1888 and lasted through May of 1890. This eruption involved powerful explosions and a thick cloud that spread white ash over a broad area. Some of the lava masses, or bombs, cast out by this eruption are said to have been gigantic. One had a reported volume of more than 300 cubic feet. There have been no major eruptions of Vulcano since the late 19th century.

Vulsini, volcano, Italy The volcano Vulsini is located near the western coast of central Italy and is part of the Roman volcanic province. The volcano is thought to have grown in three phases of activity: pyroclastic activity and extrusion of lava, producing the main portion of the Vulsini complex; large eruptions of IGNIMBRITE and the formation of a caldera by collapsing; and formation of parasitic vents after the caldera originated. Tertullian mentions a mysterious incident of "lightning" destroying Volsinii, a nearby town, around 500 B.C., and eruption of flame was reported in 104 B.C. Immediately to the east of the caldera is the Lago di Bolsena, a lake that occupies a depression believed to have originated as a GRABEN.

W

Wasatch Mountains, Utah, United States The Wasatch Mountains occupy part of the BASIN AND RANGE PROVINCE of the western United States and are noted for earthquake activity. Exposures of unweathered rock along the base of the mountains indicate that powerful earthquakes appear to have lifted the mountains considerably in recent geologic time, although before Utah was settled in the 19th century. The Wasatch Fault Zone has not generated a highly destructive earthquake since the mid-19th century, but the potential for such earthquakes exists and is cause for concern in Utah, because most of the state's population lives close to the fault zone.

Washington, United States The state of Washington, in the Pacific Northwest by the Canadian border, is among the most seismically and volcanically active portions of the United States, because of the ongoing collision between the westward-moving continental landmass and the JUAN DE FUCA CRUSTAL PLATE beneath the PACIFIC OCEAN. This region has the pattern of volcanism and deep-focus earthquake activity commonly associated with subduction zones, in this case where a plate of oceanic crust is overriden by an advancing plate of continental crust. Washington has two of the world's most famous volcanoes, Mount RAINIER near Seattle and Mount ST. HELENS near Portland, Oregon. Volcanism has been responsible for shaping much of the landscape of Washington, through such mechanisms as ashfalls and lava flows.

Earthquake activity under the state is cause for concern on the part of agencies responsible for disaster planning because the potential exists for highly destructive earthquakes in Washington, especially in the densely settled areas surrounding Puget Sound. Although the crust beneath the ocean immediately to the west of Washington is relatively quiet from a seismic standpoint, earthquake activity increases as one travels eastward across the Olympic Peninsula that separates Puget Sound from the Pacific Ocean. Puget Sound overlies an area of intense earthquake activity, where shocks in the range of 4.0 on the Richter scale occur every few months on the average and earthquakes of approximately magnitudes 2.0 to 3.0 are more commonplace. In a 1983 report to Congress, the U.S. Geological Survey (USGS) declared, "A growing body of data suggests that a great earthquake . . . could occur in the Pacific Northwest," meaning an earthquake of perhaps Richter magnitude 8.0 or above, roughly comparable to the earthquake that destroyed much of San Francisco in 1906. "In fiscal year 1983," the report concluded, "two [USGS]-university seismological teams completed a study of earthquake potential in the Pacific Northwest. They concluded that the northwestern United States may [represent] a major seismic gap which is locked and presently seismically quiet, but which may fail in great earthquakes in the future." Special conditions in the Pacific Northwest make Washington highly susceptible to damage from powerful earthquakes. Much of the Puget Sound area, for example, is settled in the form of communities built largely on hillsides, where the potential for landslides is great. Also, river deposits and other poorly consolidated materials, vulnerable to liquefaction in a major earthquake, underlie much of the developed areas in this region.

Although the state of Washington has been occupied by settlers of European descent for only about two centuries, and the record of notable earthquakes during the early years of their occupation is not always reliable, Washington already has an impressive history of major and moderate earthquakes. Considerable damage occurred in Seattle as a result of a strong earthquake on August 6, 1932. One of the most powerful and destructive earthquakes in the history of Washington occurred on April 29, 1965, near Seattle. This earthquake, estimated at Richter magnitude 6.5, caused property damage estimated at more than $12 million and affected an area of more than 100,000 square miles. The earthquake generated effects of Mercalli intensity VII over a wide area and intensity VIII in some locations in the Seattle area. An earthquake on May 18, 1980, preceded the violent eruption of Mount St. Helens by only a few moments and was part of a sequence that started in late March and led up to the catastrophic eruption. The earthquake occurred so soon before the explosive eruption that it is difficult to tell the effects of the earthquake from the acoustical effects of the explosion. (See also JUAN DE FUCA CRUSTAL PLATE; PLATE TECTONICS.)

Watchung Mountains, New Jersey, United States The Watchung Mountains are believed to represent a set of lava flows that may have extended into New England. The Watchung lavas appear to have been laid down in three episodes of volcanic activity. Lavas from these episodes make up First Watchung Mountain and Second Watchung Mountain, as well as Hook Mountain, Packanack Mountain and other ridges in the vicinity.

water Water plays an important part in earthquake and volcanic activity. Water in soil near the surface, for example, may make the soil more vulnerable to liquefaction, the condition in which earthquake waves passing through the soil increase pore-water pressure enough to make the soil lose cohesion and behave as a fluid. As a result, structures built atop the soil may undergo severe damage. This effect of water near the earth's surface has been responsible for much of the damage in some of the most destructive earthquakes in history, notably the San Francisco earthquake of 1906, which was notable for the failure of buildings constructed on landfill, and the Mexico City earthquake of 1985, where the destructive power of earthquake waves passing under the city was augmented by the unconsolidated nature of the earth below—sediments that were laid down as part of the bed of an ancient lake. Generally speaking, the soggier the soil, the greater its susceptibility to liquefaction.

Groundwater has been known to behave in strange and sometimes violent ways during earthquakes, gushing out of the earth in torrents and even spurting high into the air in the form of fountains. Sand blows, curious crater-like formations created when an earthquake forces wet sand out at the surface, are another manifestation of water's behavior in seismic events. Sand blows formed prominently, for example, during the great Charleston, South Carolina earthquake.

The levels of lakes and rivers have been seen to change markedly at the time of earthquakes, and disturbances on the surface of bodies of water during earthquakes can have devastating effects onshore. The most dramatic examples of such disturbances are of course tsunamis, the giant seismic sea waves that ripple outward from the source of an earthquake and can carry destructive energy across an entire ocean in a matter of hours. (The tsunami from the explosion of the volcano Krakatoa in the 19th century, for example, was responsible for most of the deaths from that event and circled the globe several times before finally dissipating. Almost as powerful was the tsunami that followed the Alaskan Good Friday Earthquake, devastating the shoreline of southern Alaska and carrying destruction hundreds of miles south to Crescent City, California. On occasion, tsunamis have been known to reach half a mile or more inland from the shore before receding and have left ocean-going vessels stranded on dry land hundreds of yards from the water.

In volcanic eruptions, water dissolved in magma, or molten rock rising from deep underground, affects the behavior of the magma during eruptions. This effect is commonly observed along ocean margins, where subduction zones are carrying water-bearing sediments deep into the earth's interior along with the descending slab of oceanic crust on which the sediments lie. As the descending piece of solid rock melts, the water in the sediments is dissolved in the molten rock under great pressure. When the magma rises through the mantle and the crust toward the surface, pressure on the magma is relieved, and the dissolved water in the molten rock has an opportunity to come out of solution.

This process is comparable to what happens when an ordinary can of soda is opened; as pressure on the soda (which is filled with dissolved carbon dioxide) is relieved, the gas bubbles out of solution and creates, one might say, a miniature "eruption" of gas and liquid from the can. In a volcanic eruption, much the same thing may occur; only the molten rock solidifies as the gas within it bubbles out of solution, and the result may be one of several kinds of volcanic rock, such as pumice, a finely textured, "foamy" volcanic rock that is light enough to float on water, and scoria, in which the gas bubbles, or vesicles, are larger than in pumice. As this "foamy," solidified rock breaks up and is ground to bits in the maw of a volcano, it becomes fine ash. Under certain circumstances, gases released from solution in molten rock during an eruption may build up inside the volcano and eventually produce a colossal explosion. This is what happened at Krakatoa, where the effect of water meeting molten rock at the surface was to put a damper on the volcano and allow gases released from the magma to attain tremendous pressures, which caused the island to explode with devastating effect. At Mount St. Helens in Washington state in 1980, gases built up under the north flank of the volcano until that flank burst open and sent a powerful air blast sweeping over the landscape, leveling forests. The potential for such events is relatively high where subduction zones carry water-saturated sediments into the earth's interior near chains of volcanoes, such as the Cascade Mountains, of which Mount St. Helens is part.

Mudflows are another destructive manifestation of water in volcanic eruptions. Heat released from an eruption on the summit of a volcano may melt ice or expel the contents of a crater lake, sending great flows of mud down the flanks of a mountain. These flows may accumulate to great depths and bury entire communities. Many volcanoes exhibit the phenomenon of phreatic eruptions, in which heated water burst out explosively in the form of steam. When phreatic eruptions occur in combination with eruptions of magma, the result is called a "phreato-magmatic" eruption.

Hydrothermal activity, the result of water heated underground by bodies of molten rock near the earth's surface, is a highly complex phenomenon that has had a considerable impact on human history. At some locations, the heated underground water may be tapped for generating electrical power or for heating. Elsewhere, minerals dissolved in the water may be deposited in the form of rich ores of gold, silver, copper and other metals of economic importance. Such activity is thought to have formed many of the valuable ore deposits of the western United States and western South America, for example. One especially violent and photogenic kind of hydrothermal activity is the geyser, a fountain of water and vapor that may reach heights of more than 100 feet. Some of the world's most famous geysers are located in Yellowstone National Park in the United States and in Iceland.

Sometimes the interaction of water and volcanic heat creates formations of rare beauty and complexity. At Mount Rainier in the Cascade range of the United States, for example, volcanic heat at the summit of the mountain has reached an equilibrium with the accumulation of ice and snow to produce an intricate and fragile network of caverns within the ice. These caverns are large enough to accommodate a person. (See also GEOTHERMAL ENERGY.)

Wegener, Alfred (1880–1930) A German meteorologist, Wegener is widely credited with originating the concept of continental drift, which gave rise to the theory of PLATE TECTONICS, the basis for much of modern geology. Although Wegener appears to have taken the idea of continental displacements from other authors, notably the American geologist Frank Bursley Taylor, Wegener presented a comprehensive model of continental motion and the formation of the earth's land masses in his book *The Origins of Continents and Oceans.* Wegener suggested that the continents are not fixed in position but instead float atop denser rock in the earth's mantle, just below the crust. Wegener believed the continents as arranged today were once united as a single supercontinent, which Wegener called Pangaea, meaning "all land." This supercontinent was surrounded by a world ocean called Panthalassa. The Atlantic Ocean basin, Wegener wrote, was created during the breakup of Pangaea, as the continents moved apart from one another. The energy for the continents' movements, Wegener suggested, came from a tidal motion that supposedly produced a net westward motion of landmasses. Additional energy for continental motion was believed to come from a *Polflucht,* or "flight from the poles," resulting from the earth's rotation.

Wegener's formulation had numerous flaws. He supported his account of the formation of the Atlantic Ocean basin, for example, with inaccurate data that indicated Greenland was moving westward several kilometers per year, when the actual figure is nearer several centimeters per year. For such shortcomings, Wegener's work was criticized.

Wegener graduated from Berlin University and served later on the faculty of Marburg University. He also served as an officer in the German army during World War I. Wegener himself appears to have been a colorful figure. Athletic and adventurous, he took part in the exploration of the atmosphere, setting a world record in 1902 for long-distance balloon flight (56 consecutive hours aloft) in the company of his brother Kurt. Wegener was fascinated with Greenland and joined several expeditions to the island. On his final visit in 1930, he died, presumably of a heart attack, and was buried on the icecap.

Westdahl, volcano, Alaska, United States Westdahl, in the Aleutian Islands, stands about 6,000 feet high and is located approximately 700 miles southwest of Anchorage. The ash cloud from a 1991 eruption reached an altitude greater than four miles.

West Virginia, United States The state of West Virginia lies mostly within a region of minor seismic risk, although the eastern border of the state lies at the edge of a region of higher seismic activity that extends through the western portion of Virginia. An earthquake on May 2, 1853, affected some 72,000 square miles in West Virginia (which had not yet become a separate state) and Virginia and was felt very strongly in Lynchburg, Virginia; this earthquake also alarmed residents of Wheeling and Parkersburg in West Virginia.

Whakatane Fault, North Island, New Zealand A zone of intense volcanic activity, the Whakatane Fault is the location of several active volcanoes, including NGAURUHOE, RUAPEHU, Tarawea and Tongariro. Numerous other volcanic mountains, not presently active, are also found here.

White Mountains, New Hampshire, United States The White Mountains are remarkable for the role of PLUTONS in their formation. The mountains exhibit interesting circular or crescent-shaped ranges such as the Ossipee Mountains, the Pliny Range and the Crescent Range. These ranges are thought to have formed when a magma chamber several miles below the surface moved a cylindrical mass of rock directly above the magma, possibly pushing the cylinder upward and then allowing it to fall as pressure in the magma chamber diminished. In subsequent eruptions, according to this scenario, magma flowed upward around the edges of the fallen cylinder and formed a ring-shaped pluton called a ring DIKE. (Similar structures are found in Africa, Australia, Scotland and Scandinavia.) The Rattlesnakes, a pair of mountains north of Bristol and Laconia, are thought to have originated in this manner. Ring dikes north of the Presidential Mountains indicate that a large volcano once must have stood in that location.

On a larger scale, the White Mountains are believed to have originated from a rising plume of molten rock inside the mantle that also is thought to have given rise to the hills near Montreal, Quebec as well as the New England Seamountains, a string of submarine volcanoes extending from southern New England out into the Atlantic Ocean for approximately a thousand miles. This scenario for the formation of the mountains and seamount chain has been questioned, however, because it would require a progressive increase in the age of formations as one moves westward, and estimates of age do not support this scenario. A structure similar to the ring-shaped formations of the White Mountains has been detected off the eastern coast of Massachusetts and is suspected of involvement with the two powerful earthquakes that occurred in the Boston area in 1727 and 1755. Earthquake activity in the White Mountains has resulted in damage on occasion, notably in December 1940, when two earthquakes knocked down chimneys in the vicinity of the Ossipee formation and generated cracks in the crust on snow there.

White Terrace, North Island, New Zealand A terraced formation produced by HYDROTHERMAL ACTIVITY, the geyserite White Terrace was located at Lake Rotomahana. Eruptions of a geyser supplied the mineralized waters that gave rise to the White Terrace, which covered more than seven acres. A nearby, similar formation was the Pink Terrace. The two formations were considered among the great natural treasures of New Zealand. They were destroyed during an eruption of TARAWERA in 1886.

Wisconsin, United States Wisconsin is not noted for strong earthquake activity but is affected on occasion by earthquakes in nearby Illinois, northern Michigan and Ohio.

Witori, volcano, Papua New Guinea The Witori volcano is located on New Britain Island and stands approximately 2,000 feet high. The active cone of the volcano is called Pago and erupted between 1911 and 1933, releasing a flow of lava that formed a pool on the floor of the caldera. Another eruption may have occurred around the year 1900, but this is not certain. The caldera is believed to have formed some 2,600 years ago.

Wizard Island, Oregon, United States A cinder cone, Wizard Island rises from the waters of CRATER LAKE.

Wolf, volcano, Galapagos Islands, Ecuador A shield volcano on Isabela (Albemarle) Island, Wolf has a history of unrest dating back to 1800 and possibly earlier. An explosive eruption was recorded in 1800, and lava flows were reported in 1925–26 and 1933. Other eruptions may have occurred in 1935 and 1938. A 1948 eruption involved explosive activity and lava flows, and lava

Mount Wrangell in eruption, 1902 (W. C. Mendenhall, USGS)

flowed again from the volcano in 1963. In March of 1973, an earthquake swarm occurred under the southeast flank of the volcano. Strong and steady rumbling sounds from the caldera were heard late in 1973, but no evidence of an actual eruption was found. An eruption in 1982 was remarkable in that it occurred in two hemispheres at the same time. One eruption of lava, in the caldera, took place in the northern hemisphere, while another lava flow only six miles away occurred in the southern hemisphere.

Wrangell Mountains, Alaska, United States The Wrangell Mountains include a large group of volcanoes, including Mount Drum, Mount Jarvis, Mount Sanford and Mount Wrangell. These volcanoes are thought to have grown with unusual rapidity. Two calderas, one approximately 10 miles wide and the other some 3 to 4 miles wide on average, have been described on Mount Wrangell. The summit caldera includes three craters, North Crater, East Crater and West Crater. Eruptions may have occurred at Wrangell in 1819 and 1884, although this is not certain. In 1899 Wrangell reportedly started emitting steam in great quantities following a powerful earthquake and continued smoking afterward. Steaming intensified in 1907. Melting of ice accelerated at Wrangell around 1965, and since that time much of the ice that existed in North Crater before 1965 has disappeared. Temperatures increased markedly at fumaroles in the early 1980s, and plumes were reported rising more than half a mile above the volcano in April of 1986. It has been suggested that the recent increase in heating at Wrangell had something to do with the powerful GOOD FRIDAY EARTHQUAKE in Alaska in 1964.

Wyoming, United States The state of Wyoming shows most of its seismic activity along its western border, where it touches on a zone of high seismic risk extending from southwestern Montana southward into southwestern Utah.

Wyoming has a history of abundant volcanic activity, especially in the vicinity of Yellowstone. HYDROTHERMAL ACTIVITY is found at YELLOWSTONE NATIONAL PARK in the form of numerous geysers. Volcanic and related activity at Yellowstone is believed to be the work of a giant reservoir of magma beneath the Yellowstone caldera.

Y

"year without a summer" The year 1816 is commonly called the "Year Without a Summer" because of its extremely cool temperatures, which have been attributed at least in part to the eruption of the volcano TAMBORA the previous year. Finely divided solid material ejected by Tambora is thought to have entered and lingered in the upper atmosphere, intercepting sunlight and reducing temperatures on the ground. The winter weather of 1815–16 lingered well into the spring, and summer temperatures in 1816 were so far below normal that snow reportedly fell in New England. The unusually cold winter disrupted agriculture in Europe and North America, and near-famine conditions are thought to have developed in portions of Canada. (See also CLIMATE, VOLCANOES AND.)

Yellowstone National Park, Wyoming, United States Best known for its geysers, such as the famous "Old Faithful," Yellowstone National Park occupies part of the Yellowstone Plateau, which is contiguous with the Snake River Plateau and is seen as an extension of it. The Yellowstone Plateau stands approximately 8,000 feet above sea level and is thought to have been formed by a tremendous set of volcanic eruptions as the North American crustal plate moved westward over a "hot spot" that may have resulted from the impact of a giant meteorite. According to this scenario, crustal rock melted over the "hot spot" and produced large amounts of magma that made their way to the surface in eruptions. (On the other hand, there is evidence from the chemical composition of rocks at the surface to indicate that magma involved in the Yellowstone eruptions may have originated deep in the mantle rather than by melting of continental crust near the surface.)

Although early eruptions are believed to have been small, the Yellowstone area later was the site of three great eruptions, one of which expelled more than 600 cubic miles of solid material. As huge amounts of magma were expelled, the roof of the magma chamber underlying Yellowstone collapsed and created a CALDERA occupying an area of about a thousand square miles. Vast deposits of IGNIMBRITE formed during this eruption. Another cycle of eruptions released some 70 cubic miles of magma and produced the Island Park caldera, some 17 miles wide. A third and more powerful cycle of eruptions, though not as great at the initial cycle, is thought to have deposited ash over much of North America. Deposits linked to this eruption have been found in Kansas. Again, a caldera

Fossil trees in Yellowstone National Park, Wyoming about 1890 (J. P. Iddings, USGS)

formed, this one known today as the Yellowstone caldera and measuring about 30 by 50 miles. Further eruptions at Yellowstone are possible.

Besides its numerous geysers and other evidence of HYDROTHERMAL ACTIVITY, Yellowstone is noted for its petrified forest, where eruptive activity buried and preserved more than 25 layers of forest growth believed to represent some 20,000 years of geologic time. The petrified forest has been exposed in a section more than 1,000 feet thick along the Lamar River in Yellowstone Park. The petrified forest of Yellowstone should not be confused with the petrified forest of Arizona, which appears to have formed under different conditions than that of Yellowstone. Whereas the Yellowstone forests stand where they were buried by volcanic eruptions, the trees of the Arizona petrified forest were apparently deposited and buried by floodwaters.

APPENDIX A

Chronology of Earthquakes and Eruptions

The following list represents a selection, though not a complete list, of major earthquakes and volcanic eruptions in history. In most cases, casualty figures are estimates. (There can be considerable disagreement among various sources where casualty figures are concerned, and readers should bear in mind that many of the statistics cited here on numbers of people killed in earthquakes, tsunamis and eruptions represent only informed guesses.) For earthquakes that occurred before the development of modern seismological instruments and the Richter scale of earthquake magnitude, magnitudes also represent estimates. Exact magnitude figures were not available for some earthquakes of the 20th century.

c. 1470 B.C. Santorini, Mediterranean. Eruption is thought to have killed thousands and destroyed the Minoan civilization.

1226 B.C. Etna, Sicily, volcanic eruption. Other eruptions occurred in 1170 B.C., 1149 B.C., 525 B.C., 477 B.C., 396 B.C., 140 B.C., 126 B.C., 122 B.C., A.D. 1169, 1329, 1536, 1669, 1693, 1755, 1843, 1928.

217 B.C., June. North Africa, earthquake. More than 100 cities were destroyed, and more than 50,000 people are estimated to have been killed.

A.D. 19. Syria, earthquake. More than 100,000 people believed killed.

79, August 24. Vesuvius, Italy, volcanic eruption. This eruption destroyed—and simultaneously preserved—the cities of Pompeii and Herculaneum. Other notable eruptions occurred in 203, 472, 1036, 1049, 1198, 1302, 1538, 1631, 1707, 1779, 1794, 1872, 1905, 1929 and 1944.

365, July 21. Alexandria, Egypt, earthquake. This earthquake destroyed Alexandria and the giant lighthouse for which the city was famous. Some 50,000 people are thought to have been killed.

526, May 26. Antioch, Syria, earthquake. Approximately 250,000 people are believed to have died.

856, December. Corinth, Greece, earthquake. More than 40,000 people killed.

1038, January 9. Shensi, China, earthquake. More than 20,000 killed.

1057, date unknown. Chihli, China, earthquake. 25,000 killed.

1268, date unknown. Cilicia, Asia Minor, earthquake. 60,000 killed.

1290, September 27. Chihli, China, earthquake. 100,000 killed.

1293, May 20. Kamakura, Japan, earthquake. 30,000 killed.

1362, date uncertain. Öraefajökull, Iceland, volcanic eruption. Several dozen farms destroyed.

1522, date uncertain. Masaya, Nicaragua. Volcanic eruption. Several towns destroyed.

1531, January 26. Lisbon, Portugal, earthquake. 30,000 killed.

1549, date uncertain. Nicaragua, eruption of Volcan de Agua. Thousands of people killed.

1556, January 3. Shensi, China, earthquake. 830,000 killed.

1591, date uncertain. Taal, Luzon, Philippine Islands, volcanic eruption. Thousands of people reportedly died in this eruption, the first recorded eruption of Taal. Other major eruptions occurred in 1749 and 1754, on January 28 and 30, 1911 and September 28, 1965.

1614, November 26. Echigo, Takata, Japan, earthquake. Thousands of people are thought to have been killed by an earthquake and subsequent tsunami.

1616, date uncertain. Mayon, Philippine Islands, volcanic eruption. This was the first recorded eruption of Mayon. Numerous villages were destroyed, and thousands of people were killed. Another major eruption occurred on October 23–30, 1766, killing some 2,000 people. Mayon erupted again on February 1, 1814, killing more than 2,000. Several dozen were killed in another eruption in 1853. The eruption of July 9, 1888, destroyed portions of two towns. An eruption on June 23, 1897, killed several hundred people and caused widespread destruction. Some 200 people reportedly died in an eruption on July 1, 1928.

1623, date uncertain. Irazu, Costa Rica, volcanic eruption. The death toll from this eruption is thought to be in the hundreds.

1626, July 30. Naples, Italy, earthquake. This earthquake wrecked the city, destroyed numerous villages and killed some 70,000 people. Another major earthquake in 1693 killed more than 90,000.

1650, date uncertain. Asamayama, Japan, volcanic eruption. This eruption is believed to have killed several hundred people. Another eruption in 1783 killed about 5,000 people.

1663, February 5. St. Lawrence river valley, Canada, earthquake. Maximum Mercalli intensity X. Casualty figures unavailable. This earthquake was so powerful that it caused damage to buildings in southern New England.

1692, June 7. Port Royal, Jamaica, earthquake. More than 1,000 people died in this earthquake, which submerged a large portion of the city.

1693, January 9–11. Catania, Italy, earthquake. 60,000 killed, although another estimate puts the death toll at 100,000.

1698, date uncertain. Cotopaxi, Ecuador, volcanic eruption. Hundreds of people were reported killed in this eruption. Other major eruptions occurred in 1741 and 1744 and on June 26, 1877.

1703. Tokyo (Edo), Japan, earthquake. This earthquake destroyed the city and reportedly took 200,000 lives.

1731, November 30. Peking (Beijing), China, earthquake. Some 100,000 people were reportedly killed.

1737, October 11. Calcutta, India, earthquake. 300,000 killed.

1755, June 7. Persia (in the vicinity of present-day Iran), earthquake. 40,000 killed.

1755, November 1. Lisbon, Portugal, earthquake associated with a destructive tsunami. Up to 70,000 killed.

1772, date uncertain. Papandajan, Java, Indonesia, volcanic eruption. Some 3,000 people were reported killed in this eruption, which destroyed numerous villages as well as plantations. The volcano lost more than 3,000 feet of its height in this eruption.

1779–1780, December through January. Sakurazima, Japan, volcanic eruption. This eruption destroyed several villages. Another eruption and accompanying earthquakes on January 12, 1914, destroyed many villages.

1783, February 4. Calabria, Italy, earthquake. Perhaps 40,000 killed. Another earthquake in 1797 is thought to have killed an additional 50,000 people.

1783. Iceland and Japan, volcanic eruptions. A 1783 eruption of the volcano Asamayama in Japan killed several thousand people. Between mid-June and early August, an eruption of Skaptar Jökull in Iceland caused widespread destruction of property and loss of life, killing several thousand people. Also in June, a great flow of lava occurred at Laki in Iceland, covering more than 100,000 acres of land with molten rock and killing some 10,000 people.

1792, February 10. Unzen, Japan, volcanic eruption. This eruption killed some 15,000 people and destroyed two cities.

1793, date uncertain. Miyi-Yama, Java, Indonesia, volcanic eruption. This eruption is believed to have killed more than 50,000 people.

1797, February 4. Quito, Ecuador, earthquake. 40,000 killed.

1800, date uncertain. Mount Guntur, Java, Indonesia, volcanic eruption. This eruption killed several hundred people, and a lava flow reportedly filled a valley.

1811, December 16. New Madrid, Missouri, United States, earthquake. Estimated at Mercalli intensity XI, this earthquake was one of a series of extremely powerful temblors that struck the New Madrid area (near St. Louis, Missouri and Memphis, Tennessee) in the winter of 1811–12. Other great earthquakes in this series took place on January 23 and February 7, 1812. Casualties occurred but are thought to have been few in number because the area affected was sparsely settled at the time.

1812, December 21. Near Santa Barbara, California, United States, earthquake, with possible associated tsunami. Mercalli intensity X. Several persons are thought to have been injured.

1815, April 5. Tambora, near Java, Indonesia, volcanic eruption. This volcano's 1815 eruption was one of the most powerful in history and killed tens of thousands of people. The exact final death toll from Tambora's 1815 eruption is difficult to estimate because airborne particulates from the explosion are believed to have played a part in cooling global climate over the following year, disrupting agriculture and causing the famous "Year Without a Summer."

1819, June 16. Kutch, India, earthquake. More than 1,500 killed.

1822, October. Galung Gung, Java, Indonesia, volcanic eruption. Two eruptions of Galung Gung killed several thousand people altogether and destroyed hundreds of villages.

1828, December 18. Echigo, Japan, earthquake. 30,000 killed.

1857, January 9. Fort Tejon, California, United States, earthquake. One of the most powerful earthquakes in American history, the Fort Tejon earthquake is ranked at Mercalli intensity X or above and involved a rupture of the San Andreas Fault.

1857, March 21. Tokyo, Japan, earthquake. More than 100,000 people are thought to have died in this earthquake.

1863, August 13. Bolivia and Peru, earthquake. 25,000 killed.

1868, August 16. Colombia and Ecuador, earthquake. 70,000 killed.

1872, March 26. Owens Valley, California, United States, earthquake. The Owens Valley earthquake is thought to have killed some 50 to 60 people.

1883, August 27. Krakatoa, Sunda Strait, Indonesia, volcanic eruption. The 1883 eruption of Krakatoa killed perhaps 50,000 people and generated a tsunami, or seismic sea wave, that circled the globe several times before dissipating. The wave caused tremendous destruction along shorelines near the volcano.

1886, August 31. Charleston, South Carolina, United States, earthquake. The Charleston earthquake caused tremendous property damage and killed some 60 people. Exactly what caused this earthquake is still uncertain.

1891, October 28. Mino-Owari, Japan, earthquake. 7,000 killed.

1896, June 15. Riku-Ugo, Japan, earthquake. More than 20,000 killed. A tsunami was associated with this earthquake.

1897, June 12. Assam, India, earthquake. Magnitude 8.7. More than 1,000 killed.

1902, May 6–13. Soufrière, St. Vincent, Caribbean, volcanic eruption. The death toll from this eruption of Soufrière is estimated at 3,000.

1902, May 8. Mount Pelée, Martinique, Caribbean, volcanic eruption. The 1902 eruption of Pelée is among the most spectacular and thoroughly documented eruptions. More than 30,000 people are thought to have died in this eruption, which destroyed the city of St. Pierre. Another eruption of Pelée on August 30 added to the destruction.

1902, October 24. Santa Maria, Guatemala, volcanic eruption. Some 6,000 people were killed.

1906, April 18. San Francisco, California, United States, earthquake. Magnitude approximately 8.2. The San Francisco earthquake and fire that followed destroyed the city. An estimated 700 to 2,000 people were killed.

1908, December 28. Messina, Sicily, Italy, earthquake. Magnitude 7.5. 120,000 killed, although another estimate says 160,000.

1915, January 13. Avezzano, Italy, earthquake. magnitude 7.0. 30,000 killed.

1920, December 16. Kansu, China, earthquake. An earthquake caused extensive destruction and took an estimated 180,000 lives.

1923, September 1. Tokyo, Japan, Kanto earthquake. Magnitude 7.9 to 8.3. The earthquake was accompanied by fire and destroyed much of Tokyo. More than 140,000 were killed.

1927, May 22. Kansu, China, earthquake. Estimated 100,000 people killed.

1931, December 13–28. Merapi, Java, Indonesia, volcanic eruption. More than 1,000 people are thought to have died in this eruption.

1932, December 25. Kansu, China, earthquake. Estimated 70,000 people killed.

1934, January 16. Nepal, earthquake. More than 9,000 are believed to have been killed.

1935, May 31. Quetta, India, earthquake. Magnitude 7.5. 60,000 killed.

1939, January 24. Chillán, Chile, earthquake. Magnitude approximately 7.7. Perhaps 50,000 killed.

1944, June 10. Parícutin, Mexico, volcanic eruption. Parícutin, a volcano that arose rapidly in a field beginning in 1943 and grew to be more than 1,200 feet tall, erupted violently and destroyed nearby towns.

1951, January 21. Mount Lamington, Papua New Guinea, volcanic eruption. One of the most destructive of modern times, this eruption killed some 6,000 people.

1960, February 29. Agadir, Morocco, earthquake. Magnitude 5.9. Estimated 12,000 killed.

1962, September 1. Northwestern Iran, earthquake. Magnitude 7.3. 14,000 killed.

1963, July 26. Skopje, Yugoslavia, earthquake. 1,500 to 2,000 killed; most of city destroyed.

1964, March 27. Alaska, United States, earthquake. Magnitude 8.6. The Good Friday Earthquake, as this event was known, was accompanied by a highly destructive and far-reaching tsunami. More than 100 people were killed.

1968, August 31. Iran, earthquake. Magnitude 7.4. More than 11,000 killed.

1970, May 30. Peru, earthquake. Magnitude 7.8. More than 60,000 killed.

1971, February 9. San Fernando, California, United States, earthquake. Magnitude 6.5. The San Fernando earthquake killed fewer than 100 people, but that figure could have been much higher if the earthquake had lasted slightly longer.

1972, December 23. Managua, Nicaragua, earthquake. Magnitude 6.2. Estimated 11,000 killed.

1976, February 4. Guatemala, earthquake. Magnitude 7.9. More than 20,000 killed.

1976, July 28. Tangshan, China, earthquake. Magnitude 7.6. Approximately 700,000 killed.

1978, September 16. Iran, earthquake. 25,000 killed.

1980, May 18. Mount St. Helens, Washington, United States, volcanic eruption. One of the most intensively documented and studied eruptions of all time, the eruption destroyed some 150 square miles of timber and caused $1.5 billion or more in total damage. More than 100 people were reported dead or missing.

1980, October 10. El Asnam, Algeria, earthquake. Magnitude 7.7. More than 3,000 killed.

1982, December 13. Northern Yemen, earthquake. 2,800 killed.

1983, May 2. Coalinga, California, United States, earthquake. $30 million in damage reported.

1983, May 26. Honshu, Japan, tsunami. 81 killed.

1985, September 19. Mexico City, Mexico, earthquake. Magnitude 8.1. Some 8,000 killed. Although this earthquake originated along the Pacific coast of Mexico, it caused great destruction in Mexico City because portions of the city were built on the unstable sediments of a former lake bed.

1986, August 21. Cameroon. volcanic event. An escape of volcanic gas from a lake bed killed some 1,700 people.

1987, October 1. Southern California, United States, earthquake. $200 million damage.

1988, December 7. Armenia, earthquake. 25,000 killed.

1989, October 17. San Francisco, California, United States, earthquake. 67 killed, $5 billion in damage.

1990, July 16. Luzon, Philippine Islands, earthquake. 700 killed.

1991, June 10–17. Mount Pinatubo, Philippine Islands, volcanic eruption. Hundreds killed.

1992, December 10. Island of Flores, Indonesia, earthquake. More than 1,000 killed.

1993, September 30. Maharashta, India, earthquake. 16 villages destroyed, many others damaged. Tens of thousands killed.

APPENDIX B

Eruption of Vesuvius, A.D. 79

Roman naturalist and naval officer Pliny the Elder and his nephew Pliny the Younger are remembered in connection with the catastrophic eruption of Vesuvius in A.D. 79, which destroyed the nearby cities of Pompeii and Herculaneum. Pliny the Elder was killed in the eruption, and Pliny the Younger described it, together with his uncle's death, in a pair of letters written to the historian Tacitus. Although thought to be inaccurate in some respects, Pliny's letters are worth quoting at length:

> Your request that I should send an account of my uncle's death, in order to transmit a more exact relation of it to posterity, deserves my acknowledgments . . .
>
> He was at that time with the fleet under his command at Misenum. On the 24th of August, about one in the afternoon, my mother desired him to observe a cloud which appeared of a very unusual size and shape. He had just returned from taking the benefit of the sun, and, after bathing himself in cold water, and taking a slight repast, had retired to his study. He immediately arose and went out upon an eminence, from whence he might distinctly view this very uncommon appearance. It was not at that distance discernible from what mountain the cloud issued, but it was found afterward to ascend from Mount Vesuvius. I cannot give a more exact description of its figure than by comparing it to that of a pine tree, for it shot up to a great height in the form of a trunk, which extended itself at the top into a sort of branches . . . it appeared sometimes bright, and sometimes dark and spotted, as it was more or less impregnated with earth and cinders.
>
> This extraordinary phenomenon excited my uncle's philosophical curiosity to take a nearer view of it. He ordered a light vessel to be got ready . . . As he was passing out of the house he received dispatches: the marines at Retina, terrified by the imminent peril (for the place lay beneath the mountain, and there was no retreat but by ships), entreated his aid in this extremity. He accordingly changed his first design, and what he began with a philosophical he now pursued with an heroical turn of mind.
>
> He ordered the galleys to put to sea, and went himself on board with an intention of assisting not only Retina but many other places, for the population is thick on that beautiful coast. When hastening to the place from which others fled with the utmost terror, he steered a direct course to the point of danger, and with so much calmness and presence of mind, as to be able to make and dictate his observations upon the motion and figure of that dreadful scene. He was now so nigh the mountain that the cinders, which grew thicker and hotter the nearer he approached, fell into the ships, together with pumice-stones, and black pieces of burning rock; they were in danger of not only being left around by the sudden retreat of the sea, but also from the vast fragments which rolled down from the mountain, and obstructed all the shore.
>
> Here he stopped to consider whether he should return back again; to which the pilot advised him. "Fortune," said he, "favors the brave; carry me to Pomponianus." Pomponianus was then at Stabiae, separated by a gulf, which the sea, after several insensible windings, forms upon the shore. He (Pomponianus) had already sent his baggage on board; for though he was not at that time in actual danger, yet being within view of it, and indeed extremely near, if it should in the least increase, he was determined to put to sea as soon as the wind should change. It was favorable, however, for carrying my uncle to Pomponianus, whom he found in the greatest consternation. He embraced him with tenderness, encouraging and exhorting him to keep up his spirits; and the more to dissipate his fears he ordered, with an air of unconcern, the baths to be got ready; when, after having bathed, he sat down to supper with great cheerfulness, or at least (what is equally heroic) with all the appearance of it.
>
> In the meantime, the eruption from Mount Vesuvius flamed out in several places with much violence, which the darkness of the night contributed to render still more visible and dreadful. But my uncle, in order to soothe the apprehensions of his friend, assured him it was only the burning of the villages, which the country people had abandoned to the flames; after this he retired to rest,

and it was most certain he was so little discomposed as to fall into a deep sleep; for, [as he was] pretty fat, and breathing hard, those who attended without actually heard him snore. The court which led to his apartment being now almost filled with stones and ashes, if he had continued there any longer it would have been impossible for him to have made his way out; it was thought proper, therefore, to awaken him. He got up and went to Pomponianus and the rest of his company . . . They consulted together whether it would be most prudent to trust to the houses, which now shook from side to side with frequent and violent concussions; or to fly to the open fields, where the calcined stone and cinders, though light indeed, yet fell in large showers and threatened destruction. In this distress they resolved for the fields as the less dangerous situation of the two—a resolution which, while the rest of the company were hurried into it by their fears, my uncle embraced upon cool and deliberate consideration.

They went out, then, having pillows tied upon their heads with napkins; and this was their whole defense against the storm of stones that fell around them. It was now day everywhere else, but there a deeper darkness prevailed than in the most obscure night; which, however, was in some degree dissipated by torches and other lights of various kinds. They thought proper to go down further upon the shore, to observe if they might safely put out to sea; but they found that the waves still ran extremely high and boisterous. There my uncle, having drunk a draught or two of cold water, threw himself down upon a cloth which was spread for him, when immediately the flames, and a strong smell of sulfur which was the forerunner of them, dispersed the rest of the company, and obliged him to rise. He raised himself up with the assistance of two of his servants, and instantly fell down dead, suffocated, as I conjecture, by some gross and noxious vapor, having always had weak lungs, and being frequently subject to a difficulty of breathing.

As soon as it was light again, which was not until the third day after this melancholy accident, his body was found entire, and without any marks of violence upon it, exactly in the same posture as that in which he fell, and looking more like a man asleep than dead . . .

My uncle having left us [Pliny the Younger writes], I pursued [my] studies till it was time to bathe. After which I went to suppose, and from thence to bed, where my sleep was greatly broken and disturbed. There had been, for many days before, some shocks of an earthquake, which the less surprised us as they are extremely frequent in Campania; but they were so particularly violent that night, that they not only shook everything around us, but seemed, indeed, to threaten total destruction. My mother flew to my chamber, where she found me rising in order to awaken her. We went out into a small court belonging to the house, which separated the sea from the buildings. As I was at that time but eighteen years of age, I know not whether I should call my behavior, in this dangerous juncture, courage or rashness; but I took up [the historian] Livy, and amused myself with turning over that author, and even making extracts from him, as if all about me had been in full security. While we were in this posture, a friend of my uncle's, who was just come from Spain to pay us a visit, joined us; and observing me sitting with my mother with a book in my hand, greatly condemned her calmness at the same time that he reproved me for my careless [air of] security. Nevertheless, I still went on with my author.

Though it was now morning, the light was exceedingly faint and languid; the buildings all around us tottered; and, though we stood upon open ground, yet as the place was narrow and confined, we therefore resolved to quit the town. The people followed us in the utmost consternation, and, as to a mind distracted with terror every suggestion seems more prudent than its own, pressed in great crowds about us in our way out.

Being got to a convenient distance from the houses, we stood still, in the midst of a most dangerous and dreadful scene. The chariots which we had ordered to be drawn out were so agitated backwards and forwards, though upon the most level ground, that we could not keep them steady, even by supporting them with large stones. The sea seemed to roll back upon itself, and to be driven from its banks by the convulsive motion of the earth; it is certain at least that the shore was considerably enlarged, and many sea animals were left upon it. On the other side a black and dreadful cloud, bursting with an igneous serpentine vapor, darted out a long train of fire, resembling flashes of lightning, but much larger.

Upon this the Spanish friend whom I have mentioned, addressed himself to my mother and me with great warmth and earnestness; "If your brother and your uncle," said he, "is safe, he certainly wishes you to be so too; but if he has perished, it was his desire, no doubt, that you might both survive him: why therefore do you delay your escape for a moment?" We could never think of our own safety, we said, while we were uncertain of his. Hereupon our friend left us, and withdrew with the utmost precipitation. Soon afterward, the cloud seemed to descend, and cover the whole ocean; as it certainly did the island of Capreae, and the promontory of Misenum. My mother strongly [urged] me to make my escape . . . as for herself, she said, her age and corpulency rendered all attempts of that sort impossible . . . But I absolutely refused to leave her, and taking her by the hand, I led her on; she complied with great reluctance, and not without many reproaches to herself for retarding my flight.

The ashes now began to fall upon us, though in no great quantity. I turned my head and observed behind us a thick smoke, which came rolling after us like a torrent. I proposed, while we yet had any light, to turn out of the high road lest she should be pressed to death in the dark by the crowd that followed us. We had scarce stepped out of the path when darkness overspread us, not like that of a cloudy night, or when there is no moon, but of a room when it is all shut up and all the lights are extinct. Nothing then was to be heard but the shrieks of women, the screams of children and the cries of men; some calling for their children, others for their parents, others for their husbands, and only distinguishing each other by their voices; one lamenting his own fate, another that of his family; some wishing to die from the very fear of dying . . . Among them were some who augmented the real terrors by imaginary ones, and made the frighted multitude believe that Misenum was actually in flames.

At length a glimmering light appeared, which we imagined to be rather the forerunner of an approaching burst of flames, as in truth it was, than the return of day. However, the fire fell at a distance from us; then again we were immersed in thick darkness, and a heavy shower of ashes rained upon us, which we were obliged every now and then to shake off, otherwise we should have been crushed and buried in the heap . . . At last this dreadful darkness was dissipated by degrees, like a cloud of smoke; the real day returned, and soon the sun appeared, though very faintly, and as when an eclipse is coming on. Every object that presented itself to our eyes (which were extremely weakened) seemed changed, being covered over with white ashes, as with a deep snow. We returned to Misenum, where we refreshed ourselves as well as we could, and passed an anxious night between hope and fear, for the earthquake still continued, while several greatly excited people ran up and down, heightening their own and their friends' calamities by terrible predictions . . .

Earthquake at Concepción, Chile, February 20, 1835

In his memoir *The Voyage of the* Beagle, Charles Darwin described the February 20, 1835 earthquake at Concepción, Chile and its results:

This day has been memorable in the annals of Valdavia, for the most severe earthquake experienced by the oldest inhabitant. I happened to be on shore, and was lying down in the wood to rest myself. It came on suddenly, and lasted two minutes, but the time appeared much longer. The rocking of the ground was very sensible. The undulations appeared to my companion and myself to come from due east, whilst others thought they proceeded from southwest: this shows how difficult it sometimes is to perceive the directions of the vibrations. There was no difficulty in standing upright, but the motion made me almost giddy: it was something like the movement of a vessel in a little cross-ripple, or still more like that felt by a person skating over thin ice, which bends under the weight of his body.

A bad earthquake at once destroys out oldest associations: the earth, the very emblem of solidity, has moved beneath our feet like a thin crust over a fluid;—one second of time has created in the mind a strange idea of insecurity, which hours of reflection would not have produced. In the forest, as a breeze moved the trees, I felt only the earth tremble, but saw no other effect. Captain FitzRoy and some officers were in town during the shock, and there the scene was more striking; for although the houses, from being built of wood, did not fall, they were violently shaken, and the boards creaked and rattled together. The people rushed out of doors in the greatest alarm. It is these accompaniments that create the perfect horror of earthquakes, experienced by all who have thus seen, as well as felt, their effects. Within the forest it was a deeply interesting, but by no means an awe-exciting phenomenon. The tides were very curiously affected. The great shock took place at the time of low water; and an old woman who was on the beach told me that the water flowed very quickly, but not in great waves, to high-water mark, and then as quickly returned to its proper level; this was also evident by the line of wet sand. The same kind of quick but quiet movement in the tide happened a few years since at Chiloe, during a slight earthquake, and created much causeless alarm. In the course of the evening there were many weaker shocks, which seemed to produce in the harbor the most complicated currents, and some of great strength.

March 4th.—We entered the harbor of Concepcion. While the ship was beating up to the anchorage, I landed on the island of Quiriquina. The mayor-domo of the estate quickly rode down to tell me the terrible news of the great earthquake of the 20th:—"That not a house in Concepcion or Talcahuano (the port) was standing; that seventy villages were destroyed; and that a great wave had almost washed away the ruins of Talcahuano." Of this latter statement I soon saw abundant proofs—the whole coast being strewn over with timber and furniture as if a thousand ships had been wrecked. Besides chairs, tables, book-shelves, etc., in great numbers, there were several roofs of cottages, which had been transported almost whole. The storehouses at Talcahuano had been burst open, and great bags of cotton, yerba, and other valuable merchandise were scattered on the shore. During my walk round the island, I observed that numerous fragments of rock, which, from the marine productions adhering to them, must have been lying recently in deep water, had been cast up high on the beach; one of these was six feet long, three broad, and two thick.

The island itself as plainly showed the overwhelming power of the earthquake, as the beach did that of the consequent great wave. The ground in many parts was fissured in north and south lines, perhaps caused by the yielding of the parallel and steep sides of this narrow island. Some of the fissures near the cliffs were a yard wide. Many enormous masses had already fallen on the beach; and the inhabitants thought that when the rains commenced far greater slips would happen. The effect of the vibration on the hard primary slate, which composes the foundation of the island, was still more curious: the superficial parts of some narrow ridges were as completely shivered as if they had been blasted by gunpowder. This effect, which was rendered conspicuous by the fresh fractures and the displaced soil, must be confined to near the surface, for otherwise there would not exist a block of solid rock throughout Chile; nor is this improbable, as it is known that the surface of a vibrating body is affected differently from the central part. It is, perhaps, owing to this same reason, that earthquakes do not cause quite such terrific havoc within deep mines as would be expected. I believe this convulsion has been more effectual in lessening the size of the island of Quiriquina, than the ordinary wear-and-tear of the sea and weather during the course of a whole century.

The next day I landed at Talcahuano, and afterwards rode to Concepcion. Both towns presented the most awful yet interesting spectacle I ever beheld. To a person who had formerly known them, it possibly might have been still more impressive; for the ruins were so mingled together, and the whole scene possessed so little the air of a hospitable place, that it was scarcely possible to imagine its former condition. The earthquake commenced at half-past eleven o'clock in the forenoon. If it had happened in the middle of the night, the greater number of the inhabitants (which in this one province must amount to many thousands) must have perished, instead of less than a hundred; as it was, the invariable practice of running out of doors at the first trembling of the ground, alone saved them. In Concepcion each house, or row of houses, stood by itself, a heap or line of ruins; but in Talcahuano, owing to the great wave, little more than one layer of bricks, tiles and timber, with here and there part of a wall left standing, could be distinguished. From this circumstance Concepcion, although not so completely desolated, was a more terrible, and if I may so call it, picturesque sight. The first shock was very sudden. The mayor-domo at Quiriquina told me, that the first notice he received of it, was finding both the horse he rode and himself, rolling together on the ground. Rising up, he was again thrown down. He also told me that some cows which were standing on the steep side of the island were rolled into the sea. The great wave caused the destruction of many cattle; on one low island, near the head of the bay, seventy animals were washed off and drowned. It is generally thought that this has been the worst earthquake ever recorded in Chile; but as the very severe ones occur only after long intervals, this cannot easily be known; nor indeed would a much worse shock have made any difference, for the ruin was now complete. Innumerable small tremblings followed the great earthquake, and within the first twelve days no less than three hundred were counted.

After viewing Concepcion, I cannot understand how the greater number of inhabitants escaped unhurt. The houses in many parts fell outwards; thus forming in the middle of the streets little hillocks of brickwork and rubbish. Mr. Rouse, the English consul, told us that he was at breakfast when the first movement warned him to run out. He had scarcely reached the middle of the courtyard, when one side of his house came thundering down. He retained presence of mind to remember, that if he once got on top of that part which had already fallen, he would be safe. Not being able from the motion of the ground to stand, he crawled up on his hands and knees; and no sooner had he ascended this little eminence, than the other side of the house fell in, the great beams sweeping close in front of his head. With his eyes blinded, and his mouth choked with the cloud of dust which darkened the sky, at last he gained the street. As shock succeeded shock, at the interval of a few minutes, no one dared approach the shattered ruins; and no one knew whether his dearest friends and relations were not perishing from the want of help. Those who had saved any property were obliged to keep a constant watch, for thieves prowled about, and at each little trembling of the ground, with one hand they beat their breast and cried "misericorda!" and then with the other filched what they could from the ruins. The thatched roofs fell over the fires, and flames burst forth in all parts. Hundreds knew themselves ruined, and few had the means of providing food for the day.

Earthquakes alone are sufficient to destroy the prosperity of any country. If beneath England the now inert subterranean forces should exert those powers, which most assuredly in former geological ages they have exerted, how completely would the entire condition of the country be changed! What would become of the lofty houses, thickly packed cities, great manufactories, the beautiful public and private edifices? If the new period of disturbance were first to commence by some great earthquake in the dead of the night, how terrific would be the carnage! England would at once be bankrupt; all papers, records, and accounts would from that moment be lost. Government being unable to collect the taxes, and failing to maintain its authority, the hand of violence and rapine would remain uncontrolled. In every large town famine would go forth, pestilence and death following in its train.

Shortly after the shock, a great wave was seen from the distance of three or four miles, approaching in the middle of the bay with a smooth outline; but along the shore it tore up cottages and trees, as it swept onwards with irresistible force. At the head of the bay it broke in a fearful line of white breakers, which rushed up to a height of 23 vertical feet above the highest spring-tides. Their force must have been prodigious; for at the Fort a cannon with its carriage, estimated at four tons in weight, was moved 15 feet inwards. A schooner was left in the midst of the ruins, 200 yards from the beach. The first wave was followed by two others, which in their retreat carried away a vast wreck of floating objects. In one part of the bay, a ship was pitched high and dry on shore, was carried off, again driven on shore, and again carried off. In another part, two large vessels anchored near together were whirled about, and their cables were thrice wound round each other; though anchored at a depth of 36 feet, they were for some minutes aground. The great wave must have traveled slowly, for the inhabitants of Talcahuano had time to run up the hills behind the town; and some sailors pulled out seaward, trusting successfully to their boat riding securely over the swell, if they could reach it before it broke. One old woman with a little boy, four or five years old, ran into a boat, but there was nobody to row it out: the boat was consequently dashed against an anchor and cut in twain; the old woman was drowned, but the child was picked up some hours afterwards clinging to the wreck. Pools of salt-water were still standing amidst the ruins of the houses, and children, making boats with old tables and chairs, appeared as happy as their parents were miserable. It was, however, exceedingly interesting to observe, how much more active and cheerful all appeared than could have been expected. It was remarked with much truth, that from the destruction being universal, no one individual was humbled more than another, or could suspect his friends of coldness—that most grievous result of the loss of wealth. Mr. Rouse, and a large party whom he kindly took under his protection, lived for the first week in a garden beneath some apple-trees. At first they were as merry as if it had been a picnic; but soon afterward heavy rain caused much discomfort, for they were absolutely without shelter.

In Captain Fitzroy's excellent account of the earthquake, it is said that two explosions, one like a column of smoke and another like the blowing of a great whale, were seen in the bay. The water also appeared everywhere to be boiling; and it "became black, and exhaled a most disagreeable sulphurous smell." These latter circumstances were observed in the Bay of Valparaiso during the earthquake of 1822; they may, I think, be accounted for, by the disturbance of the mud at the bottom of the bay containing organic matter in decay. In the Bay of Callao, during a calm day, I noticed, that as the ship dragged her cable over the bottom, its course was marked by a line of bubbles. The lower orders in Talcahuano thought that the earthquake was caused by some old Indian women, who two years ago, being offended, stopped the volcano of Antuco. This silly belief is curious, because it shows that experience has taught them to observe, that there exists a relation between the suppressed action of the volcanos, and the trembling of the ground. It was necessary to apply the witchcraft to the point where their perception of cause and effect failed; and this was the closing of the volcanic vent. This belief is the more singular in this particular instance, because, according to Captain FitzRoy, there is reason to believe that Antuco was noways affected.

The town of Concepcion was built in the usual Spanish fashion, with all the streets running at right angles to each other; one set ranging S.W. by W, and the other set N.W. by N. The walls in the former direction certainly stood better than those in the latter; the greater number of the masses of brickwork were thrown down towards the N.E. Both these circumstances perfectly agree with the general idea, of the undulations having come from the S.W., in which quarter subterranean noises were also heard; for it is evident that the walls running S.W. and N.E. which presented their ends to the point whence the undulations came, would be much less likely to fall than those walls which, running N.W. and S.E., must in their whole lengths have been at the same instant thrown out of the perpendicular; for the undulations, coming from the S.W., must have extended in N.W. and S.E. waves, as they passed under the foundations. This may be illustrated by putting books edgeways on a carpet, and then . . . imitating the undulations of an earthquake; it will be found that they fall with more or less readiness, according as their direction more or less coincides with the line of the waves. The fissures in the ground generally, though not uniformly, extended in a S.E. and N.W. direction, and therefore corresponded to the lines of undulation or of principal flexure. Bearing in mind all these circumstances, which so clearly point to the S.W. as the chief focus of disturbance, it is a very interesting fact that the island of S. Maria, situated in that quarter, was, during the general uplifting of the land, raised to nearly three times the height of any other part of the coast.

The different resistance offered by the walls, according to their direction, was well exemplified in the case of the Cathedral. The side which fronted the N.E. presented a grand pile of ruins, in the midst of which door-cases and masses of timber stood up, as if floating in a stream. Some of the angular blocks of brickwork were of great dimensions; and they were rolled to a distance on the level plaza, like fragments of rock at the base of some high mountain. The side walls (running S.W. and N.E.), though exceedingly fractured, yet remained standing; but the vast buttresses (at right angles to them, and therefore parallel to the walls that fell) were in many cases cut clean off, as if by a chisel, and hurled to the ground. Some square ornaments on the coping of these same walls, were moved by the earthquake into a diagonal position. A similar circumstance was observed after an earthquake at Valparaiso, Calabria, and other places, including some of the ancient Greek temples. This twisting displacement, at first appears to indicate a vorticose movement beneath each point thus affected; but this is highly improbable. May it not be caused by a tendency in each stone to arrange itself in some particular position, with respect to the lines of vibration,—in a manner somewhat similar to pins on a sheet of paper when shaken? Generally speaking, arched doorways or windows stood much better than any other part of the buildings. Nevertheless, a poor, lame old man, who had been in the habit, during trifling shocks, of crawling to a certain doorway, was this time crushed to pieces.

I have not attempted to give any detailed description of the appearance of Concepcion, for I feel that it is quite impossible to convey the mingled feelings which I experienced. Several of the officers visited it before me, but their strongest language failed to give a just idea of the scene of desolation. It is a bitter and humiliating thing to see works, which have cost man so much time and labour, overthrown in one minute; yet compassion for the inhabitants was almost instantly banished, by the surprise in seeing a state of things produced in a moment of time, which one was accustomed to attribute to a succession of ages. In my opinion, we have scarcely beheld, since leaving England, any sight so deeply interesting.

In almost every severe earthquake, the neighbouring waters of the sea are said to have been greatly agitated. The disturbance seems generally, as in the case of Concepcion, to have been of two kinds: first, at the instant of the shock, the water swells up high on the beach with a gentle motion, and then as quietly retreats; secondly, some time afterwards, the whole body of the sea retires from the coast, and then returns in waves of overwhelming force. The first movement seems to be an immediate consequence of the earthquake affecting differently a fluid and a solid, so that their respective levels are slightly deranged: but the second case is a far more important phenomenon. During most earthquakes, and especially during those on the west coast of America, it is certain that the first great movement of the waters has been a retirement. Some authors have attempted to explain this, by supposing that the water retains its level, whilst the land oscillates upward; but surely the waters close to the land, even on a rather steep coast, would partake of the motion of the bottom: moreover . . . similar movements of the sea have occurred at islands far distant from the chief line of disturbance, as was the case with Juan Fernandez during this earthquake, and with Madeira during the famous Lisbon shock. I suspect (but the subject is a very obscure one) that a wave, however produced, first draws the water from the shore, on which it is advancing to break: I have observed that this happens with the little waves from the paddles of a steam-boat. It is remarkable that whilst Talcahuano and Callao (near Lima), both situated at the head of large shallow bays, have suffered during every severe earthquake from great waves, Valparaiso, seated close to the edge of profoundly deep water, has never been overwhelmed, though so often shaken by the severest shocks. From the great wave not immediately following the earthquake, but sometimes after the interval of even half an hour, and from distant lands being affected similarly with the coasts near the focus of the disturbance, it appears that the first wave rises in the offing; and as this is of general occurrence, the cause must be general: I suspect we must look to the line, where the less disturbed waters of the deep ocean join the water nearer the coast, which has partaken of the movements of the land, as the place where the great wave is first generated; it would also appear that the wave is larger or smaller, according to the extent of shoal water which has been agitated together with the bottom on which it rested.

The most remarkable effect of this earthquake was the permanent elevation of the land; it would probably be far more correct to speak of it as the cause. There can be no doubt that the land round the Bay of Concepcion was upraised two or three feet; but it deserves notice, that owing to the wave having obliterated the old lines of tidal action on the sloping sandy shores, I could discover no evidence of this fact, except in the united testimony of the inhabitants, that one little rocky shoal, now exposed, was formerly covered with water. At the island of S. Maria (about thirty miles distant) the elevation was greater on one part, Captain

FitzRoy found beds of putrid mussel-shells still adhering to the rocks, ten feet above highwater mark: the inhabitants had formerly dived at lower-water spring-tides for these shells. The elevation of this province is particularly interesting, from its having been the theater of several other violent earthquakes, and from the vast numbers of sea-shells scattered over the land, up to a height of certainly 600, and I believe, of 1,000 feet. At Valparaiso, as I have remarked, similar shells are found at the height of 1,300 feet: it is hardly possible to doubt that this great elevation has been effected by successive small uprisings, such as that which accompanied or caused the earthquake of this year, and likewise by an insensibly slow rise, which is certainly in progress on some parts of this coast.

The island of Juan Fernandez, 360 miles to the N.E., was, at the time of the great shock of the 20th, violently shaken, so that the trees beat against each other, and a volcano burst forth under water close to the shore: these facts are remarkable because this island, during the earthquake of 1751, was then also affected more violently than other places at an equal distance from Concepcion, and this seems to show some subterranean connection between these two points. Chiloe, about 340 miles southward of Concepcion, appears to have been shaken more strongly than the intermediate district of Valdivia, where the volcano of Villarica was noways affected, whilst in the Cordillera in front of Chiloe, two of the volcanoes burst forth at the same instant into violent action. These two volcanos, and some neighboring ones, continued for a long time in eruption, and ten months afterwards were again influenced by an earthquake at Concepcion. Some men, cutting wood near the base of one of these volcanos, did not perceive the shock of the 20th, although the whole surrounding Province was then trembling; here we have an eruption relieving and taking the place of an earthquake . . . Two years and three-quarters afterwards, Valdivia and Chiloe were again shaken, more violently than on the 20th, and an island in the Chonos Archipelago was permanently elevated more than eight feet. It will give a better idea of the scale of these phenomena, if . . . we suppose them to have taken place at corresponding distances in Europe:—then would the land from the North Sea to the Mediterranean have been violently shaken, and at the same instant of time a large tract of the eastern coast of England would have been permanently elevated, together with some outlying islands,—a train of volcanos on the coast of Holland would have burst forth in action, and an eruption taken place at the bottom of the sea, near the northern extremity of Ireland—and lastly, the ancient vents of Auvergne, Cantal, and Mont d'Or would each have sent up to the sky a dark column of smoke, and have long remained in fierce action. Two years and three-quarters afterwards, France, from its center to the English Channel, would have been again desolated by an earthquake, and an island permanently upraised in the Mediterranean.

The space, from under which volcanic matter on the 20th was actually erupted, is 720 miles in one line, and 400 miles in another line at right angles to the first: hence, in all probability, a subterranean lake of lava is here stretched out, of nearly double the area of the Black Sea. From the intimate and complicated manner in which the elevatory and eruptive forces were shown to be connected during this train of phenomena, we may confidently come to the conclusion, that the forces which slowly and by little starts uplift continents, and those which at successive periods pour forth volcanic matter from open orifices, are identical. From many reasons, I believe that the frequent quakings of the earth on this line of coast are caused by the rending of the strata, necessarily consequent on the tension of the land when upraised, and their injection by fluidified rock. This rending and injection would, if repeated often enough (and we know that earthquakes repeatedly affect the same areas in the same manner), form a chain of hills;—and the linear island of S. Mary, which was upraised thrice the height of the neighbouring country, seems to be undergoing this process. I believe that the solid axis of a mountain differs in its manner of formation from a volcanic hill only in the molten stone having been repeatedly injected, instead of having been repeatedly ejected. . . .

Eruption of Mount Pelée, Martinique, May 8, 1902

Many firsthand accounts of the eruption of Mount Pelée and the destruction of St. Pierre and vicinity were recorded, despite the virtually complete loss of life on shore. One of the most comprehensive and vivid accounts of the catastrophe comes from Comte de Fitz-James, a French traveler, who with his companion Baron Fontenilliat witnessed the destruction of St. Pierre from the relative safety of a boat in the harbor:

> From the depths of the earth came rumblings, an awful music which cannot be described. I called my companion's name, and my voice echoed back at me from a score of angles. All the air was filled with the acrid vapors that had belched from the mouth of the volcano . . .
>
> From a boat in the roadstead . . . I witnessed the cataclysm that came upon the city. We saw the

shipping destroyed by a breath of fire. We saw the cable ship *Grappler* keel over under the whirlwind, and sink as through drawn down into the waters of the harbor by some force from below. The *Roraima* was overcome and burned at anchor. The *Roddam* . . . was able to escape like a stricken moth which crawls from a flame that has burned its wings . . .

Our own danger was great, and had it not been for the bravery and courage of the Baron I would have perished . . . I was stunned, unable to lift a hand to assist myself. Baron de Fontenilliat dragged me from the boat into the water, where he supported me . . .

[Before the eruption, it] was such a morning as . . . is almost impossible to describe. Low hanging clouds gave the scene a dismal appearance, and this was heightened by the fine volcanic dust which filled the atmosphere, making respiration difficult. This dust was next to impalpable. It could not be seen as it floated in the air, but it settled so rapidly that my hand, resting upon the edge of the boat, was covered completely in less than three minutes.

As we made our way across the water we more than half faced Mont Pelee, which was throwing off a heavy cloud of smoke, steam and ashes. No flames were to be seen. On shore the inhabitants were making their way about the waterfront. The city was to our right. Small craft plied about the harbor, some trading with the ships that were at anchor, while in some fishermen were going out to the fishing grounds, just off Carbet . . .

[The] calm of that morning was almost abnormal. Not a ripple was to be seen upon the face of the sea. Not a breath of air was stirring, which made it more difficult for us to breathe . . .

The rumblings from the bowels of the mountain were majestic in tone. I cannot tell you just how they sounded, but perhaps you can imagine a mighty hand playing upon the strings of a harp greater than all the world. The notes produced were deep and full of threatenings. There was a jarring sensation, and every now and then there was a commotion of the waters that caused a swell without making the surface break.

Out from the shore put a small launch carrying the pennant of Governor Mouttet. The Governor at the last moment had realized that the situation was filled with a terrible danger. He was attempting to escape with his family and a few friends. I had commented to Baron de Fontenilliat upon the appearance of the Governor's craft . . . [The Governor, as later evidence proved, was too late in his attempt at flight.]

While we were talking there came an explosion that . . . I can liken only to a shot from a mammoth cannon. The breath of fire swept down upon the city and waterfront with all of the force that could have been given to it by such a cannon. Of this comparison I shall have more to say later. For the present it will do to add that the explosion was without warning and that the effect was instantaneous. Cinders were shot into our [faces] with stinging effect.

The air was filled with flame. Involuntarily we raised our hands to protect our faces. I noted the same gesture when I saw the bodies of victims on shore; arms had been raised and the hands were extended with palms outward, a gesture that in a peculiar manner indicated dread and horror.

[When] the frightful explosion came, our two boatmen were either thrown from the boat or with a quick impulse they sprang overboard. It was the one thing [they could] do to save their lives; but . . . they lost their presence of mind and, instead of staying by the side of the boat, they swam away in the direction of Precheur, which we were approaching when the [explosion] came. It was impossible for them to land at Precheur, so they were compelled to put back. They then struck out across the bay, evidently hoping to reach Carbet. We saw neither of them again, and I have no doubt they were drowned.

My brave companion . . . sprang into the water, and when he saw that I did not move he reached up and catching me by the shoulder, dragged me from the boat. I was stunned at first, and, though it was not a physical injury, I could not move of my own volition until the cold water restored my senses. It was thus that we could see all that happened about us.

The *Grappler* rushed through the water as far as her anchor-cable would permit. Then she seemed to rise by the bow, and when she settled back she sank almost before the force of the explosion had spent itself.

The *Roraima* was all a mass of flames for several seconds. We could see the poor wretches aboard . . . her rushing about in a vain attempt to escape from the fire that enveloped them. Captain Muggah—or, at least, I suppose that it was he—made an attempt to give orders to the . . . crew. Then he staggered to the railing and fell overboard.

The *Roddam* was also overcome. Her gangway was over the side. Her upper works were wrecked, but by heroic effort those on board were able to let slip the anchor chain, and, after many attempts, the ship began to move. She . . . crawled away. It was a splendid display of courage. At least three hours elapsed after the explosion before the *Roddam* cleared the harbor.

On shore all was aflame. The city burned with a terrible roar. We realized that the inhabitants had all died, as not one was to be seen making an attempt to escape. Not a cry was heard save from the ships that were in the harbor.

Our own condition was desperate in the extreme. The heat was intense. We were able to keep our faces above the surface of the water for a second at a time at the most. We would take a mouthful of air and then sink into the water to stay there until forced to come to the surface again. This lasted only about three minutes. After that we were able to float by the side of the boat, dipping only occasionally.

The water began to get so warm that I feared we had escaped roasting only to be boiled to death. In reality the water did not get so warm as to be uncomfortable. [The water] at the surface was many degrees warmer than that a foot below.

. . . St. Pierre was mantled by a dense black cloud. Our eyes could not penetrate it, but it lifted a few seconds, revealing below it a second cloud, absolutely distinct from it. The second cloud was yellow, apparently made up of sulphurous gases. It lifted as did the first, both rising like blankets, and in a similar manner they floated away. Then, as the yellow cloud lifted from the earth, we saw the flames devouring the city, from which all show of life had disappeared. . .

When we could sustain the heat that filled the air we clambered into the boat and rowed back to Carbet. The *Roddam* had just gone out from the harbor, the *Roraima* was a smoking wreck, the *Grappler* had disappeared entirely, and little was to be seen of the other craft.

At Carbet we found the village absolutely deserted. Two portions of it had been ruined. [The portion by] the water's edge had been swept by the great wave which followed the explosion . . . [The] wave . . . was of terrific force, and it added to the confusion all along the shore. Part of Carbet had been struck by the wave of fire from the volcano, but the greater portion of the village was left uninjured.

When we got ashore we called aloud, and only the echo of our voices answered us. Our fear was great, but we did not know which way to turn, and had it been our one thought to escape we would not have known how to do so. It was about one o'clock in the afternoon when we reached shore. Our weariness was beyond description. Sleep was the one thing that I wanted, but I overcame the desire and, with Baron de Fontenilliat, set off to make our way to St. Pierre, hoping that we might still render some assistance to the injured.

Not knowing the paths, we attempted to enter the city from the direction traveled by the blast of the volcano. That brought us to the flames and we were driven back. Then we went further into the country, and so happened to meet two soldiers who . . . had been in camp at Colson, far back from St. Pierre, but, on leave, had wandered in toward the city. They heard the explosion and rushed down from the hills to give aid where it was needed. When they went in through the streets it was at the risk of their lives. They were the only ones who ventured into St. Pierre that afternoon. They came upon a sailor so injured that he could not move. Picking him up, they carried him back out of the danger zone . . .

Again entering the city, [the soldiers] found five women in a hut. They were much injured, but were not dead. The soldiers gave them drink and put food within their reach, and then left them, promising to return with assistance as soon as possible . . .

Now, to show the folly of those upon whom responsibility fell in that hour of terrible disaster, I may say that when those two soldiers reached their camp they were sent to the guard-house for having remained away after hours. They told of the five suffering women, and their officer insisted that the tale had been arranged by them for the purpose of escaping punishment. They were kept under guard all Thursday night and all of the next day and the following night.

During those thirty-six hours the two soldiers made no complaint of their own treatment, but they continued to beg that assistance be sent to the women whom they had left so badly injured. Finally their plea prevailed, and on Saturday they were permitted to lead a rescue party to St. Pierre. Then their story was fully verified. One of the women was still alive. She told how . . . her four friends . . . had died late Friday night. She was taken to Fort de France, where she died a few days later . . .

Our shoes were burned to a crisp, but we plodded about those hills as long as we were able to move. Then we returned to Carbet, and remained there that night . . .

It is impossible to describe even in the most faint manner the horrors of St. Pierre. There were some things that can be made clear, but many more that cannot be explained by anything known to human reason.

It happened that one of the first bodies found . . . [when] we entered St. Pierre on Monday was that of a pretty little girl about four years old. She sat in a lifelike position by the side of a box containing her toys. But how shall we explain the fact that the house in which she was found was in absolute ruins, and, instead of being under the debris, the body was on top of it all? It was as though the little girl and her box of toys had been lifted into the air, and, after the building had fallen into ruins, had been dropped back to earth.

So it was in the streets. The explosion happened just before eight o'clock. It was a feast day. Mass was called for eight o'clock, and many were on their way to the cathedral. All of these had been lifted into the air, and after the ruins had fallen the bodies dropped back . . .

We saw great stones that seemed to be marvels of strength, but when touched with the toe of a boot they crumbled into impalpable dust. I picked up a bar of iron. It was about an inch and a half thick and three feet long. It had been manufactured square and then twisted so as to give it greater strength. The fire that came down from Mont Pelee had taken from the iron all of its strength and had left it so that when I twisted it, it fell into filaments, like so much broom straw.

Back of the cathedral was the savannah. Great trees had been torn up by the roots, leaving holes twenty feet deep and thirty to forty feet across. Then these holes had been filled by the ashes that poured down from the volcano. Trees were cut off as though by a mighty knife in the hands of a giant reaper. Everywhere were banks of cinders and ashes.

When the Baron and I first went into the ruined city we were too awe-struck to speak. Then . . . I called to him. His name echoed back to us from a score of standing walls. All about us were bodies. On few faces was to be seen the peace which I have seen mentioned by others. I believe that almost all had time to realize what was upon them, but they did not have time to suffer. Their arms were outstretched . . . The hands were open and the fingers were spread. It was a common gesture, and I believe that it was the act of men and women who threw up their arms to ward off a blow which they knew was descending upon them . . .

I know that the explosion of Mont Pelee was not accompanied by anything like an earthquake, for . . . when we entered St. Pierre we found the fountains all flowing, just as though nothing had happened. They continued to flow, and are flowing still, unless destroyed by the later explosions.

There was no flow of lava. It was all ashes, dust, gas and mud . . .

One vessel caught in the eruption was the steamship *Roraima,* of the Quebec Steamship Company. The ship arrived in St. Pierre around 7 A.M. on the morning of the eruption and had trouble making its way into the port because of darkness and a heavy ashfall. Assistant Purser Thompson's account of the explosion of St. Pierre and the destruction of St. Pierre follows:

I saw St. Pierre destroyed. It was blotted out by one great flash of fire. Nearly 40,000 persons were all killed at once. Out of 18 vessels lying in the roads [that is, the harbor] only one, the British steamer *Roddam,* escaped, and she, I hear, lost more than half on board. It was a dying crew that took her out.

Our boat . . . arrived at St. Pierre early Thursday morning. For hours before we entered the roadstead we could see flames and smoke rising from Mont Pelee. No one on board had any idea of danger. Captain G. T. Muggah was on the bridge, and all hands got on deck to see the show.

The spectacle was magnificent. As we approached St. Pierre we could distinguish the rolling and leaping of the red flames that belched from the mountain in huge volumes and gushed high into the sky. Enormous clouds of black smoke hung over the volcano.

When we anchored at St. Pierre I noticed the cable steamer *Grappler,* the *Roddam,* three or four American schooners and a number of Italian and Norwegian barks. The flames were then spurting straight up in the air, now and then waving to one side or the other for a moment and again leaping suddenly higher up.

There was a constant muffled roar. It was like the biggest oil refinery in the world burning up on the mountain top. There was a tremendous explosion about 7.45 o'clock, soon after we got in. The mountain was blown to pieces. There was no warning. The side of the mountain was ripped out, and there was hurled straight toward us a solid wall of flames. It sounded like thousands of cannon.

The wave of fire was on us and over us like a lightning flash. It was like a hurricane of fire. It saw it strike the . . . *Grappler* broadside on and capsize her. From end to end she burst into flames and then sank. The fire rolled in mass straight down upon St. Pierre and the shipping. The town vanished before our eyes and the air grew stifling hot, and we were in the thick of it.

Wherever the mass of fire struck the sea the water boiled and sent up vast clouds of steam. The sea was torn into huge whirlpools that careened toward the open sea. One of these horrible hot whirlpools swung under the *Roraima* and pulled her down on her beam ends with the suction. She careened way over to port, and then the fire hurricane from the volcano smashed her, and over she went on the opposite side. The fire wave swept off the masts and smokestack as if they were cut with a knife.

Captain Muggah was the only one on deck not killed outright. He was caught by the fire wave and terribly burned. He yelled to get up the anchor, but, before two fathoms were heaved in the *Roraima* was almost upset by the boiling whirlpool, and the fire wave had thrown her down on her beam ends to starboard. Captain Muggah was overcome by the flames. He fell unconscious from the bridge and toppled overboard.

The blast of fire from the volcano lasted only a few minutes. It shriveled and set fire to everything it touched. Thousands of casks of rum were stored in St. Pierre, and these were exploded by the terrific heat. The burning rum ran in streams down every street and out to the sea. This blazing rum set fire to the *Roraima* several times. Before the volcano burst the landings of St. Pierre were crowded with people. After the explosion not one living being was seen on land. Only 25 of those on the *Roraima* out of 68 were left after the first flash.

The French cruiser *Suchet* came in and took us off at 2 p.m. She remained nearby, helping all she could, until 5 o'clock, then went to Fort de France with all the people she had rescued. At that time it looked as if the entire north end of the island was on fire.

Another crew member of the *Roraima* witnessed a horrible spectacle on deck:

Hearing a tremendous report and seeing the ashes falling thicker, I dived into a room, dragging with me Samuel Thomas, a gangway man and . . . [shut] the door tightly. Shortly after I heard a voice, which I recognized as that of the chief mate, Mr. Scott. Opening the door with great caution, I drew him in. The nose of Thomas was burned by the intense heat.

We three and Thompson, the assistant purser . . . were the only persons who escaped practically uninjured. The heat being unbearable, I emerged in a few moments, and the scene that presented itself to my eyes baffles description. All around on the deck were the dead and dying covered with boiling mud. There they lay, men, women and little children, and the appeals of the latter for water were heart-rending. When water was given them they could not swallow it, owing to their throats being filled with ashes or burnt with the heated air.

The ship was burning aft, and I jumped overboard, the sea being intensely hot. I was at once swept seaward by a tidal wave, but, the sea receding a considerable distance, the return wave washed me against an upturned sloop to which I clung. I was joined by a man so dreadfully burned and disfigured as to be unrecognizable. Afterwards I found he was the captain of the *Roraima,* Captain Muggah. He was in dreadful agony, begging piteously to be put on board his ship.

Picking up some wreckage which contained bedding and a tool chest, I, with the help of five others who had joined me on the wreck, constructed a rude raft, on which we placed the captain . . . Seeing the *Roddam,* which arrived in port shortly after we anchored, making for the Roraima, I said goodbye to the captain and swam back to the *Roraima.* [The captain's body was recovered later.]

The *Roddam,* however, burst into flames and put to sea. I reached the *Roraima* at about half-past 2, and was afterwards taken off by a boat from the French warship *Suchet.* Twenty-four others with myself were taken on to Fort de France. Three of these died before reaching port. A number of others have since died.

Ellery Scott, mate of the *Roraima,* recounted his view of the eruption and of the last moments of Captain Muggah:

All hands had had breakfast. I was standing on the fo'c's'l head trying to make out the marks on the pipes of a ship 'way out and heading for St. Lucia. I wasn't looking at the mountain at all. But I guess the captain was, for he was on the bridge, and the last time I heard him speak was when he shouted, "Heave up, Mr. Scott; heave up [raise the anchor]." I gave the order to the men, and I think some of them did jump to get the anchor up, but nobody knows what really happened for the next fifteen minutes. I turned around toward the captain and then I saw the mountain.

Did you ever see the tide come into the Bay of Fundy? It doesn't sneak in a little at a time as it does 'round here. It rolls in, in waves. That's the way the cloud of fire and mud and white-hot stones rolled down from that volcano over the town and over the ships. It was on us in almost no time, but I saw it, and in the same glance I saw our captain bracing to meet it on the bridge. He was facing the fire cloud with both hands gripped hard to the bridge rail, his legs apart and his knees braced back stiff. I've seen him brace himself that same way many a time in a rough sea with the spray going mast-head high and green water pouring along the decks.

I saw the captain . . . at the same time I saw that ruin coming down on us. I don't know why, but that last glimpse of poor Muggah on his bridge will stay with me just as long as I remember St. Pierre, and that will be long enough.

In another instant it was all over for him. As I was looking at him he was all ablaze. He reeled and fell on the bridge with his face toward me. His mustache and eyebrows were gone in a jiffy. His hat had gone, and his hair was aflame, and so were his clothes from head to foot. I knew he was conscious when he fell, by the look in his eyes, but he didn't make a sound.

That all happened a long way inside of half a minute; then something new happened. When the wave of fire was going over us, a tidal wave of the sea came out from the shore and did the rest. That wall of rushing water was so high and so solid that it seemed to rise up and join the smoke and flame above. For an instant we could see nothing but the water and the flame.

That tidal wave picked the ship up like a canoe and then smashed her. After one list to starboard the ship righted, but the masts, the bridge, the funnel and all the upper works had gone overboard.

I had saved myself from fire by jamming a metal ventilator cover over my head and jumping from the fo'c's'l head. Two St. Kitts [natives] saved me from the water by grabbing my legs and pulling me down into the fo'c's'l after them. Before I could get up, three men tumbled in on top of me. Two of them were dead.

Captain Muggah went overboard, still clinging to the fragments of his wrecked bridge. Daniel Taylor, the ship's cooper, and a Kitts native jumped overboard to save him. Taylor managed to push the captain on to a hatch that had floated off from us, and then they swam back to the ship for more assistance, but nothing could be done for the captain. Taylor wasn't sure he was alive. The last we saw of him or his dead body, it was drifting shoreward on that hatch . . .

[After] staying in the fo'c's'l about twenty minutes, I went out on deck. There were just four of us left aboard who could do anything . . . It was still raining fire and hot rocks, and you could hardly see a ship's length for dust and ashes, but we could stand that. There were burning men and some women and two or three children lying around the deck. Not just burned, but burning, then, when we got to them. More than half the ship's company had been killed in that first rush of flame. Some had rolled overboard when the tidal wave came, and we never saw so much as their bodies. The cook was burned to death in his galley. He had been paring potatoes for dinner, and what was left of his right hand held the shank of his potato knife. The wooden handle was in ashes. All that happened to [the] man in less than a minute. The donkey engineman was killed on deck sitting in front of his boiler. We found parts of some bodies—a hand, or an arm or a leg. Below decks there were some twenty alive.

The ship was on fire, of course, what was left of it. The stumps of both masts were blazing. Aft she was like a furnace, but forward the flames had not got below deck, so we four carried those who were still alive on deck into the fo'c's'l. All of them were burned and most of them were half strangled . . .

My own son's gone, too. It had been his trick at lookout ahead during the dog watch that morning, when we were making for St. Pierre, so I supposed at first when the fire struck us that he was asleep in his bunk and safe. But he wasn't. Nobody could tell me where he was. I don't know whether he was burned to death or rolled overboard and drowned . . .

After getting all hands that had any life left in them below and [attending] to the best we could, the four of us that were left halfway ship-shape started in to fight the fire . . . Thanks to the tidal wave that cleared our decks, there wasn't much left to burn, so we got the fire down so's we could live on board with it several hours more, and then [we] turned to knock a raft together out of what timber and truck we could find below. Our boats had gone overboard with the masts and funnel.

We made that raft for something over thirty that were alive . . . But we did not have to risk the raft, for . . . the *Suchet* came along and took us all off. We thought for a minute just after we were wrecked that we were to get help from a ship that passed us . . . but she kept on. We learned afterward that she was the *Roddam.*

One of the strangest experiences associated with this eruption was reported by Captain Eric Lillienskjold of the Danish steamship *Nordby,* which was several hundred miles from the island:

We were plodding along slowly that day. About noon I took the bridge to make an observation. It seemed to be hotter than ordinary. I shed my coat and vest and got into what little shade there was. As I worked it grew hotter and hotter, I didn't know what to make of it. Along about 2 o'clock in the afternoon it was so hot that all hands got to talking about it. We reckoned that something queer was coming off, but none of us could explain what it was. You could almost see the pitch softening in the seams.

Then, as quick as you could toss a biscuit over its rail, the *Nordby* dropped—regularly dropped—three or four feet down into the sea. No sooner did it do this than big waves, that looked like they were coming from all directions at once, began to smash against our sides. This was queerer yet, because the water a minute before was as smooth as I ever saw it. I had all hands piped on deck, and we battened down everything loose to make ready for a storm. And we got it all right—the strangest storm you ever heard tell of.

There was something wrong with the sun that afternoon. It grew red and then dark red and then, about a quarter after 2, it went out of sight altogether. The day got so dark that you couldn't see half a ship's length ahead of you. We got our lamps going, and put on our oilskins, ready for a hurricane. All of a sudden there came a sheet of lightning that showed up the whole tumbling sea for miles and miles. We sort of ducked, expecting an awful crash of thunder, but it didn't come. There was no sound except the big waves pounding against our sides. There wasn't a breath of wind.

Well, sir, at that minute there began [the] most exciting time I've ever been through, and I've been on every sea on the map for twenty-five years. Every second there'd be waves 15 or 20 feet high, belting us head-on, stern-on and broadside, all at once. We could see them coming, for without any stop at all, flash after flash of lightning was blazing all about us.

Something else we could see, too. Sharks! There were hundreds of them on all sides, jumping up and down in the water. And sea birds! A flock of them, squawking and crying, made for our rigging and perched there. They seemed like they were scared to death. But the queerest part of it all was the water itself. It was hot—not so hot that our feet could not stand it when it washed over the deck, but hot enough to make us think that it had been heated by some kind of a fire.

Well, that sort of thing went on hour after hour. The waves, the lightning, the hot water and the sharks, and all the rest of the odd things happening, frightened the crew out of their wits . . . Mighty strange things happen on the sea, but this topped them all.

I kept to the bridge all night. When the first hour of morning came, the storm was still going on. We were all pretty much tired out by that time, but there was no such thing as trying to sleep. The waves were still batting us around, and we didn't know whether we were one mile or a thousand miles from shore. At two o'clock in the morning all the queer goings on stopped just the way they began—all of a sudden. We lay to until daylight; then we took our reckonings and started off again. We were about 700 miles off Cape Henlopen . . . None of us was hurt, and the old *Nordby* herself pulled through all right, but I'd sooner stay ashore than see waves without wind and lightning without thunder.

Captain Freeman of the *Roddam* recorded his experiences of the eruption and its aftermath:

I went to anchorage between 7 and 8 and had hardly moored when the side of the volcano opened out with a terrible explosion. A wall of fire swept over the town and the bay. The *Roddam* was struck boardside by the burning mass. The shock to the ship was terrible, nearly capsizing her.

Hearing the awful report of the explosion and seeing the great wall of flames approaching the steamer, those on deck sought shelter wherever it was possible, jumping into the cabin, the forecastle and even into the hold. I was in the chart room, but the burning embers were borne by so swift a movement of the air that they were swept in through the door and port holes, suffocating and scorching me badly. I was terribly burned by these embers about the face and hands, but managed to reach the deck. Then, as soon as it was possible, I mustered the few survivors who seemed able to move, ordered them to slip the anchor, leaped for the bridge and [rang] the engine for full speed astern. The second and the third engineer and a fireman were on watch below and so escaped injury . . . but the men on deck could not work the steering gear because it was jammed by the debris from the volcano. We accordingly went ahead and astern until the gear was free, but in this running backward and forward it was two hours after the first shock before we were clear of the bay.

One of the most terrifying conditions was that, the atmosphere being charged with ashes, it was totally dark. The sun was completely obscured, and the air was only illuminated by the flames from the volcano and those of the burning town and shipping. It seems small to say that the scene was terrifying in the extreme. As we backed out we passed close to the *Roraima,* which was one mass of blaze. The steam was rushing from the engine room, and the screams of those on board were terrible to hear. The cries for help were all in vain, for I could do nothing but save my own ship. When I last saw the *Roraima* she was settling down by the stern . . .

When the *Roddam* was safely out of the harbor of St. Pierre, with its desolations and horrors, I made for St. Lucia. Arriving there, and when the ship was safe, I mustered the survivors as well as I was able and searched for the dead and injured. Some I found in the saloon where they had . . . sought for safety, but the cabins were full of burning embers that had blown in through the port holes. Through these the fire swept as through funnels and burned the victims where they lay or stood, leaving a circular imprint of scorched and burned flesh. I brought ten on deck who were thus burned; two of them were dead, the others survived, although in a dreadful state of torture from their burns. Their screams of agony were heartrending . . . The ship was covered from stem to stern with powdered lava, which retained its heat for hours after it had fallen. In many cases it was practically incandescent, and to move about the deck in this burning mass was not only difficult but absolutely perilous. I am only now able to begin thoroughly to clear and search the ship for any damage done by this volcanic rain, and to see if there are any corpses in out-of-the-way places. For instance, this morning, I found one body in the peak of the forecastle. The body was horribly burned and the sailor had evidently crept in there in his agony to die.

On the arrival of the *Roddam* at St. Lucia the ship presented an appalling appearance. Dead and calcined bodies lay about the deck, which was also crowded with injured, helpless and suffering people . . . The woodwork of the cabins and bridge and everything inflammable on deck were constantly igniting, and it was with great difficulty that we few survivors managed to keep the flames down. My ropes, awnings, tarpaulins were completely burned up.

I witnessed the entire destruction of St. Pierre. The flames enveloped the town in every quarter with such rapidity that it was impossible that any person could be saved. As I have said, the day was suddenly turned to night, but I could distinguish by the light of the burning town people distractedly running about on the beach. The burning buildings stood out from the surrounding darkness like black shadows. All this time the mountain was roaring and shaking, and in the intervals between these terrifying sounds I could hear the cries of despair and agony from the thousands who were perishing. These cries added to the terror of the scene, but it is impossible to describe its horror or the dreadful sensations it produced. It was like witnessing the end of the world.

Captain Cantell of the British steamship *Etona* visited the *Roddam* at St. Lucia on May 11 and described the damage to the vessel afterward:

The *Roddam* was covered with a mass of fine bluish gray dust or ashes of cementlike appearance. In some places it lay two feet deep on the decks. This matter had fallen in a red-hot state all over the steamer, setting fire to everything it struck that was burnable, and, when it fell on the men on board, burning off limbs and large pieces of flesh. This was shown by finding portions of human flesh when the decks were cleared of the debris. The rigging, ropes, tarpaulins, sails, awnings, etc., were charred or burned, and most of the upper stanchions and spars were swept overboard or destroyed by fire. Skylights were smashed and cabins were filled with volcanic dust. The scene of ruin was deplorable.

The captain, though suffering the greatest agony, succeeded in navigating his vessel to the port of Castries, St. Lucia, with 18 dead bodies on the deck and human limbs scattered about. A sailor stood by constantly wiping the captain's injured eyes . . .

Captain Cantell had witnessed the eruption from a relatively safe distance:

The weather was clear and we had a fine view, but the old outlines of St. Pierre were not recognizable. Everything was a mass of blue lava, and the formation of the land itself seemed to have changed. When we were about eight miles off the northern end of the island Mount Pelee began to belch a second time. Clouds of smoke and lava shot into the air and spread all over the sea, darkening the sun. Our decks in a few minutes were covered with a substance that looked like sand dyed a bluish tint, and which smelled like phosphorus . . .

We were about four miles off the northern end of the island when suddenly there shot up in the air to a tremendous height a column of smoke. The sky darkened and the smoke seemed to swirl down upon us. In fact, it spread all around, darkening the atmosphere as far as we could see. I called Chief Engineer Farrish to the deck.

"Do you see that over there?" I asked, pointing to the eruption . . .

"Well, Farrish, rush your engines as they have never been rushed before," I said to him . . .

We began to cut through the water at almost twelve knots. Ordinarily we make ten knots. We could see no more of the land contour, but everything seemed to have been enveloped in a great cloud. There was no fire visible, but the lava dust rained down upon us steadily. In less than an hour there were two inches of it upon our deck.

The air smelled like phosphorus. No one dared look up to try to locate the sun, because one's eyes would fill with lava dust. Some of the blue lava dust is sticking to our mast yet, although we have swabbed decks and rigging again and again to be clear of it.

After little more than an hour's fast running we saw daylight ahead and began to breathe easier . . .

Eruption of Soufrière, St. Vincent, May 7, 1902

Following are several eyewitness accounts:

I was fishing at some distance from the shore when my boatman said to me, "Look at the Soufriere, sir. It is smoking!"

From the top of the cone, reaching far up into the heavens, a dark column of smoke arose, while the mouth of the crater itself glowed like a gigantic forge belching a huge jet of yellow flame. The mass of smoke spread out into branches extending for miles, and clouds of sulphurous vapor, overflowing, as it were, the bowl of the crater, began to roll down the mountain slopes.

We reached shore and started to run for our lives. We were soon enveloped in impenetrable darkness, and I was unable to distinguish the white shirt of my boatman at a yard's distance. But as he knew every inch of the ground, I held on to a stick he had, and so we stumbled on until we reached a place of safety. The incessant roar of the volcano, the rumbling of the thunder, the flashes of the lightning added to the terrific grandeur of the scene. At last we emerged from the pall of death, half suffocated, and with our temples throbbing as if they were going to burst.

***

We [a group of several persons who were rescued from a house] heard the mountain roaring the whole morning, but we thought it would pass off, and we did not like to abandon our homes, so we chanced it. About half-past one it began to rain pebbles and stones, some of which were alight; but then, although we were afraid, we could not leave. The big explosion must have taken place at half-past two o'clock. There was fire all around me, and I could not breathe. My hands and feet got burned, but I managed to reach the house where the others were.

In two hours everything was over, although pebbles and dust fell for a long time after. My burns got so painful and stiff that I could not move. We remained until Sunday morning without food or water. Five persons died, and as none of us could throw the bodies out, or even move, we had to lie alongside the bodies until we were rescued.

***

From Sunday night, May 4, the heat had been oppressive. Never had I experienced such heat before. It was with the utmost difficulty one could breathe, and to sleep was impossible . . .

On Tuesday I learned . . . that the Chateau Belaire side of the mountain was showing signs of activity. On Wednesday morning, between nine and ten o'clock, the lightning and thunder began. Such lightning and such thunder! Oh, it is terrible to remember, and thrice terrible it was to behold! Blinding flashes that zigzagged with hissing fury and a lurid light ominous of destruction . . .

In the meantime some fisher girls who came down from the mountain said they had observed the water in the mountain lake to be boiling rapidly and the grass in the vicinity to be torn up. Then, you will understand, I got anxious. The storm grew in fury. The thunder became louder and louder . . .

Amid the crashing thunder peals and the dreadful lightning there began to fall a shower of small pebbles, and later on there were stones as big as your fist. Meanwhile dismal rumblings were heard, as though the mountain groaned under the weight of accumulated fury, and the earth swayed in deep sympathy.

At half-past two the explosion occurred and darkness fell upon the land . . . The sounds were weird and abysmal, and caused our hearts to quiver with fear.

The rain of big stones continued up to about eleven o'clock at night, when sand began to fall. From where we were, we could see the reflection of the fire in the sky, but could not see the blaze . . . But at last morning broke . . . a dull, dismal, dreary day came, not much distinguishable from the preceding night . . . But it was day, and that fact afforded some measure of relief. We could see and hear others in the town . . . Among those who came into [Georgetown] or were brought in were many who had been stricken by lightning and were paralyzed, or who had been scorched by the burning hot sand and were blistered and sore.

***

In company with several gentlemen [a clergyman wrote], on Wednesday at noon I left in a small rowboat to go to Cheateau Belaire, where we hoped to get a better view of the eruption. As we passed Layou, the first town on the leeward coast, the smell of sulphuretted hydrogen was very perceptible. Before we got halfway on our journey a vast column of steam, smoke and ashes ascended to a prodigious elevation, falling apparently in the vicinity of Georgetown . . . We were about eight miles from the crater, as the crow flies, and . . . I judge that the awful pillar [the eruption column] was fully eight miles in height.

We were proceeding rapidly to our point of observation, when an immense cloud, dark, dense and apparently thick with volcanic material, descended over our pathway . . . This mighty bank of sulphurous vapor and smoke assumed at one time the shape of a gigantic promontory, then appeared as a collection of twirling, revolving cloud whorls, turning with rapid velocity; now assuming the shape of gigantic cauliflowers, then efflorescing into beautiful flower shapes, some dark, some effulgent, some bronze, others pearly white and all brilliantly illuminated by electric flashes.

Darkness, however, soon fell upon us. The sulphurous air was laden with fine dust that fell thickly upon and around us, discoloring the sea. A black rain began to fall, followed by another rain of [volcanic material]. The electric flashes were marvelously rapid in their motions, and numerous beyond all computation. These, with the thundering roar of the mountain, mingled with the dismal roar of the lava, the shocks of earthquakes, the falling stones, the enormous quantity of material ejected from the belching crater, producing a darkness as dense as a starless midnight . . . combined to make up a scene of horror.

An unnamed press correspondent who visited the island provided some idea of the destruction visited upon St. Vincent by the volcano:

The entire northern portion of the island is covered with ashes to an average depth of eighteen inches, varying from a thin layer at Kingstown to two feet or more at Georgetown. The crops are ruined, nothing green can be seen, the streets of Georgetown are cumbered with snowdriftlike heaps of ashes, and ashes rest so heavily on the roofs that in several cases they have caused them to fall in. There will soon be 5,000 destitute persons in need of assistance from the government, which is already doing everything possible to relieve the sufferers. There are a hundred injured people in the hospital at Georgetown, gangs of men are searching for the dead or rapidly burying them in trenches, and all that can be done under the circumstances is being accomplished.

The arrival here of the first detachment of the Ambulance Corps, which brought sufferers from

Georgetown, caused a sensation. This batch consisted of a hundred persons, whose charred bodies exhaled fetid odors, and whose loathsome faces made even the hospital attendants shudder. All these burned persons were suffering fearfully from thirst and uttering, when strong enough to do so, agonizing cries for water. It is doubtful whether any of the whole party will recover.

While the outbreak of the volcano on the island of Martinique killed more people outright, more territory has been ruined in St. Vincent, hence there is greater destitution here. The injured persons were horribly burned by the hot grit, which was driven along with tremendous velocity. Twenty-six persons who sought refuge in a room ten feet by twelve were all killed. One person was brained by a huge stone some nine miles from the crater.

Rough coffins are being made to receive the bodies of the victims. The hospital here is filled with dying people. Fifty injured persons are lying on the floor of that building, as there are no beds for their accommodation, though cots are being rapidly constructed of boards . . .

Since midnight on Tuesday the subterranean detonations here have ceased, and the Soufriere on Wednesday relapsed, apparently, into perfect repose, no smoke rising from the crater, and the fissures emitting no vapor. The stunted vegetation that formerly adorned the slopes of the mountain has disappeared, having given place to gray-colored lava, which greets the eye on every side. The atmosphere is dry. Rain would be welcome, as there is a great deal of dust in the air, which is disagreeable and irritating to the throats and eyes, and is causing the merchants to put all their dry-goods under cover . . . [People] who have remained on the estates are half-starved, and the few Carib survivors are leaving their caves and pillaging abandoned dwelling houses and shops . . .

The report that the volcanic lake which occupied the top of the mountain has disappeared, now appears to be confirmed. A sea of lava, emitting sulfurous fumes, now apparently occupies the place, and several new craters have been formed. The last time the volcano showed activity, on Tuesday last, the craters, old and new, and numerous fissures in the mountain sides discharged hot vapor, deep subterranean murmurings were heard, the ground trembled at times, from the center of the volcano huge volumes of steam rose like gigantic pine trees toward the sky, and a dense black smoke, mingling with the steam, issued from a new and active crater, forming an immense pall over the northern hills, lowering into the valleys and then rising and spreading until it enveloped the whole island in a peculiar gray mist . . .

The sulphurous vapors, which still exhale all over the island, are increasing the sickness and mortality among the surviving inhabitants, and are causing suffering among the new arrivals . . .

The stench in the afflicted districts is terrible beyond description. Nearly all the huts left standing are filled with dead bodies. In some cases disinfectants and the usual means of disposing of the dead are useless, and cremation has been resorted to. When it is possible the bodies are dragged with ropes to the trenches and are there hastily covered up, quicklime being used when available. Many of the dead bodies were so covered with dust that they were not discovered until walked upon by visitors, or by the relieving officers or their assistants . . .

The volcano resumed activity on the night of May 18. Earthquakes were felt on the island, and smoke emanated from fissures and craters on Soufriere. As churchgoers returned from services around 8:30 p.m., one account of the eruption says, "an alarming luminous cloud suddenly ascended many miles high in the north of the island, and drifted sluggishly to the northeast. Incessant lightning fell on the mountain, and one severe flash seemed to strike about three miles from Kingstown. The thunderous rumblings in the craters lasted for two hours and then diminished until they became mere murmurings. During the remainder of the night the volcano was quiet, though ashes fell from 10 o'clock until midnight. The inhabitants were frenzied with fear at the time of the outbreak, dreading a repetition of the catastrophe which had caused such terrible loss of life on the island. They ran from the streets into the open country, crying and praying for preservation from another calamity. No one on the island of St.. Vincent slept that night . . .

The continuous agitation of the volcano and the absence of rain caused the vicinity of the afflicted villages to look like portions of the Desert of Sahara. A thick, smoky cloud overspread the island, all business was suspended, the streets were empty and everyone was terror-stricken. The feeling of suspense grew painful. People passed their time gazing at the northern sky, where the thunder clouds gathered and the mournful roaring of the volcano was heard. Ashes and pumice fell slowly in the [outlying] districts, and a new reign of terror existed in the island. But during the next day the volcanic disturbances moderated, and some degree of calm returned to the afflicted islanders.

BIBLIOGRAPHY

Because the literature on earthquakes and volcanoes is so vast, the following list represents only a selected bibliography on the subject.

Aaronson, S. "The Social Cost of Earthquake Prediction." *New Scientist,* March 17, 1977, 634–6.

Adams, W., ed. *Tsunamis in the Pacific Ocean.* Honolulu: East-West Center Press, 1970.

Aitken, F., and E. Hilton. *A History of the Earthquake and Fire.* San Francisco: Hilton, 1906.

Anderson, C. "Animals, Earthquakes and Eruptions." *Field Museum of Natural History Bulletin* 44, no. 5 (1973): 9–11.

Andrews, A. *Earthquake.* London: Angus and Robertson, 1963.

Ballard, R., and J. Grassle. "Return to Oases of the Deep." *National Geographic,* November 1979, 686–705.

Barnea, J. "Geothermal Power." *Scientific American,* January 1972, 70–77.

Bird, J., and B. Isacks, eds. *Plate Tectonics: Selected Papers from the Journal of Geophysical Research.* Washington, D.C.: American Geophysical Union, 1972.

Blong, R. J. *Volcanic Hazards.* Orlando, Fla.: Academic Press, 1984.

Bolt, B. *Earthquakes: A Primer.* New York: Freeman, 1978.

———. *Inside the Earth.* New York: Freeman, 1982.

Bolt, B., W. Horn, G. Macdonald and R. Scott. *Geological Hazards.* Berlin: Springer-Verlag, 1975.

Booth, B. "Predicting Eruptions." *New Scientist,* September 9, 1976, 526–28.

———. "Volcanic Hazards." *New Scientist,* August 26, 1976, 432–35.

Boyd, F., and H. Meyer, eds. *Kimberlites, Diatremes and Diamonds: Their Geology, Petrology and Geochemistry.* Washington, D.C.: American Geophysical Union, 1979.

Brander, J. "The Iranian Earthquake: What Next?" *New Scientist,* September 28, 1978, 930–31.

Brander, J., and R. Lewin. "Another Theory Shakes the Experts." *New Scientist,* October 12, 1978, 91.

Briggs, P. *Will California Fall Into the Sea?* New York: McKay, 1972.

Bullard, F. *Volcanoes in History, in Theory, in Eruption.* Austin: University of Texas Press, 1962.

———. *Volcanoes of the Earth.* Austin: University of Texas Press, 1984.

Burke, K., and J. Wilson. "Hot Spots on the Earth's Surface." *Scientific American,* August 1976, 46–57.

Clague, D., and C. Darymple. "The Hawaiian-Emperor Volcanic Chain." *Volcanism in Hawaii.* U.S. Geological Survey, Professional Paper 1350. 1987, 5–54.

Cox, A., ed. *Plate Tectonics and Geomagnetic Reversals.* New York: Freeman, 1973.

Crandell, D., and D. Mullineaux. *Potential Hazards from Future Eruptions of Mount St. Helens Volcano.* U.S. Geological Survey, Bulletin 1383-C. 1978.

Decker, R., and B. Decker. *Volcanoes.* Rev. ed. New York: Freeman, 1989.

Decker, R., T. Wright and P. Stauffer, eds. *Volcanism in Hawaii.* U.S. Geological Survey, Professional Paper 1350. 2 vols. 1987.

Douglas, J. "Letter from Tokyo (6): Waiting for the 'Great Tokai Quake.' " *Science News,* April 29, 1978, 282–3.

Eiby, G. *Earthquakes.* Auckland, New Zealand: Heineman, 1980.

Fisher, R., and H.-U. Schmincke. *Pyroclastic Rocks.* Berlin: Springer-Verlag, 1984.

Francis, P. *Volcanoes.* Middlesex, England: Penguin Books, 1976.

Fried, J. *Life Along the San Andreas Fault.* New York: Saturday Review Press, 1973.

Gold, T., and S. Sofer. "The Deep-Earth-Gas Hypothesis." *Scientific American,* June 1980, 154–61.

Green, J., and N. Short, eds. *Volcanic Landforms and Surface Features.* New York: Springer-Verlag, 1971.

Gutenberg, B., and C. Richter. *Seismicity of the Earth*

and Associated Phenomena. Princeton, N.J.: Princeton University Press, 1954.

Haas, J., and D. Mileti. *Socioeconomic Impact of Earthquake Prediction on Government, Business and Industry*. Boulder: Institute of Behavioral Sciences, University of Colorado, 1976.

Halacy, D. *Earthquakes: A Natural History*. Indianapolis: Bobbs-Merrill, 1974.

Harris, S. *Fire Mountains of the West*. Missoula, Mont.: Mountain Press, 1987.

Heezen, B., and C. Hollister. *The Face of the Deep*. New York: Oxford University Press, 1971.

Iacopi, R. *Earthquake Country*. Menlo Park, Calif.: Lane, 1973.

International Association of Volcanology and Chemistry of the Earth's Interior. *Catalog of Active Volcanoes of the World*. Parts 1–22. Rome, Italy: Int'l. Assn. of Volcanology, 1951–75.

Jeffreys, H. *Earthquakes and Mountains*. London: Methuen, 1950.

Kerr, R. "Earthquakes: Prediction Proving Elusive." *Science,* April 28, 1978, 419–21.

———. "Prospects for Earthquake Prediction Wane." *Science,* November 2, 1979, 542–5.

Kruger, P., and C. Otte, eds. *Geothermal Energy*. Stanford, Calif.: Stanford University Press, 1973.

Kummel, B. *History of the Earth*. 2d ed. San Francisco: Freeman, 1970.

Lawson, A. *The California Earthquake of April 18, 1906*. Report of the State Earthquake Investigation Commission. Washington, D.C.: Carnegie Institution, 1908.

Leet, L., and F. Leet. *Earthquake: Discoveries in Seismology*. New York: Dell, 1964.

Lipman, P., and D. Mullineaux, eds. *The 1980 Eruptions of Mount St. Helens*. U.S. Geological Survey, Professional Paper 1250. 1981.

Lynch, J. *Our Trembling Earth*. New York: Dodd, 1940.

Macdonald, G. *Volcanoes*. Englewood Cliffs, N.J.: Prentice-Hall, 1972.

Macdonald, G., A. Abbott and F. Peterson. *Volcanoes in the Sea*. Honolulu: University of Hawaii Press, 1983.

Macelwane, J. *When the Earth Quakes*. Milwaukee, Wis.: Bruce, 1947.

Marvin, U. *Continental Drift: The Evolution of a Concept*. Washington, D.C.: Smithsonian Institution, 1973.

Menard, H. *Marine Geology of the Pacific*. New York: McGraw-Hill, 1964.

Morris, C. *The Volcano's Deadly Work*. New York: Scull, 1902.

Muffler, L., ed. *Assessment of Geothermal Resources of the United States—1978*. U.S. Geological Survey, Circular 790. 1979.

Newhall, C., and D. Dzurisin. *Historical Unrest at Large Calderas of the World*. U.S. Geological Survey, Bulletin 1855. 1988.

Oakeshott, G. *California's Changing Landscapes*. New York: McGraw-Hill, 1978.

———. *Volcanoes and Earthquakes*. New York: McGraw-Hill, 1976.

Panel on Earthquake Prediction of the Committee on Seismology. *Predicting Earthquakes*. Washington, D.C.: National Academy of Sciences, 1976.

Park, C., and R. MacDiarmid. *Ore Deposits*. New York: Freeman, 1975.

Paterson, D. "Methane from the Bowels of the Earth." *New Scientist,* June 29, 1978, 896–897.

Press, F., and R. Siever. *Earth*. New York: Freeman, 1986.

Richter, C. *Elementary Seismology*. San Francisco: Freeman, 1958.

Rikitake, T. *Earthquake Prediction*. Amsterdam: Elsevier, 1976.

Ritchie, D. *The Ring of Fire*. New York: Atheneum, 1981.

———. *Superquake!* New York: Crown, 1988.

Rona, P. "Plate Tectonics and Mineral Resources." *Scientific American,* July 1973, 86–95.

Rybach, L., and L. Muffler. *Geothermal Systems: Principles and Case Histories*. New York: Wylie, 1981.

Smith, F. *The Seas in Motion*. New York: Crowell, 1973.

Smith, R. "Waiting for the Other Plate to Drop in California." *Science,* February 24, 1978, 797.

Stearns, H. *Geology of the State of Hawaii*. 2d ed. Palo Alto, Calif.: Pacific Books, 1985.

Steinbrugge, K. *Earthquakes, Volcanoes and Tsunamis*. New York: Scandia America Group, 1982.

Stommel, H., and E. Stommel. "The Year Without a Summer." *Scientific American,* June 1979, 176–80.

Sugimura, A., and S. Uyeda. *Island Arcs: Japan and its Environs*. Amsterdam: Elsevier, 1973.

Sullivan, R., with D. Mustart and J. Galehouse. "Living in Earthquake Country: A Survey of Residents Living Along the San Andreas Fault." *California Geology,* January 1977, 3–8.

Sullivan, W. *Continents in Motion*. New York: McGraw-Hill, 1974.

———. *Landprints*. New York: Times Books, 1984.

Tarling, D., and M. Tarling. *Continental Drift*. New York: Pelican, 1974.

Tazieff, H. *When the Earth Trembles.* New York: Harcourt, Brace and World, 1964.

Thomas, G., and M. Witts. *The San Francisco Earthquake.* New York: Stein and Day, 1971.

Thorarinsson, S. *Surtsey.* New York: Viking, 1967.

U.S. Geological Survey. *The San Fernando, California, Earthquake of February 9, 1971.* Professional Paper 733. 1971.

Uyeda, S. *The New View of the Earth.* New York: Freeman, 1978.

Verney, P. *The Earthquake Handbook.* London: Paddington, 1979.

Wegener, A. *The Origins of the Oceans and Continents.* New York: Methuen, 1967.

Wertenbaker, W. *The Floor of the Sea.* Boston: Little, Brown, 1974.

Wexler, H. "Volcanoes and World Climate." *Scientific American,* April 1952, 74–81.

Wilcoxson, K. *Chains of Fire.* Radnor, Pa.: Chilton, 1966.

Williams, H., and A. McBirney. *Volcanology.* San Francisco: Freeman, Cooper, 1979.

Wyllie, P. "The Earth's Mantle." *Scientific American,* March 1975, 50–57.

Ziony, J., ed. *Evaluating Earthquake Hazards in the Los Angeles Region—An Earth-Science Perspective.* U.S. Geological Survey, Professional Paper 1360. 1985.

INDEX

Main topics have page numbers in **boldface** type. Illustration locators are in *italics*. Maps are indicated by the letter *"m"*

D

N

R

S

T

U

V